Communications in Computer and Information Science 1913

Rationale

The CCIS series is devoted to the publication of proceedings of computer science conferences. Its aim is to efficiently disseminate original research results in informatics in printed and electronic form. While the focus is on publication of peer-reviewed full papers presenting mature work, inclusion of reviewed short papers reporting on work in progress is welcome, too. Besides globally relevant meetings with internationally representative program committees guaranteeing a strict peer-reviewing and paper selection process, conferences run by societies or of high regional or national relevance are also considered for publication.

Topics

The topical scope of CCIS spans the entire spectrum of informatics ranging from foundational topics in the theory of computing to information and communications science and technology and a broad variety of interdisciplinary application fields.

Information for Volume Editors and Authors

Publication in CCIS is free of charge. No royalties are paid, however, we offer registered conference participants temporary free access to the online version of the conference proceedings on SpringerLink (http://link.springer.com) by means of an http referrer from the conference website and/or a number of complimentary printed copies, as specified in the official acceptance email of the event.

CCIS proceedings can be published in time for distribution at conferences or as post-proceedings, and delivered in the form of printed books and/or electronically as USBs and/or e-content licenses for accessing proceedings at SpringerLink. Furthermore, CCIS proceedings are included in the CCIS electronic book series hosted in the SpringerLink digital library at http://link.springer.com/bookseries/7899. Conferences publishing in CCIS are allowed to use Online Conference Service (OCS) for managing the whole proceedings lifecycle (from submission and reviewing to preparing for publication) free of charge.

Publication process

The language of publication is exclusively English. Authors publishing in CCIS have to sign the Springer CCIS copyright transfer form, however, they are free to use their material published in CCIS for substantially changed, more elaborate subsequent publications elsewhere. For the preparation of the camera-ready papers/files, authors have to strictly adhere to the Springer CCIS Authors' Instructions and are strongly encouraged to use the CCIS LaTeX style files or templates.

Abstracting/Indexing

CCIS is abstracted/indexed in DBLP, Google Scholar, EI-Compendex, Mathematical Reviews, SCImago, Scopus. CCIS volumes are also submitted for the inclusion in ISI Proceedings.

How to start

To start the evaluation of your proposal for inclusion in the CCIS series, please send an e-mail to ccis@springer.com.

Nicholas Olenev · Yuri Evtushenko · Milojica Jaćimović · Michael Khachay · Vlasta Malkova
Editors

Advances in Optimization and Applications

14th International Conference, OPTIMA 2023 Petrovac, Montenegro, September 18–22, 2023 Revised Selected Papers

Editors
Nicholas Olenev
FRC CSC RAS
Moscow, Russia

Yuri Evtushenko
FRC CSC RAS
Moscow, Russia

Milojica Jaćimović
University of Montenegro
Podgorica, Montenegro

Michael Khachay
Krasovsky Institute of Mathematics
and Mechanics
Ekaterinburg, Russia

Vlasta Malkova
FRC CSC RAS
Moscow, Russia

ISSN 1865-0929 ISSN 1865-0937 (electronic)
Communications in Computer and Information Science
ISBN 978-3-031-48750-7 ISBN 978-3-031-48751-4 (eBook)
https://doi.org/10.1007/978-3-031-48751-4

This Springer imprint is published by the registered company Springer Nature Switzerland AG
The registered company address is: Gewerbestrasse 11, 6330 Cham, Switzerland

Paper in this product is recyclable.

Preface

This volume contains the second part of the refereed proceedings of the XIV International Conference on Optimization and Applications (OPTIMA 2023)[1].

Organized annually since 2009, the conference has attracted a significant number of researchers, academics, and specialists in many fields of optimization, operations research, optimal control, game theory, and their numerous applications in practical problems of data analysis and software development.

The broad scope of OPTIMA has made it an event where researchers involved in different domains of optimization theory and numerical methods, investigating continuous and discrete extremal problems, designing heuristics and algorithms with theoretical bounds, developing optimization software, and applying optimization techniques to highly relevant practical problems can meet together and discuss their approaches and results. We strongly believe that this facilitates collaboration between researchers working in optimization theory, methods, and applications, to advance optimization theory and methods and employ them on valuable practical problems.

The conference was held during September 18–22, 2023, in Petrovac, Montenegro, on the picturesque Budvanian Riviera on the azure Adriatic coast. For those who were not able to come to Montenegro this year, an online session was organized. The main organizers of the conference were the Montenegrin Academy of Sciences and Arts, Montenegro, FRC CSC RAS, Russia, and the University of Évora, Portugal. This year, the key topics of OPTIMA were grouped into eight tracks:

(i) Mathematical Programming
(ii) Global Optimization
(iii) Continuous Optimization
(iv) Discrete and Combinatorial Optimization
(v) Optimal Control
(vi) Game Theory and Mathematical Economics
(vii) Optimization in Economics and Finance
(viii) Applications

The Program Committee and the reviewers of the conference included more than one hundred well-known experts in continuous and discrete optimization, optimal control and game theory, data analysis, mathematical economics, and related areas from leading institutions of 26 countries including Argentina, Australia, Austria, Belgium, China, Finland, France, Germany, Greece, India, Israel, Italy, Lithuania, Kazakhstan, Mexico, Montenegro, The Netherlands, Poland, Portugal, Russia, Serbia, Sweden, Taiwan, the UAE, the UK, and the USA. This year we received 107 submissions mostly from Russia but also from Algeria, Armenia, Azerbaijan, Belarus, China, Egypt, France, Germany, Iran, Kazakhstan, Montenegro, Poland, Turkey, United Arab Emirates, the UK and the USA. Each submission was single-blind reviewed by at least three PC members or invited

[1] http://agora.guru.ru/display.php?conf=OPTIMA-2023.

reviewers, experts in their fields, to supply detailed and helpful comments. Out of 68 qualified submissions, the Program Committee decided to accept 27 papers to the first volume of the proceedings. Thus the acceptance rate for this volume was about 40%. In addition, after a short presentation of the candidate submissions, discussion at the conference, and subsequent revision, the Program Committee proposed 21 papers out the remaining 41 papers to be included in this second volume of the proceedings.

The conference featured three invited lecturers, and several plenary and keynote talks. The invited lectures included:

- Alexey Tret'yakov, Systems Research Institute, Polish Academy of Sciences, Warsaw, Poland, "The pth-Order Karush-Kuhn-Tucker Type Optimality Conditions for Nonregular Inequality Constrained Optimization Problems"
- Panos M. Pardalos, University of Florida, USA, "Diffusion in Networks"
- Yurii Nesterov, UCLouvain, Belgium, "Universality, the New Trend in Development of Optimization Schemes"

We would like to thank all the authors for submitting their papers and the members of the PC for their efforts in providing exhaustive reviews. We would also like to express special gratitude to all the invited lecturers and plenary speakers.

October 2023

Nicholas Olenev
Yuri Evtushenko
Milojica Jaćimović
Michael Khachay
Vlasta Malkova

Organization

Program Committee Chairs

Milojica Jaćimović	Montenegrin Academy of Sciences and Arts, Montenegro
Yuri G. Evtushenko	FRC CSC RAS, Russia
Michael Yu. Khachay	Krasovsky Institute of Mathematics and Mechanics, Russia
Vlasta U. Malkova	FRC CSC RAS, Russia
Nicholas N. Olenev	CEDIMES-Russie, FRC CSC RAS, Russia

Program Committee

Majid Abbasov	St. Petersburg State University, Russia
Samir Adly	University of Limoges, France
Kamil Aida-Zade	Institute of Control Systems of ANAS, Azerbaijan
Alexander P. Afanasiev	Institute for Information Transmission Problems, RAS, Russia
Yedilkhan Amirgaliyev	Suleyman Demirel University, Kazakhstan
Anatoly S. Antipin	FRC CSC RAS, Russia
Adil Bagirov	Federation University, Australia
Artem Baklanov	International Institute for Applied Systems Analysis, Austria
Evripidis Bampis	LIP6 UPMC, France
Olga Battaïa	ISAE-SUPAERO, France
Armen Beklaryan	National Research University Higher School of Economics, Russia
Nikolay Belotelov	FRC CSC RAS, Russia
Vladimir Beresnev	Sobolev Institute of Mathematics, Russia
Anton Bondarev	Xi'an Jiaotong-Liverpool University, China
Sergiy Butenko	Texas A&M University, USA
Vladimir Bushenkov	University of Évora, Portugal
Igor A. Bykadorov	Sobolev Institute of Mathematics, Russia
Alexander Chichurin	John Paul II Catholic University of Lublin, Poland
Duc-Cuong Dang	INESC TEC, Portugal
Tatjana Davidovic	Mathematical Institute of Serbian Academy of Sciences and Arts, Serbia
Stephan Dempe	TU Bergakademie Freiberg, Germany
Alexandre Dolgui	IMT Atlantique, LS2N, CNRS, France

Yuri Nesterov	Université Catholique de Louvain, Belgium
Yuri Nikulin	University of Turku, Finland
Evgeni Nurminski	Far Eastern Federal University, Russia
Natalia K. Obrosova	FRC CSC RAS, Russia
Victor Orlov	Moscow State University of Civil Engineering, Russia
Panos Pardalos	University of Florida, USA
Dmitry Pasechnyuk	Mohammed bin Zayed University of Artificial Intelligence, United Arab Emirates
Alexander V. Pesterev	V.A. Trapeznikov Institute of Control Sciences, Russia
Alexander Petunin	Ural Federal University, Russia
Stefan Pickl	University of the Bundeswehr Munich, Germany
Leonid Popov	IMM UB RAS, Russia
Mikhail A. Posypkin	FRC CSC RAS, Russia
Alexander N. Prokopenya	Warsaw University of Life Sciences, Poland
Artem Pyatkin	Novosibirsk State University; Sobolev Institute of Mathematics, Russia
Ioan Bot Radu	University of Vienna, Austria
Soumyendu Raha	Indian Institute of Science, India
Leonidas Sakalauskas	Institute of Mathematics and Informatics, Lithuania
Sergei Semakov	FRC CSC RAS, Russia
Yaroslav D. Sergeyev	University of Calabria, Italy
Natalia Shakhlevich	University of Leeds, UK
Alexander A. Shananin	Moscow Institute of Physics and Technology, Russia
Bismark Singh	University of Southampton, UK
Angelo Sifaleras	University of Macedonia, Greece
Mathias Staudigl	Maastricht University, The Netherlands
Fedor Stonyakin	V. I. Vernadsky Crimean Federal University, Russia
Alexander Strekalovskiy	Matrosov Institute for System Dynamics & Control Theory, SB RAS, Russia
Vitaly Strusevich	University of Greenwich, UK
Michel Thera	University of Limoges, France
Tatiana Tchemisova	University of Aveiro, Portugal
Anna Tatarczak	Maria Curie-Skłodowska University, Poland
Alexey A. Tretyakov	Siedlce University of Natural Sciences and Humanities, Poland
Stan Uryasev	University of Florida, USA
Frank Werner	Otto von Guericke University Magdeburg, Germany

Adrian Will	National Technological University, Argentina
Anatoly A. Zhigljavsky	Cardiff University, UK
Aleksandra Zhukova	FRC CSC RAS, Russia
Julius Žilinskas	Vilnius University, Lithuania
Yakov Zinder	University of Technology, Australia
Tatiana V. Zolotova	Financial University under the Government of the Russian Federation, Russia
Vladimir I. Zubov	FRC CSC RAS, Russia
Anna V. Zykina	Omsk State Technical University, Russia

Organizing Committee Chairs

Milojica Jaćimović	Montenegrin Academy of Sciences and Arts, Montenegro
Yuri G. Evtushenko	FRC CSC RAS, Russia
Nicholas N. Olenev	FRC CSC RAS, Russia

Organizing Committee

Anna Flerova	FRC CSC RAS, Russia
Alexander Gasnikov	National Research University Higher School of Economics, Russia
Alexander Gornov	Institute of System Dynamics and Control Theory, SB RAS, Russia
Vesna Dragović	Montenegrin Academy of Sciences and Arts, Montenegro
Vladimir Jaćimović	University of Montenegro, Montenegro
Michael Khachay	Krasovsky Institute of Mathematics and Mechanics, Russia
Yury Kochetov	Sobolev Institute of Mathematics, Russia
Vlasta Malkova	FRC CSC RAS, Russia
Anton Medennikov	FRC CSC RAS, Russia
Oleg Obradovic	University of Montenegro, Montenegro
Natalia Obrosova	FRC CSC RAS, Russia
Mikhail Posypkin	FRC CSC RAS, Russia
Kirill Teymurazov	FRC CSC RAS, Russia
Yulia Trusova	FRC CSC RAS, Russia
Svetlana Vladimirova	FRC CSC RAS, Russia
Ivetta Zonn	FRC CSC RAS, Russia
Vladimir Zubov	FRC CSC RAS, Russia

Contents

Game Theory and Mathematical Economics

Optimization in Economics and Finance

Applications

Mathematical Programming

A Randomised Non-descent Method for Global Optimisation

Dmitry A. Pasechnyuk[1,2,3(✉)] and Alexander Gornov[4]

[1] Moscow Institute of Physics and Technology, Dolgoprudny, Russia
dmitry.vilensky@mbzuai.ac.ae
[2] Mohamed bin Zayed University of Artificial Intelligence, Abu Dhabi, UAE
[3] Institute for Information Transmission Problems RAS, Moscow, Russia
[4] Matrosov Institute for System Dynamics and Control Theory, Irkutsk, Russia
gornov@icc.ru

Abstract. This paper proposes novel algorithm for non-convex multimodal constrained optimisation problems. It is based on sequential solving restrictions of problem to sections of feasible set by random subspaces (in general, manifolds) of low dimensionality. This approach varies in a way to draw subspaces, dimensionality of subspaces, and method to solve restricted problems. We provide empirical study of algorithm on convex, unimodal and multimodal optimisation problems and compare it with efficient algorithms intended for each class of problems.

Keywords: Random subspace · Subspace sampling · Zeroth-order optimization

1 Introduction

Speaking of optimisation as a computational discipline, practitioner and therefore a theorist face with a great collection of functions with no structure sufficient to guarantee that their optimum can be found in reasonable time. The only properties of these functions which a designer of algorithms can rely on are continuity and boundedness of domain. This is what we mean by global optimisation. Most of natural problems in a non-simplified form satisfy this description, textbook example is intermolecular potential optimisation. We hope that mentioned properties are natural enough to cover most of real-life variational phenomena. Thus, solving global optimisation problems is essential purpose of applied mathematics.

Many practical general-purpose global optimisation methods represent a heuristic, which has proved the efficiency in applications. Great part of global optimisation as a science consist of systematic collection of such heuristics and computational facts about them. Classic works reflecting the described methodology include [8,21]. To systematise numerical tests, the benchmarks are developed, for example [3]. Despite the growth of another, theoretical part of global optimisation science, this paper belongs to former one, and describes and tests a heuristic, which first appeared in preprint [15].

N. Olenev et al. (Eds.): OPTIMA 2023, CCIS 1913, pp. 3–14, 2024.
https://doi.org/10.1007/978-3-031-48751-4_1

Global optimisation makes severe demands on a general-purpose heuristic. Continuous functions challenge us with expensive-to-evaluate function's value and gradient, curse of dimensionality, numerous spurious optima (local extrema and saddle points). Further moreover, descent direction given by anti-gradient ceases to carry global information about function if no growth conditions introduced. This is the reason why local descent-type methods are not widely used in practice of global optimisation as standalone algorithms, but usually serve as an auxiliary local minimisation procedure for basin-hopping-type algorithms [22].

The philosophy behind the proposed approach is following. Since the monograph [12], the framework of local oracles is the dominant point of view on optimisation algorithms. In broad terms, in the beginning of algorithm's operation it has no information about the particular function to optimise and cannot classify it to point the optimum. Every next iteration of the algorithm calls an oracle, which gives some information to classify function more precisely. For example, if class of functions consist of two functions with no common values, one evaluation of function will fully characterise the function and its optimum will be known from pre-knowledge. When enough information is received to narrow the class of possible function so that their optima are close, optimisation completes. Further development of this concept for the case of local oracles is a foundation of convex optimisation. We cannot rely on local information, so we receive the information about the function through *hypotheses*. Every next iteration of our algorithm test the hypothesis of some form. Hypothesis may be unsuccessful, or *unproductive*, and give no information for optimisation (in this case, algorithm go to next step), or *productive* and give some information to construct a step. When algorithm cannot come to a productive hypothesis for a long time, it indicates that good enough approximation of the optimum is already found. The particular instance, which is called Solar method, of the meta-algorithmic scheme described above is proposed in this paper. Its hypotheses are of the form: "Does the point with optimal function's value in section of feasible set by a given subspace improves current best function's value?" — and its step is a jump to that point if hypothesis is verified.

The paper is organised as follows. In Sect. 2 we summarise the development of algorithms allied to one described in this paper and provide a literature review. The algorithm itself is described in Sect. 3, in Sect. 3.1 this description is extended with description of ways to draw a random subspace. Section 4 consists of empirical evaluation of algorithm's efficiency on test functions representing convex, unimodal and global optimisation problems in Sects. 4.1, 4.2 and 4.3 correspondingly. There we compare proposed algorithm with algorithms known to be efficient on corresponding class, and test the dependencies of algorithm's convergence on parameters of problem and algorithm's hyperparameters.

2 Related Work

One forerunner of the algorithm we propose is coordinate descent method. It is known to be especially efficient in large-scale optimisation problems. In convex optimisation setting it provably converges to the optimum with known convergence rates. Classic works devoted to the analysis of coordinate descent methods are [14,20]. At the same time, this method is not widely used in non-convex and global optimisation. Attempts to go beyond the class of convex problems were made [16,23], but corresponding results are as well of a local nature.

Another branch related to optimisation over random subspaces is sketching. This technique consists in implicit choice of a subspace by proper preconditioning of the gradient. Unlike coordinate descent method, this approach is more general in variants of subspace to choose. It became popular in community developing randomised optimisation algorithms [6,7,9] and has acquired extensions for second-order oracle [5] and non-convex problems [1]. Unlike the algorithm proposed in this paper, efficacy of these methods is based on Johnson–Lindenstrauss lemma or similar theoretical principles.

Second closest predecessor of our algorithm is classical steepest gradient descent with line-search. Properties of related methods in global optimisation environment is reflected in [25] to some extent. However, algorithms with global line-search along descent or random directions are out of mainstream and it is hard to find modern works on this topic, as is easily seen if trying to browse in Google Scholar.

3 The Algorithm

One iteration of the algorithm, which is called Solar method, consists of three principal steps: construction of set to restrict the problem (random drawn subspace or manifold), solving restricted optimisation problem, and jump to the new point if it improves current record.

In general case, random manifold can be constructed using cloud of $< p$ reference points to interpolate or approximate through them. In dependence on method, one should store p points in some data structure with cheap insertion and extraction of minimum of finding closest neighbours. In variants we consider it is necessary to maintain p minimums, which can be done by heap or simple sorted list. If closest neighbours need to be found for construction, one should use k-d tree.

Pseudocode of Solar method is listed in Algorithm 1. Naming of the randomly generated inclusion r explains the name of method: the metaphor is that we move along random rays from current sun-like point.

3.1 Subspace, or Manifold, Sampling

Substantive part of the algorithm is drawing random ray based on available points and gradient vectors. We considered two variants of the algorithms differing in method to draw a subspace: vanilla, in which subspace is fully random, and cone-restricted, in which there is a dominant direction such that angle between random subspace and that direction is bounded. In both variants, r is

Algorithm 1: Solar method

Data: Number of outer iterations $K \in \mathbb{N}$, number of total inner iterations $N \in \mathbb{N}$, total dimensionality $n \in \mathbb{N}$, number of base variables $b \in \mathbb{N} : 1 \leq b < n$, number of probes $p \in \mathbb{N}$, convex indicator function $\chi : \mathbb{R}^n \to \{0, +\infty\}$ of feasible set $Q \in \mathbb{R}^n$, initial point $x \in Q$, zeroth-order oracle $f : \mathbb{R}^n \to \mathbb{R}[$, first-order oracle $\nabla f : \mathbb{R}^n \to \mathbb{R}^n]$

Result: $x_{\text{best}} = T.\text{extract\_min}() \in \mathbb{R}^n$

Initialise data structure T;
$T.\text{insert}((f(x), x))$;
for $i = 1, ..., K$ **do**
  Choose $|B| = b$ random unique indices from set $\{1, ..., n\}$;
  for $j = 1, ..., \lfloor N/K \rfloor$ **do**
    for $k = 1, ..., p$ **do**
      $(f_k, x_k) = T.\text{extract\_min}()$;
    end
    Construct the ray $r : \mathbb{R}^b \hookrightarrow \mathbb{R}^n$ using points $(x_1, ..., x_p)[$ and gradients $(\nabla f(x_1), ..., \nabla f(x_p))]$;
    $x_{\text{candidate}} \in \arg\min_{x \in \text{im } r} \{f(x) + \chi(x)\}$;
    $T.\text{insert}((f(x_{\text{candidate}}), x_{\text{candidate}}))$;
    for $k = 1, ..., p - 1$ **do**
      $T.\text{insert}((f_k, x_k))$;
    end
  end
end

constructed as follows:

$$r : b \to x_1 + A(b - x_1[B]),$$

where $x[B]$ means vector from $\mathbb{R}^b$ composed of values of x by indices B, and $A \in \mathbb{R}^{n \times b}$ is randomised matrix containing linear factors defining the affine subspace, which is different for variants under consideration.

In vanilla variant, A is generated as follows:

$$\begin{aligned} c_{ij} &\propto \mathcal{U}(-1, 1) \\ a_{ij} &= \tan(\pi/2 \cdot C) \\ A[B] &= I_b, \end{aligned}$$

where I_b is $b \times b$ identity matrix, and $A[B]$ means submatrix of A containing rows by indices B, analogously.

In cone-restricted variant, procedure is more complicated. Independently on dominant direction, angles under tan must be in $(-\pi/2, \pi/2)$. A, therefore, is generated as follows:

$$
\begin{aligned}
c_{ij} &\propto \mathcal{U}(-1,1)\\
\alpha_{ij} &= \arctan \frac{g_i}{g_j}\\
\underline{\alpha}_{ij} &= \max\{-\pi/2+\epsilon, \alpha_{ij}-\beta\cdot a\}\\
\overline{\alpha}_{ij} &= \min\{\alpha_{ij}+\beta\cdot a, \pi/2-\epsilon\}\\
A &= \tan\left(\frac{\overline{\alpha}+\underline{\alpha}}{2}+\frac{\overline{\alpha}-\underline{\alpha}}{2}\circ C\right)\\
A[B] &= I_b,
\end{aligned}
$$

where $\epsilon = 10^{-16}$ is machine zero, β is coefficient dependent on iteration of algorithm (can be used to increase angle), a is initial angle of cone from which subspaces are drawn. $\circ$ denotes Hadamard product. $g \in \mathbb{R}^n$ is vector of dominant direction, it may be equal to $\nabla f(x_1)$ (first-order variant) or $x_2 - x_1$ (secant variant).

4 Numerical Experiments

To assess the practical performance of proposed algorithm and compare it with existing efficient algorithms, numerical experiments on quadratic (representing class of convex problems), Rosenbrock–Skokov (typical example of non-convex unimodal problem) and Rastrigin and DeVilliersGlasser02 functions (both are from benchmark [3]) are carried out.

Common details of further experiments are following. To solve the restricted problems of low dimensionality, Solar method use Nelder–Mead method [11] set to 10 iterations in all the cases. Each curve or point on presented plots is equipped with shadow: for classic randomised algorithms, sizes of lower and upper shadow are determined by standard deviation of function value measured in 3 or 5 runs with different random seeds; for Solar method, upper shadow is given by function value in run with worst final function value, and lower bound – with best final function value, correspondingly.

The implementation of algorithms used in the experiments and reference implementation of Solar method are in Python 3 and available as open source[1].

4.1 Convex Problem: Quadratic Function

Let's consider the problem of quadratic function optimisation with uniformly random positive-definite matrix $AA^\top$, some vector b and scalar c:

$$
\begin{gathered}
f(x) = x^\top AA^\top x + b^\top x + c\\
a_{ij} \propto \mathcal{U}(0,a),\ b_i \propto \mathcal{U}(0,1),\ c_i \propto \mathcal{U}(0,1)
\end{gathered}
$$

with three particular instances of different dimensionality and conditioning:

[1] https://github.com/dmivilensky/Solar-method-non-convex-optimisation.

1. $a = 1$, $\kappa \approx 3.78 \cdot 10^3$
   $x \in \mathbb{R}^{10}, -20 \le x \le 20$
   $\|x^0 - x^*\|_2 \approx 23.16$, $f(x^*) \approx -4.77$
2. $a = 5\sqrt{2}$, $\kappa \approx 5.44 \cdot 10^4$
   $x \in \mathbb{R}^{25}, -5 \le x \le 5$
   $\|x^0 - x^*\|_2 \approx 8.40$, $f(x^*) \approx -1.42$
3. $a = \sqrt{10}$, $\kappa \approx 3.70 \cdot 10^5$
   $x \in \mathbb{R}^{50}, -5 \le x \le 5$
   $\|x^0 - x^*\|_2 \approx 1.23$, $f(x^*) \approx 0.71$

where $\kappa = \lambda_{\max}(AA^\top)/\lambda_{\min}(AA^\top)$ is a condition number. In all the cases $\kappa \gg 1$, which makes chosen quadratic functions hard to optimise using standard gradient descent.

For this example, we apply Solar method in zeroth-order variant and compare it with zeroth-order methods of convex optimisation for fairness. We consider zeroth-order gradient descent with two-point feedback [13] and its variant with line-search. To represent zeroth-order accelerated algorithms, we take Momentum three point method [4] in deterministic setting.

The description of the results presented on Fig. 1 follows.

Comparison of Algorithms. As is easily seen from Fig. 1a, Solar method takes second place after steepest gradient descent. This is expectable that Solar method is not leading method for convex optimisation, because it puts much effort into exploration by choosing random direction instead of exploitation of given knowledge that gradient direction is descent direction leading to optimum. Unexpectedly, it performs better than Momentum three point method with theoretical convergence rate corresponding to accelerated methods.

Dependency of Convergence Rate of Problem's Properties. It is important to choose proper dimensionality B of random subspaces, but it is expectable that optimal choice depends on properties of the problem, main of which are total dimensionality N and conditioning $\mu/L \equiv 1/\kappa$. On Fig. 1b convergence rate is estimated by relative function suboptimality after 5000 inner iterations of Solar method. One point represents convergence rate of method for given dimensionality, condition number and variable proportion of base variables B/N. It can be seen that form of dependence of convergence rate on dimensionality of subspaces varies, but as a rule low dimensionality is better. It can be explained by the uniform choice of number of iterations for method solving restricted problems, while the complexity of restricted problems grow with their dimensionality and more iterations are required. In some sense, curves reflect the dependency of error accumulated due to insufficient iterations of auxiliary method on dimensionality of subspaces. However, guided by limitation of budget, this plot should guide in choice of B.

Dependency of Convergence Rate on Algorithm's Hyperparameters. Parameter for number of outer iterations K of Solar method, or derived parameter for number of inner iterations per chosen set of base variables N/K, where N is total number of iterations, does not seem to be essential. If random manifold is used for restricting the problem, this parameter can affect on

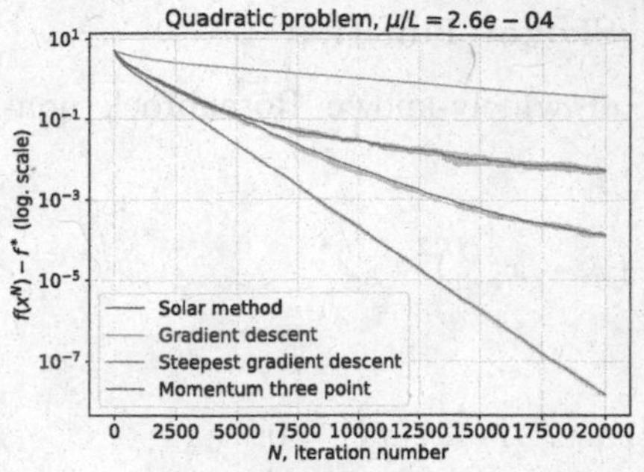

(a) Convergence curves Solar method, two-point zeroth-order gradient descent with or without line-search and Momentum three point method

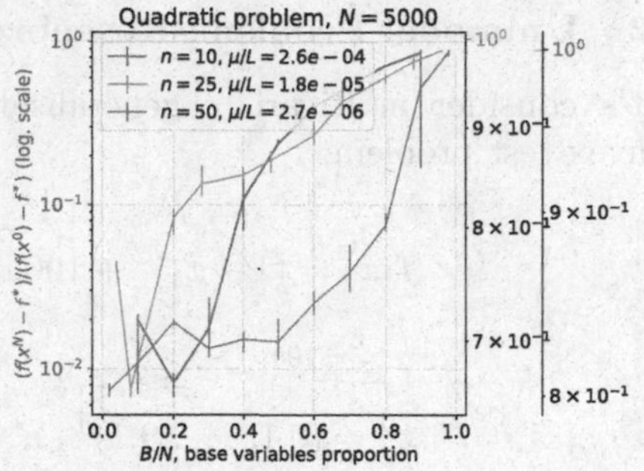

(b) Function suboptimality achieved by Solar method on different quadratic functions with different proportion of base variables B/N

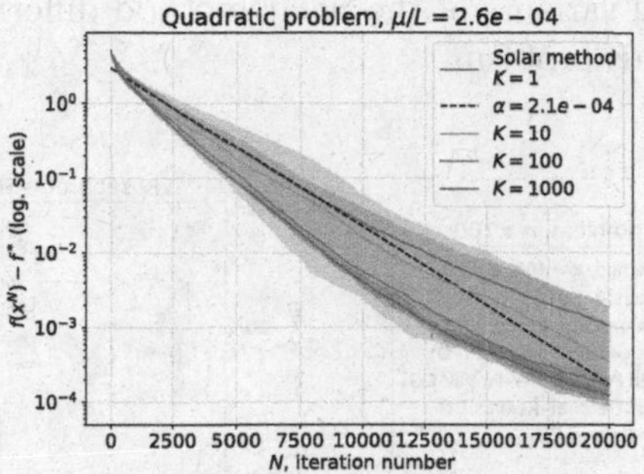

(c) Convergence curves of Solar method for different $K = 1, 10, 100, 1000$ and estimation of order of linear convergence

Fig. 1. Practical efficiency of Solar method on quadratic functions

variability of restricted problems, but linear subspaces are invariant on choice of base variables. Nevertheless, due to artefacts of random sampling (even in spherically symmetric sampling which we used) parameter K affects practical performance. Figure 1c shows that the lower is number of inner iterations per chosen set of base variables, the less is dispersion of trajectories and the better resulting function value is on average.

Linear Convergence in the Beginning. Another phenomenon Fig. 1c shows is that in the beginning the convergence rate of Solar method is almost linear. We estimate order of convergence α by linear fit to show that it is close to $1/\kappa$, as for classic gradient descent.

4.2 Unimodal Problem: Rosenbrock–Skokov Function

Let's consider multivariate generalisation of widely-known Rosenbrock non-convex test problem:

$$f(x) = (1 - x_1)^2 + 100 \cdot \sum_{i=2}^{100} \left(x_i - x_{i-1}^2\right)^2$$

$$x \in \mathbb{R}^{100}, -3 \leq x \leq 3$$

$$x^0 = (0.1, \ldots, 0.1)^\top, x^* = (1, \ldots, 1)^\top, f(x^*) = 0$$

For this example, we consider variant of Solar method which uses gradient, together zeroth-order variant and compare them with first-order algorithms of conjugate gradient type. In particular, we run Fletcher–Reeves [2] and Polak–Ribiere–Polyak [17,18] conjugate gradient methods with or without restarts [19]. Besides, we test several variants of the Solar method differing in a way to draw random subspace on this problem.

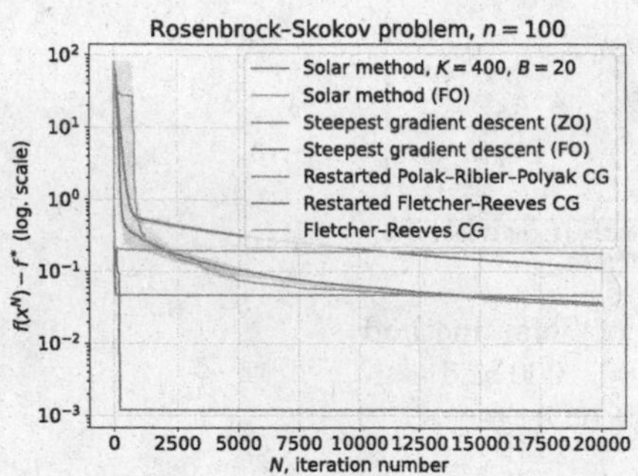

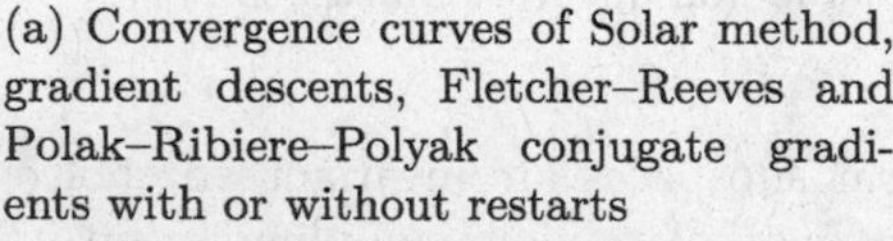

(a) Convergence curves of Solar method, gradient descents, Fletcher–Reeves and Polak–Ribiere–Polyak conjugate gradients with or without restarts

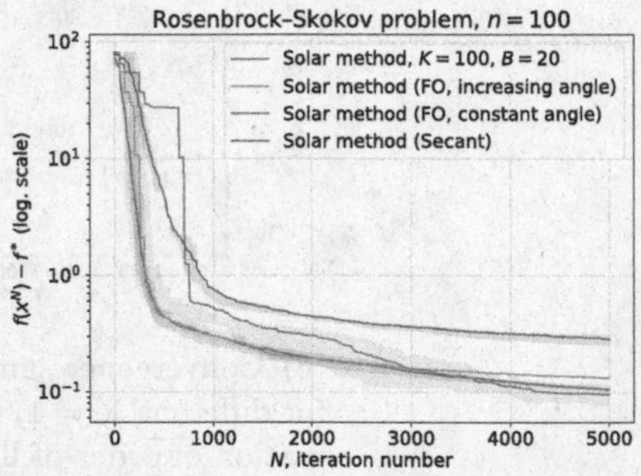

(b) Convergence curves of several variants of Solar method with different ways to choose random subspace: sampling in cone around gradient or secant direction through two points

Fig. 2. Practical efficiency of Solar method on Rosenbrock–Skokov function

The description of the results presented on Fig. 2 follows.

Comparison of Algorithms. As one can see from the Fig. 2a, both variants of Solar method overtake Polak–Ribiere–Polyak algorithm and unrestarted Fletcher–Reeves algorithm (restarted Fletcher–Reeves algorithm is the only algorithm overtaking Solar method, and far ahead all other methods). Despite the instant convergence of conjugate gradient methods in the beginning, they stuck in a valley, while Solar method preserves steady convergence rate.

Comparison of Variants of Solar Method. There were tested four variants of Solar method: vanilla, in which subspace is defined by random linear dependency between base and the rest of variables, first-order, in which linear dependencies are in cone with constant angle around gradient, similar but with angle increasing over time and secant, in which two best points are maintained and cone is taken around segment connecting them. Briefly, the fastest variant is first-order one with constant angle, but the performance of all the variants is only slightly different, so in practice one should choose variant according to its efficiency on particular problem.

4.3 Global Problems: Rastrigin and DeVilliersGlasser02 Functions

Let's consider two global optimisation problems from benchmark [3] to assess the performance of Solar method in multi-extrema setting. First problem is for classical Rastrigin function, with 200 variables:

$$f(x) = 10 \cdot 200 \cdot \sum_{i=1}^{200} (x_i^2 - 10\cos(2\pi x_i))$$

$$x \in \mathbb{R}^{200}, -5.12 \le x \le 5.12$$

$$x^0 = (5, \ldots, 5)^\top, \; x^* = (0, \ldots, 0)^\top, \; f(x^*) = 0$$

and second is the hardest problem in [3], for DeVilliersGlasser02 function of only 5 variables:

$$f(x) = \sum_{i=1}^{24} \left(x_1 x_2^{t_i} \tanh(x_3 t_i + \sin(x_4 t_i)) \cos(t_i e^{x_5}) - y_i\right)^2,$$

where $t_i = 0.1(i-1)$, $y_i = 53.81 \cdot 1.27^{t_i} \tanh(3.012 t_i + \sin(2.13 t_i)) \cos(t_i e^{0.507})$

$x \in \mathbb{R}^5, 1 \le x \le 60$

$x^0 = (30, \ldots, 30)^\top, x^* = (53.81, 1.27, 3.012, 2.13, 0.507)^\top, f(x^*) = 0$

For this example, we compare Solar method with two classic and practically efficient algorithms of global optimisation: Simulated Annealing [24], and Monotonic Sequence Basin Hopping [10].

The results presented on Fig. 3 can be described together as both of experiments compare algorithms' performance. In both cases Solar method turns out to be more efficient in search of deeper local optimum. Note, that for better performance dimensionality of subspaces should be chosen not very small. The explanation of the computational fact that Solar method is more efficient than standard hopping algorithms is that Solar method is less sensitive to choice of hyperparameters. There is no doubt that after grid search of jump length and proper temperature decreasing policy classic algorithms will achieve at least not worse function value in the same time. But this search of hyperparameters is time-consuming. In turn, Solar method has a great advantage that allows it jump farther: instead of localisation of the jump by limiting its length, Solar method restricts it on low-dimensional subspace, which does not take away the opportunity to find a better local optimum far from current best point.

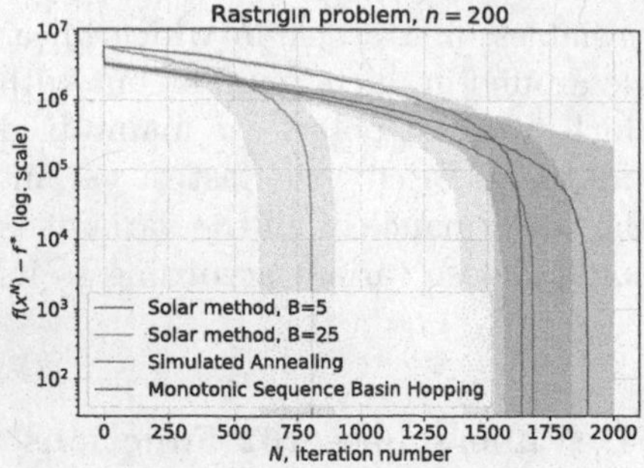

(a) Convergence curves of Solar method, Simulated Annealing and MSBH on Rastrigin problem

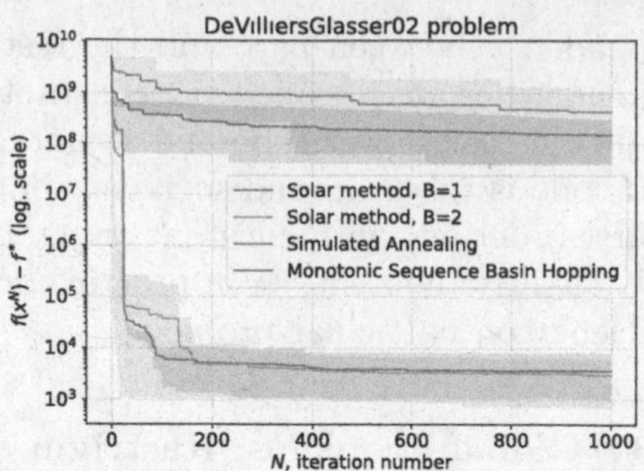

(b) Convergence curves of Solar method, Simulated Annealing and MSBH on DeVilliersGlasser02 problem

Fig. 3. Practical efficiency of Solar method on global optimisation test functions

5 Discussion

This paper proposes heuristic global optimisation algorithm called Solar method, which is based on optimisation over randomly drawn subspaces to search for a point step to. The algorithm demonstrated competitive or leading convergence rate on convex, unimodal and global test functions in comparison with algorithms known as the best in practice general-purpose methods in corresponding classes of problems.

The proposed algorithm is an instance of more general scheme, so its further development shall be aimed at more delicate exploitation of advantages of the scheme itself. Theoretical and empirical study of this scheme is one barely permeable but, we believe, very perspective direction for the future work.

From the practical point of view, the usability of the algorithm itself shall be improved. Firstly, more detailed study of how the inexactness in solving the auxiliary problems accumulate and affect on overall performance is required. Secondly, current technical solution of maintaining the k best points is too time-consuming, so one needs to find more convenient data structure for this purpose.

Finally, algorithm can be extended in several directions. Firstly, it is needed to explore more options of choice of random subspace and consider random manifold generality. Secondly, the algorithm can use itself as an algorithm to solve restricted problems, which gives complex multi-level algorithm with unknown performance and properties.

References

1. Cartis, C., Fowkes, J., Shao, Z.: Randomised subspace methods for non-convex optimization, with applications to nonlinear least-squares. arXiv preprint arXiv:2211.09873 (2022)

2. Fletcher, R., Reeves, C.M.: Function minimization by conjugate gradients. Comput. J. **7**(2), 149–154 (1964)
3. Gavana, A.: Global optimization benchmarks and ampgo (2018)
4. Gorbunov, E., Bibi, A., Sener, O., Bergou, E.H., Richtárik, P.: A stochastic derivative free optimization method with momentum. arXiv preprint arXiv:1905.13278 (2019)
5. Gower, R., Kovalev, D., Lieder, F., Richtárik, P.: RSN: randomized subspace Newton. In: Advances in Neural Information Processing Systems 32 (2019)
6. Gower, R.M., Richtárik, P., Bach, F.: Stochastic quasi-gradient methods: Variance reduction via Jacobian sketching. Math. Program. **188**, 135–192 (2021)
7. Grishchenko, D., Iutzeler, F., Malick, J.: Proximal gradient methods with adaptive subspace sampling. Math. Oper. Res. **46**(4), 1303–1323 (2021)
8. Jones, D.R.: A taxonomy of global optimization methods based on response surfaces. J. Global Optim. **21**, 345–383 (2001)
9. Kozak, D., Becker, S., Doostan, A., Tenorio, L.: A stochastic subspace approach to gradient-free optimization in high dimensions. Comput. Optim. Appl. **79**(2), 339–368 (2021)
10. Leary, R.H.: Global optimization on funneling landscapes. J. Global Optim. **18**(4), 367 (2000)
11. Nelder, J.A., Mead, R.: A simplex method for function minimization. Comput. J. **7**(4), 308–313 (1965)
12. Nemirovskij, A.S., Yudin, D.B.: Problem complexity and method efficiency in optimization. Wiley-Interscience (1983)
13. Nesterov, Y., Spokoiny, V.: Random gradient-free minimization of convex functions. Found. Comput. Math. **17**, 527–566 (2017)
14. Nesterov, Y.E.: Efficiency of coordinate descent methods on huge-scale optimization problems. SIAM J. Optim. **22**(2), 341–362 (2012)
15. Pasechnyuk, D.A., Gornov, A.: Solar method for non-convex problems: hypothesizing approach to an optimization (2022). http://dmivilensky.ru/preprints/Solar%20method%20for%20non-convex%20problems.pdf
16. Patrascu, A., Necoara, I.: Efficient random coordinate descent algorithms for large-scale structured nonconvex optimization. J. Global Optim. **61**(1), 19–46 (2015)
17. Polak, E., Ribiere, G.: Note sur la convergence de méthodes de directions conjuguées. Revue française d'informatique et de recherche opérationnelle. Série rouge **3**(16), 35–43 (1969)
18. Polyak, B.T.: The conjugate gradient method in extremal problems. USSR Comput. Math. Math. Phys. **9**(4), 94–112 (1969)
19. Powell, M.J.D.: Restart procedures for the conjugate gradient method. Math. Program. **12**, 241–254 (1977)
20. Richtárik, P., Takáč, M.: Iteration complexity of randomized block-coordinate descent methods for minimizing a composite function. Math. Program. **144**(1–2), 1–38 (2014)
21. Törn, A., Zilinskas, A.: Global optimization, vol. 350. Springer (1989)
22. Wales, D.J., Doye, J.P.: Global optimization by basin-hopping and the lowest energy structures of lennard-jones clusters containing up to 110 atoms. J. Phys. Chem. A **101**(28), 5111–5116 (1997)
23. Xu, Y., Yin, W.: A globally convergent algorithm for nonconvex optimization based on block coordinate update. J. Sci. Comput. **72**(2), 700–734 (2017)

24. Xue, G.L.: Parallel two-level simulated annealing. In: Proceedings of the 7th International Conference on Supercomputing, pp. 357–366 (1993)
25. Zhigljavsky, A.A.: Theory of global random search, vol. 65. Springer Science & Business Media (2012)

Global Optimization

Optimizing Parallelization Strategies for the Big-Means Clustering Algorithm

Ravil Mussabayev[1,2,3(✉)] and Rustam Mussabayev[3]

[1] Department of Mathematics, University of Washington, Padelford Hall C-138, Seattle, WA 98195-4350, USA
ravmus@uw.edu

[2] Huawei Russian Research Institute, Smolenskaya Square 5, 121099 Moscow, Russia

[3] Institute of Information and Computational Technologies, Laboratory for Analysis and Modeling of Information Processes, Pushkin str. 125, 050010 Almaty, Kazakhstan
rustam@iict.kz

Abstract. This study focuses on the optimization of the Big-means algorithm for clustering large-scale datasets, exploring three distinct parallelization strategies. We conducted extensive experiments to assess the computational efficiency, scalability, and clustering performance of each approach, revealing their benefits and limitations. The paper also delves into the trade-offs between computational efficiency and clustering quality, examining the impacts of various factors. Our insights provide practical guidance on selecting the best parallelization strategy based on available resources and dataset characteristics, contributing to a deeper understanding of parallelization techniques for the Big-means algorithm.

Keywords: Big data · Clustering · Minimum sum-of-squares · Divide and conquer algorithm · Decomposition · K-means · K-means++ · Global optimization · Unsupervised learning

1 Introduction

Clustering is a fundamental task that involves identifying groups of similar objects within a given set. This problem is challenged by the rapid growth of digital data and has applications in many domains, such as image and video analysis [19], customer segmentation [5], information retrieval [8], anomaly detection [18], pattern recognition and classification [16], vector quantization and data compression [20], natural language processing [1], bioinformatics [16], gene expression analysis [10], network and traffic analysis [7], time series analysis [15], medical diagnosis [13], social media analysis [21].

There are many models for cluster analysis. The most basic and extensively studied one is the minimum sum-of-squares clustering (MSSC) [2]. Given a set of m data points $X = \{x_1, \ldots, x_m\}$ in the Euclidean space $\mathbb{R}^n$, it solves the problem of finding k cluster centers (centroids) $C = (c_1, \ldots, c_k) \in \mathbb{R}^{n \times k}$ that

N. Olenev et al. (Eds.): OPTIMA 2023, CCIS 1913, pp. 17–32, 2024.
https://doi.org/10.1007/978-3-031-48751-4_2

minimize the sum of squared distances from each data point x_i to its nearest cluster center c_j:

$$\min_{C} \quad f(C, X) = \sum_{i=1}^{m} \min_{j=1,\ldots,k} \|x_i - c_j\|^2 \tag{1}$$

where $\|\cdot\|$ stands for the Euclidean norm. Equation (1) is the objective function, which is called the sum-of-squared distances. Each solution C uniquely defines the corresponding partition $X = X_1 \cup \ldots \cup X_k$, where each X_j stands for the set of points closest to centroid c_j than to any other one. For general k and m, the MSSC is known to be an NP-hard problem [2].

MSSC can be considered as a global optimization problem whose objective is to partition the dataset into subsets, called clusters. The major advantage of MSSC is that minimizing the single objective, as expressed in (1), leads to the simultaneous minimization of similarity between objects within the same cluster, while maximizing the dissimilarity between objects in different clusters. Consequently, the resulting clustering accuracy, as measured by (1), is a crucial criterion for evaluating clustering algorithms.

It has been demonstrated that global minimizers more accurately reflect the clustering structure of a given dataset [9]. The pursuit of global minimizers in MSSC is complicated by the high non-convexity of the objective function, which leads most MSSC-solving algorithms to many locally minimizing traps. Although various optimization methods have been proposed to address this high non-convexity challenge, each approach has its strengths and weaknesses, and there is no one-size-fits-all solution. As a result, further research is needed to develop more efficient and robust techniques for locating global minimizers in the context of MSSC and big data.

As of today, the Big-means algorithm is considered state-of-the-art among MSSC algorithms for clustering big data [14]. The decomposition principle is at the heart of the Big-means algorithm [14]. This principle not only serves as the algorithm's cornerstone but also facilitates its efficient parallelization. Parallelization is one of the core approaches employed for big data clustering. However, the original paper [14] did not discuss this critical aspect of the proposed algorithm in detail. In the current work, we endeavor to explore this dimension comprehensively in order to significantly boost the performance of the Big-means algorithm for big data clustering.

Our paper has the following roadmap. Section 2 reviews the main developments and strategies related to parallelization of clustering algorithms. Section 3 describes the main theses and contributions of this study, including the proposed parallelization strategies, as well as discusses the nuances involved in the implementation of modern high-performance techniques in practice. Section 4 describes the experimental setting used in our study. Finally, Sect. 5 brings a detailed explanation and interpretation of the obtained experimental results, while containing some practical guidelines for choosing the optimal parallelization strategy, which will be useful for practitioners in the field of big data clustering. Also, the final section concludes our work and determines promising future research directions.

2 Related Works

In the realm of big data clustering, numerous methods have been developed to tackle the challenges posed by the sheer volume, high dimensionality, and streaming nature of the data.

Traditional clustering methods such as K-means and its improved version, K-means++ [3], are widely used for their simplicity and effectiveness. However, their direct application to big data can be problematic due to their high time complexity and the requirement to store all data in memory, making them less feasible for big data applications.

To overcome these challenges, several parallel and distributed clustering algorithms have been proposed. The MapReduce framework is commonly employed to enable scalable and efficient computation [6]. The K-means algorithm was parallelized using MapReduce by Zhao et al. [22], with the process being significantly sped up by distributing the computation across multiple nodes.

For dealing with data that cannot be stored in memory, one popular approach is the mini-batch K-means algorithm, an online version of K-means which uses random subsets (mini-batches) of the data at each iteration, providing significant computation speedups with a minor sacrifice in clustering quality [17].

A step forward was made by Bahmani et al. [4], who proposed the scalable K-means++ algorithm, combining the advantages of K-means++ and mini-batch K-means to offer a solution that is both fast and provides high-quality clusters.

Big data is often seen as a challenge to be overcome by most standard algorithms and alternative heuristics, rather than an advantage to be leveraged to improve clustering results. Therefore, there is a significant need for new big data clustering algorithms that are relatively simple, effective for big data processing and able to use big data as an advantage to enhance clustering results. These algorithms should balance simplicity, result quality, and convergence speed, and perform global search for the optimal solution without relying on known global optimization metaheuristics. Our proposed Big-means algorithm [14], utilizing parallel processing and intelligent sample selection, seeks to fill this gap.

3 Methodology

3.1 Big-Means Algorithm

The Big-means algorithm is designed for solving the large-scale Minimum Sum-of-Squares Clustering (MSSC) problem. It uses a heuristic approach and focuses on computational efficiency and solution quality by working with a subset of the data in each iteration, instead of the entire dataset. The pseudocode of Big-means is shown in Algorithm 1.

Algorithm 1: Big-means Clustering

Result: Compute the final centroids C and cluster assignments Y for a dataset X using the Big-means algorithm.

Initialization:
Initialize all k centroids C as degenerate;
Set iteration counter (or CPU time) $t = 0$;
while $t < T$ **do**
    Draw a random sample S of size s from X;
    for *each centroid c in C* **do**
        if *c is a degenerate cluster* **then**
            Reinitialize c using K-means++ on S;
        end
    end
    Compute new centroids C_{new} using K-means on S with initial centroids C;
    if $f(C_{new}, S) < f(C, S)$ **then**
        $C \leftarrow C_{\text{new}}$;
    end
    $t \leftarrow t + 1$;
end
$Y \leftarrow$ Assign each point in X to nearest centroid in C;

To start, the algorithm randomly creates samples of size s from the given dataset X, where s is much smaller than the total number of feature vectors m. The first sample is clustered using the K-means++ algorithm, which sets the initial configuration of the centroids denoted as C. As the algorithm progresses, the centroids are updated with each new sample that improves the clustering solution based on the objective function evaluated on the sample. This approach follows a "keep the best" principle, ensuring that the best solution found so far is always prioritized. Also, it is important to note that the current best (incumbent) solution C is replaced or kept intact in each iteration, without growing in size as the number of iterations increases.

Big-means addresses degenerate clusters (also called empty clusters) differently than traditional approaches. When all data points initially associated with a cluster are reassigned to other clusters during the K-means process, Big-means reinitializes those empty clusters using the K-means++ algorithm. This introduces new potential cluster centers and enhances the overall clustering solution by providing more opportunities to minimize the objective function.

The shaking procedure is a crucial element of the Big-means algorithm. It refers to creating new samples in each iteration, which perturbs the incumbent solution and introduces variability into the clustering results. By considering the full dataset as a cloud of points in an n-dimensional space, each sample represents a sparse approximation of this cloud. This brings diversity and adaptability into the clustering process. Also, it prevents the algorithm from being trapped in suboptimal solutions.

Once a stop condition is met, such as reaching a limit on computational resources (e.g., the elapsed CPU time) or processing a maximum number of samples, the algorithm distributes the data points in the entire dataset into clusters based on their proximity to the final set of centroids C. In some cases, this final step may be skipped if only the centroids or a limited sample of object assignments are needed. The time complexity of each Big-means iteration is $\mathcal{O}(s \cdot n \cdot k)$ (where k is the number of clusters), making Big-means much faster than traditional K-means or K-means++ when working with a small sample from a larger dataset.

In essence, the Big-means algorithm attempts to balance the efficiency of the K-means algorithm with the need for quality clustering results. It achieves this by applying the K-means algorithm iteratively on random subsets of the data, progressively refining the location of the centroids, and effectively handling degenerate clusters.

The Big-means algorithm is a promising approach for clustering large datasets, providing scalability, efficiency, and robustness.

3.2 Parallelization Strategies for the Big-Means Algorithm

Implementing parallelization is a crucial factor for algorithms dealing with the Minimum Sum-of-Squares Clustering (MSSC) problem, given its NP-hard nature and its typical application to big data. High-Performance Computing (HPC) technologies, which leverage the computational power of supercomputers and computer clusters, offer a robust platform for such algorithms. The scalability of these HPC algorithms is largely facilitated by the parallel processing of the data across multiple computing nodes, computers, or processors.

With regards to the Big-means algorithm, its structure makes it well-suited for effective parallelization, which can be achieved through several strategies:

1. Inner parallelism (Big-means-inner): In this approach, incoming data samples are processed in a sequence, while the K-means and K-means++ processes involved in clustering are parallelized. This means that the main loop of the Big-means algorithm remains sequential, but all the internal loops of K-means and K-means++ are executed in parallel;
2. Competitive parallelism (Big-means-competitive): In this strategy, 'worker' processes start simultaneously on each available processor by initializing their own data samples from the big dataset. Each worker keeps clustering its individual stream of data samples completely independently from other workers, employing the sequential versions of K-means and K-means++. For every iteration, workers only use their own preceding best centroids for initialization. This parallelization mode is termed competitive because each worker operates independently from the others. When the stopping criterion for every worker is met, the best solution among all workers is selected;
3. Collective parallelism (Big-means-collective): Similar to competitive parallelism, workers begin clustering their individual data samples in parallel on each available processor. However, after independently initializing their first

sample, each worker initializes every subsequent sample using the best set of centroids observed so far from all previous iterations across all workers. This parallelization mode is termed collective since the workers share information about the best solutions. When the stopping criterion for every worker is met, the best solution among all workers is chosen.

Algorithm 2: Competitive Big-means Clustering

Result: Compute the final centroids C and cluster assignments Y for a dataset X using the competitive Big-means algorithm.

```
Initialization:
C_w ← Mark all k centroids as degenerate for each worker w;
t_w ← 0 for each worker w;
while t_w < T for any worker w do
    for each parallel worker w do
        S_new,w ← Random sample of size s from X;
        for each c ∈ C_w do
            if c is a degenerate cluster then
                Reinitialize c using K-means++ on S_w;
            end
        end
        C_new,w ← K-means clustering on S_w with initial centroids C_w;
        if f(C_new,w, S_new,w) < f(C_w, S_w) then
            C_w ← C_new,w;
        end
        t_w ← t_w + 1;
    end
end
C_best ← Centroids of the worker with the smallest objective function value;
Y ← Assign each point in X to nearest centroid in C_best;
```

The source code for the Big-means algorithm, which includes implementations of various parallelization strategies, is available at https://github.com/R-Mussabayev/bigmeans/.

In this article, we focus solely on researching the efficiency of various strategies of parallel interaction between individual workers. We assume that each worker has equal access to the full-sized dataset and can independently draw samples from it. For the sake of simplicity, in this study we are not exploring various available opportunities for further optimization of the algorithm, particularly those concerning distributed data storage across different nodes of the computing system. Such optimizations merit a separate study.

3.3 Parallelization Nuances in the Big-Means Algorithm

The Big-means algorithm, which is partitioning-based clustering method, features steps that are inherently suited for parallelization. Specifically, two main components lend themselves to concurrent execution: distance computation and centroid update.

Algorithm 3: Collective Big-means Clustering

Result: Compute the final centroids C and cluster assignments Y for a dataset X using the collective Big-means algorithm.

```
Initialization:
C_w ← Mark all k centroids as degenerate for each worker w;
t_w ← 0 for each worker w;
while t_w < T for any worker w do
    for each parallel worker w do
        S_w ← Random sample of size s from X;
        C_best ←
        Centroids of the worker with the smallest objective function value;
        for each c ∈ C_best do
            if c is a degenerate cluster then
                Reinitialize c using K-means++ on S_w;
            end
        end
        C_new,w ← K-means clustering on S_w with initial centroids C_best;
        if f(C_new,w, S_w) < f(C_best, S_best) then
            C_w ← C_new,w;
        end
        t_w ← t_w + 1;
    end
end
C_best ← Centroids of the worker with the smallest objective function value;
Y ← Assign each point in X to nearest centroid in C_best;
```

In the distance computation phase, the algorithm applies K-means++ to a sample, computing the distances from data points to centroids. The independence across data points in this operation offers the potential for simultaneous execution. The potential for parallelism is not just confined to the initial K-means++ computation, but extends to the computations on the subsequent data samples.

The second component, the centroid update, occurs in each subsequent iteration following the computation of new centroids C_{new}. During this step, the algorithm calculates the distance from all sample points to the new centroids to redefine the clusters. Just as the distance computation phase, this step is also highly parallelizable.

However, while parallel computing can significantly accelerate these computations, it also introduces certain overheads. Specifically, balancing the load across different cores or processors is crucial to mitigate communication and synchronization overheads. For instance, if the number of samples s is substantially less than the number of available processors, idle processors could arise during the sample clustering step, resulting in inefficient resource utilization.

Aside from computational efficiency and scalability, the correct and robust implementation of parallel strategies in the Big-means algorithm necessitates careful consideration of concurrency control. Specifically, race conditions pose significant challenges. Race conditions are situations where a program's outcome varies depending on the sequence or timing of the execution of threads. In the context of Big-means, multiple threads might simultaneously read and write to shared memory locations. For instance, this can happen when updating centroid information or membership of data points. If not properly handled, these concurrent operations can lead to inconsistent and erroneous clustering results.

To prevent such situations, synchronization mechanisms such as locks, semaphores, or atomic operations could be used. These constructs ensure that only one thread accesses shared data at a time, preserving data consistency and integrity. Furthermore, it might be beneficial to design the algorithm in such a way that reduces the need for shared memory access, thus minimizing the potential for race conditions.

However, these measures should be implemented judiciously, as excessive synchronization can lead to thread contention, reducing parallel efficiency and potentially negating the benefits of parallelization. Striking the right balance between data protection and computational efficiency is a critical aspect of optimizing parallel strategies for Big-means. This underscores the interplay between software design and algorithmic considerations in the era of parallel computing.

Moreover, it is essential to underscore that while parallelization can enhance the performance of Big-means, the final clustering results should remain consistent, regardless of the number of processors deployed. That is, parallelization should primarily affect the computational speed rather than the resulting configuration of clusters. Therefore, our study underlines the importance of effective parallelization strategies in harnessing the full potential of the Big-means algorithm without compromising the accuracy of clustering results. However, it is worth noting that unlike other parallelized algorithms, this requirement is not mandatory for Big-means. This is because the more iterations it performs within a fixed time interval, the potentially higher accuracy it can achieve.

Most of the parallelization strategies used in Big-means provide a degree of robustness to the initialization of centroids. By allowing each worker to independently initialize the centroids at the starting point of the algorithm, the algorithm mitigates the impact of poor initial centroid selection, which is a common issue for the K-means clustering.

4 Experiments

The experiments were performed on a system running Ubuntu 22.04 64-bit, powered by an AMD EPYC 7663 56-Core Processor, with up to 16 cores used in our experiments. The system was equipped with 1.46 TB of RAM. The software stack consisted of Python 3.10.11 along with NumPy 1.24.3 and Numba 0.57.0. The Numba [12] package was used to accelerate Python code execution and facilitate parallelism. The use of Numba is particularly advantageous for these purposes due to its ability to compile Python code into machine code at runtime and its capabilities for executing code on multiple processors.

The performance of various parallelization versions of Big-means was evaluated on 19 publicly available datasets. Descriptions of these datasets are provided in Table 1, with more details available on their corresponding webpages listed in Table 2. In addition to these, four datasets were normalized, bringing the total to 23 datasets.

Table 1. Brief description of the datasets

Datasets	No. instances m	No. attributes n	Size $m \times n$	File size
CORD-19 Embeddings	599616	768	460505088	8.84 GB
HEPMASS	10500000	28	294000000	7.5 GB
US Census Data 1990	2458285	68	167163380	361 MB
Gisette	13500	5000	67500000	152.5 MB
Music Analysis	106574	518	55205332	951 MB
Protein Homology	145751	74	10785574	69.6 MB
MiniBooNE Particle Identification	130064	50	6503200	91.2 MB
MFCCs for Speech Emotion Recognition	85134	58	4937772	95.2 MB
ISOLET	7797	617	4810749	40.5 MB
Sensorless Drive Diagnosis	58509	48	2808432	25.6 MB
Online News Popularity	39644	58	2299352	24.3 MB
Gas Sensor Array Drift	13910	128	1780480	23.54 MB
3D Road Network	434874	3	1304622	20.7 MB
KEGG Metabolic Relation Network (Directed)	53413	20	1068260	7.34 MB
Skin Segmentation	245057	3	735171	3.4 MB
Shuttle Control	58000	9	522000	1.55 MB
EEG Eye State	14980	14	209720	1.7 MB
Pla85900	85900	2	171800	1.79 MB
D15112	15112	2	30224	247 kB

Table 2. Information about the used datasets

Datasets	URLs
CORD-19 Embeddings	https://www.kaggle.com/allen-institute-for-ai/CORD-19-research-challenge
HEPMASS	https://archive.ics.uci.edu/ml/datasets/HEPMASS
US Census Data 1990	https://archive.ics.uci.edu/ml/datasets/US+Census+Data+(1990)
Gisette	https://archive.ics.uci.edu/ml/datasets/Gisette
Music Analysis	https://archive.ics.uci.edu/ml/datasets/FMA%3A+A+Dataset+For+Music+Analysis
Protein Homology	https://www.kdd.org/kdd-cup/view/kdd-cup-2004/Data
MiniBooNE Particle Identification	https://archive.ics.uci.edu/ml/datasets/MiniBooNE+particle+identification
MFCCs for Speech Emotion Recognition	https://www.kaggle.com/cracc97/features
ISOLET	https://archive.ics.uci.edu/ml/datasets/isolet
Sensorless Drive Diagnosis	https://archive.ics.uci.edu/ml/datasets/dataset+for+sensorless+drive+diagnosis
Online News Popularity	https://archive.ics.uci.edu/ml/datasets/online+news+popularity
Gas Sensor Array Drift	https://archive.ics.uci.edu/ml/datasets/gas+sensor+array+drift+dataset
3D Road Network	https://archive.ics.uci.edu/ml/datasets/3D+Road+Network+(North+Jutland,+Denmark)
KEGG Metabolic Relation Network (Directed)	https://archive.ics.uci.edu/ml/datasets/KEGG+Metabolic+Relation+Network+(Directed)
Skin Segmentation	https://archive.ics.uci.edu/ml/datasets/skin+segmentation
Shuttle Control	https://archive.ics.uci.edu/ml/datasets/Statlog+(Shuttle)
EEG Eye State	https://archive.ics.uci.edu/ml/datasets/EEG+Eye+State
Pla85900	http://softlib.rice.edu/pub/tsplib/tsp/pla85900.tsp.gz
D15112	https://github.com/mastqe/tsplib/blob/master/d15112.tsp

All datasets comprised exclusively numeric features, without any missing values. They ranged from having as few as two attributes to as many as 5,000, and the number of instances varied from thousands (smallest being 7,797) to tens of millions (largest being 10,500,000). The datasets were deliberately unbalanced, with significant variance in their number of objects and features, as this allowed us to evaluate Big-means as a universal algorithm suitable for various dataset sizes. Furthermore, the use of the same methodology and datasets as those in Karmitsa et al. [11] enabled additional comparisons of our results.

Each of the 23 datasets was clustered using each algorithm n_{exec} times into clusters of sizes: 2, 3, 5, 10, 15, 20, 25. Following the methodology of the experiments in [11], some datasets ("Shuttle Control", "EEG Eye State" and their normalized versions) underwent additional clustering into 4 clusters. The total number of conducted experiments was 14,732. The results of each experiment were analyzed for relative error ε, spent CPU time t, and baseline time $\bar{t}$. For each algorithm and choice of a dataset and k, relative error (error gap) ε is defined as:

$$\varepsilon(\%) = \frac{100 \times (f - f^*)}{f^*}$$

where $f^* = f^*(X, k)$ is the best value of the objective function observed on the whole dataset X using k clusters from the available past experiments and history records. Throughout the experiments, we noted instances where certain algorithms attained accuracies surpassing the existing best-known benchmarks. Consequently, encountering negative error gaps (ε) in some cases is not unusual. In fact, this underscores the exceptional performance of an algorithm that has surpassed our initial expectations.

For the Big-means algorithm, CPU time t is considered to be the time of the last change of the incumbent solution C. The CPU time t is measured in seconds. For the parallel multi-worker Big-means versions, t is defined to be the time of the last change of the incumbent solution C_w for the worker w that achieved the best value of the objective function $f(C_w, S_w)$ on the sample S_w. For a fixed pair (X, k), baseline time $\bar{t}$ of algorithm A_i is defined to be the time of achieving the baseline sample objective value $\bar{f}_s$:

$$\bar{f}_s = \max_{j \in \{1,2,3,4\}} med\,(f_s^*(A_j)) \tag{2}$$

where $f_s^*(A_j)$ is the best value of the objective function evaluated on a sample using algorithm A_j. The median $med(\cdot)$ is calculated across n_{exec} executions of algorithm A_j for the fixed pair (X, k). This allows to cancel adverse random noise in the computations. Then, to capture the performance of the worst algorithm, the maximum of the medians is calculated. In the parallel multi-worker versions of Big-means, $\bar{t}$ represents the time t in seconds taken by the worker w that achieves the baseline sample objective value f_s^* before any other worker. For each k, the minimum, median, and maximum values of relative error ε and CPU time t were calculated relative to n_{exec} runs.

The maximum CPU time of Big-means was capped at t_{max} seconds. The clustering process on each sample was stopped when the number of iterations exceeded 300, or the relative tolerance between two consecutive objective function values was less than 10^{-4}. For K-means++, three candidate points were considered when generating the next centroid, choosing only the best one.

The rule of thumb was used to determine the sample size s of Big-means in our experiments. We adjusted s until neither increasing nor decreasing it improved the objective function values. For each pair (X, k), the choice of parameters t_{max} and n_{exec} precisely matched the values specified in the original Big-means paper [14].

Two preliminary experiments were conducted to establish necessary baselines. First, the number of used CPUs (workers) varied across the values 2, 4, 8, 12, 16, and the resulting relative error ε and CPU time t were measured. For this preliminary experiment, the number of executions n_{exec} of each algorithm A_i was set to 3 for every pair (X, k). We observed that the results were quite robust with respect to the CPU number, so we believed that this choice of n_{exec} was enough for obtaining statistically significant results. The values for n_{exec} parameter for the main experiment were much larger and match the values provided in [14]. The outcome of the first preliminary experiment allowed us to determine the optimal number of workers for the subsequent experiments. We established that having 8 CPUs would be the optimal value for all subsequent experiments. In this context, optimal selection means that this value achieves the best balance between solution quality and execution time simultaneously for all algorithms, allowing for fair comparisons under equal conditions.

In the second preliminary experiment, three parallelized versions of Big-means, along with the fully sequential version, were run over all datasets according to the methodology described above. Then, the baseline sample objective values $\overline{f}_s$ were computed according to definition (2). These values served as baselines in the main experiment.

The main experiment was conducted using the baselines established in the preliminary experiments. Tables 3 and 4 display the summarized results. Each row represents a dataset, while the 'MinGap', 'MedianGap', and 'MaxGap' columns represent the minimum, median, and maximum values of the metric across different runs for the current algorithm on the given dataset. These scores were summarized in the last rows using mean values, an approach justified due to its sensitivity to outliers, providing a comprehensive measure of our algorithms' performance across all datasets. In both tables, the highest resulting values for each experiment (algorithm, data pair (X, k)) are displayed in bold. In Table 3, a success is counted when the performance of an algorithm aligns with the top outcome among all algorithms in the present experiment.

Table 3. Relative clustering accuracies ϵ (%) for different algorithms. The highest accuracies for each experiment (algorithm, data pair (X, k)) are displayed in bold. Success is indicated when an algorithm's performance matches the best result among all algorithms for the current experiment.

Dataset	Big-means-sequential				Big-means-inner				Big-means-competitive				Big-means-collective			
	#Succ	MinGap	MedianGap	MaxGap	#Succ	MinGap	MedianGap	MaxGap	#Succ	MinGap	MedianGap	MaxGap	#Succ	MinGap	MedianGap	MaxGap
CORD-19 Embeddings	5/49	0.01	0.32	2.7	9/49	0.0	0.14	4.03	9/49	0.01	**0.05**	**0.43**	11/49	0.01	0.09	0.49
HEPMASS	5/49	0.0	0.33	1.7	4/49	0.0	0.27	1.27	19/49	**−0.06**	0.12	0.34	10/49	0.0	0.17	1.25
US Census Data 1990	22/140	0.01	3.75	163.36	15/140	0.03	4.05	163.52	35/140	0.05	**1.94**	**5.62**	24/140	0.03	2.74	8.41
Gisette	14/105	−1.72	**0.01**	0.43	18/105	−1.65	**0.01**	0.55	15/105	−1.68	**0.01**	0.28	25/105	**−1.76**	**0.01**	**0.19**
Music Analysis	22/140	0.02	1.16	26.17	40/140	**0.01**	1.02	9.24	31/140	0.04	**0.62**	**3.29**	26/140	0.03	0.87	4.01
Protein Homology	24/105	0.03	0.96	18.83	20/105	**0.01**	**0.64**	18.53	31/105	0.07	0.84	3.12	15/105	0.06	1.01	2.89
MiniBooNE Particle Identification	14/105	−0.49	0.18	116.8	22/105	−0.45	0.06	21.68	17/105	−0.51	**0.01**	4031563.99	31/105	−0.48	**0.01**	**1.0**
MiniBooNE Particle Identification (normalized)	29/140	**0.0**	0.8	7.12	32/140	**0.0**	0.66	7.06	27/140	0.01	0.68	3.64	32/140	**0.0**	**0.52**	**3.16**
MFCCs for Speech Emotion Recognition	21/140	**0.01**	1.21	5.62	30/140	**0.01**	0.94	5.03	24/140	0.02	0.13	2.29	33/140	0.02	**0.11**	**2.14**
ISOLET	10/105	−0.1	0.82	2.59	14/105	−0.01	0.59	3.39	29/105	**−0.15**	0.23	**1.44**	25/105	−0.1	0.32	1.89
Sensorless Drive Diagnosis	42/280	−2.41	1.45	100.19	40/280	**−2.42**	1.12	100.2	63/280	−2.41	0.03	21.95	75/280	**−2.42**	**−0.0**	**8.31**
Sensorless Drive Diagnosis (normalized)	39/280	0.02	3.59	14.45	52/280	**0.01**	3.47	13.83	77/280	0.02	**1.9**	7.74	58/280	0.02	2.46	**7.24**
Online News Popularity	36/140	0.0	2.96	29.08	31/140	**−0.15**	2.57	31.05	25/140	0.01	**0.87**	**9.16**	27/140	0.0	0.97	20.25
Gas Sensor Array Drift	31/210	−0.08	2.83	33.08	31/210	−0.79	3.42	33.04	56/210	**−0.85**	0.15	8.46	34/210	−0.77	0.4	12.48
3D Road Network	53/280	**0.0**	0.24	5.87	67/280	**0.0**	**0.19**	5.49	65/280	**0.0**	**0.19**	2.85	49/280	**0.0**	0.21	**2.67**
Skin Segmentation	29/210	−1.18	4.83	18.67	33/210	**−1.22**	3.95	22.34	40/210	−1.15	**0.2**	**9.12**	53/210	−1.06	0.23	12.36
KEGG Metabolic Relation Network (Directed)	24/140	−0.97	2.55	124.91	19/140	−1.09	2.61	124.83	24/140	−1.27	0.04	157.01	34/140	**−1.29**	0.03	**17.47**
Shuttle Control	18/120	−1.87	5.03	78.35	19/120	−2.81	4.99	91.65	23/120	**−3.62**	1.41	**24.3**	27/120	−3.11	**0.48**	152.42
Shuttle Control (normalized)	28/160	0.03	2.09	31.05	24/160	**0.02**	1.97	31.98	43/160	0.07	1.5	**9.26**	27/160	0.07	1.55	16.75
EEG Eye State	34/160	−0.01	0.57	29.91	39/160	−0.01	0.24	29.91	34/160	**−0.1**	0.02	29.91	23/160	−0.01	0.02	**4.25**
EEG Eye State (normalized)	42/240	−0.33	0.42	65.96	34/240	−0.34	0.43	185.07	56/240	−0.33	**0.0**	65.96	41/240	−0.34	**0.0**	**0.75**
Pla85900	45/280	**0.0**	0.36	2.85	34/280	**0.0**	0.42	2.8	71/280	**0.0**	**0.12**	**1.46**	55/280	**0.0**	0.24	2.01
D15112	19/105	**0.0**	0.71	4.66	13/105	**0.0**	0.16	16.71	30/105	**0.0**	**0.1**	1.78	23/105	**0.0**	0.13	2.12
Overall Results	606/3683	−0.39	1.82	38.45	640/3683	−0.47	1.66	40.14	844/3683	**−0.52**	**0.69**	175301.45	758/3683	−0.48	0.72	**12.37**

Table 4. Resulting clustering times $\bar{t}$ (sec.) with respect to baseline sample objective values $\overline{f}_s$. The lowest clustering times for each experiment (algorithm, data pair (X, k)) are displayed in bold.

Dataset	Big-means-sequential			Big-means-inner			Big-means-competitive			Big-means-collective		
	Min	Median	Max	Min	Median	Max	Min	Median	Max	Min	Median	Max
CORD-19 Embeddings	**0.76**	20.26	36.59	**0.76**	**10.66**	**36.19**	2.87	18.93	41.22	4.25	14.53	36.71
HEPMASS	0.89	8.63	29.73	0.8	2.89	29.79	0.91	3.32	27.94	0.78	**2.76**	19.33
US Census Data 1990	0.11	0.67	2.94	0.1	**0.42**	3.07	**0.07**	0.82	3.06	**0.07**	0.46	**2.32**
Gisette	3.53	21.1	40.85	**1.15**	**5.93**	**9.26**	4.94	32.24	53.5	7.68	25.46	53.28
Music Analysis	0.22	4.09	9.49	**0.18**	**2.49**	**8.51**	0.6	3.37	9.27	0.36	3.4	10.43
Protein Homology	0.17	2.72	5.56	**0.07**	**1.1**	**3.17**	0.42	3.17	9.07	0.38	3.28	11.35
MiniBooNE Particle Identification	**0.16**	2.45	10.56	0.29	**1.07**	**2.82**	0.46	4.75	14.68	0.48	4.75	16.45
MiniBooNE Particle Identification (normalized)	0.02	0.46	1.32	**0.01**	**0.25**	**1.01**	0.04	0.77	1.6	0.02	0.82	1.57
MFCCs for Speech Emotion Recognition	0.08	0.62	1.05	**0.05**	**0.25**	**0.96**	0.08	0.86	1.88	0.07	0.7	1.74
ISOLET	0.15	1.84	4.91	**0.13**	**0.59**	**4.69**	0.4	2.85	5.49	0.35	2.32	5.72
Sensorless Drive Diagnosis	0.06	0.6	2.93	**0.03**	**0.26**	**1.09**	0.06	1.63	4.61	0.13	1.64	5.16
Sensorless Drive Diagnosis (normalized)	**0.01**	0.11	0.32	**0.01**	**0.1**	**0.29**	**0.01**	0.25	0.72	**0.01**	0.28	0.63
Online News Popularity	**0.02**	0.35	0.84	**0.02**	**0.23**	**0.71**	0.09	0.48	1.22	0.04	0.49	1.45
Gas Sensor Array Drift	0.05	0.74	2.06	**0.03**	**0.36**	**1.95**	0.29	1.24	2.34	0.26	1.09	2.89
3D Road Network	0.03	0.32	1.34	0.03	**0.28**	**0.71**	**0.02**	0.77	2.28	0.04	0.76	2.12
Skin Segmentation	**0.01**	0.08	**0.21**	**0.01**	**0.07**	**0.21**	0.02	0.18	0.5	**0.01**	0.1	0.41
KEGG Metabolic Relation Network (Directed)	**0.02**	0.38	1.19	0.06	**0.21**	**0.97**	0.07	0.7	2.06	0.15	0.67	1.92
Shuttle Control	0.05	0.43	1.41	0.03	**0.22**	**1.4**	0.18	0.6	1.49	**0.02**	0.48	**1.4**
Shuttle Control (normalized)	**0.0**	**0.16**	**0.4**	**0.0**	**0.16**	**0.4**	0.01	0.28	0.41	0.02	0.18	**0.4**
EEG Eye State	**0.02**	0.48	**1.44**	**0.02**	**0.44**	1.48	0.12	0.55	**1.44**	0.03	0.46	1.46
EEG Eye State (normalized)	**0.0**	0.03	0.99	**0.0**	**0.02**	0.99	**0.0**	0.18	0.71	**0.0**	0.19	**0.68**
Pla85900	**0.0**	0.53	1.49	0.01	**0.35**	1.5	0.02	0.54	**1.43**	0.01	0.47	1.45
D15112	0.03	**0.33**	1.5	**0.01**	0.36	1.41	**0.01**	0.42	**1.0**	0.02	0.4	1.44
Overall Results	0.28	2.69	6.92	**0.17**	**1.44**	**4.89**	0.51	2.98	8.17	0.66	3.0	7.84

5 Conclusion and Future Works

In this research, we compared and evaluated three parallelization strategies of the Big-means algorithm on a variety of datasets. Our comparative analysis was based on two key metrics: the resulting relative clustering accuracy ε and the runtime with respect to the baseline $\bar{t}$. These metrics allowed us to assess both the effectiveness and the efficiency of each parallelization strategy.

Based on the experimental results, it was observed that the Big-means-sequential strategy performed consistently worse than the Big-means-inner parallelization strategy for all datasets, as measured by both metrics. This indicates that utilizing a parallel version of Big-means is justified whenever feasible. Parallelization of Big-means offers a significant boost of the resulting clustering accuracy and convergence time.

The analysis of clustering times across different strategies provided valuable insights. Notably, the Big-means-inner strategy consistently demonstrated faster convergence to baselines compared to other strategies across most datasets. This effect was particularly pronounced for the largest datasets, as indicated in the first rows of Table 4. To achieve satisfactory clustering quality, significant sample sizes were employed for certain datasets, relative to the sizes of the datasets themselves. By leveraging parallelized K-means++ and K-means on each new sample, Big-means-inner achieved accelerated processing times compared to the sequential worker in other parallel Big-means strategies. These findings highlight the substantial impact of dataset characteristics on the efficiency-accuracy trade-off. These results also reinforce the importance of balancing sample size and quality of clusters. Larger sample sizes typically resulted in better approximation of the overall data distribution but also led to slower algorithmic performance.

Additionally, when considering the Big-means-competitive and Big-means-collective strategies, an interesting balance between parallelism and associated time overhead was observed. While employing these strategies resulted in improved final solutions, the coordination of multiple processors and the complexities introduced by the Numba library led to increased convergence times. On average, using 8 CPUs, the convergence times were up to twice as long compared to the Big-means-inner version.

However, if the convergence time is not a critical factor, both the Big-means-competitive and Big-means-collective strategies exhibit considerably improved global clustering quality compared to other versions of Big-means. On average, using 8 CPUs, the resulting quality is up to three times better.

The Big-means-competitive scheme demonstrates slightly better clustering quality compared to the Big-means-collective scheme. This improvement can be attributed to multiple initializations at the beginning. More precisely, the K-means algorithm is highly sensitive to the quality of initial initialization. There exist two ways to clustering of samples by multiple workers: either spending a significant amount of time on local search with a single initialization or conducting multiple different initializations. Our experiments suggest that the latter approach seems to be more advantageous than the former. At some point, Big-means-collective transitions to processing the results of a single initialization, although it is not guaranteed to be the best choice (it may only be good at the beginning). On the other hand, the alternative approach continues to cluster a multitude of different K-means++ initializations, from which the best one is selected at the end.

In conclusion, this study reveals that there is no one-size-fits-all parallelization strategy for the Big-means algorithm. Instead, the optimal strategy appears to be data-dependent, suggesting the need for adaptive techniques that can select the most suitable strategy based on the characteristics of the dataset. Nevertheless, in the majority of cases, we advise practitioners to utilize the competitive parallelization strategy of the Big-means algorithm.

For future work, we plan to investigate adaptive techniques that can dynamically select the optimal parallelization strategy based on the dataset at hand. We also aim to delve deeper into the trade-offs observed in this study to gain a better understanding of their impacts on algorithmic performance and accuracy.

The insights and observations gleaned from this study pave the way for further research into efficient and adaptive parallelization strategies for the Big-means algorithm and beyond. It is our hope that this research contributes significantly to the fields of data clustering and high-performance computing.

Acknowledgements. This research was funded by the Science Committee of the Ministry of Science and Higher Education of the Republic of Kazakhstan (Grants No. AP09259324 and AP09058174).

References

1. Alguliyev, R.M., Aliguliyev, R.M., Isazade, N.R., Abdi, A., Idris, N.: Cosum: text summarization based on clustering and optimization. Expert Syst. **36**(1) (2019). https://doi.org/10.1111/exsy.12340
2. Aloise, D., Deshpande, A., Hansen, P., et al.: Np-hardness of euclidean sum-of-squares clustering. Mach. Learn. (2009). https://doi.org/10.1007/s10994-009-5103-0
3. Arthur, D., Vassilvitskii, S.: K-means++: The advantages of careful seeding. In: Proceedings of the Eighteenth Annual ACM-SIAM Symposium on Discrete Algorithms, SODA 2007, pp. 1027–1035. Society for Industrial and Applied Mathematics, USA (2007)
4. Bahmani, B., Moseley, B., Vattani, A., Kumar, R., Vassilvitskii, S.: Scalable k-means++. In: Proceedings of the VLDB Endowment, vol. 5, pp. 622–633 (2012)
5. Chen, X., Fang, Y., Yang, M., Nie, F., Zhao, Z., Huang, J.Z.: Purtreeclust: A clustering algorithm for customer segmentation from massive customer transaction data. IEEE Trans. Knowl. Data Eng. **30**(3), 559–572 (2018). https://doi.org/10.1109/TKDE.2017.2763620
6. Dean, J., Ghemawat, S.: Mapreduce: simplified data processing on large clusters. Commun. ACM **51**, 107–113 (2008)
7. Depaire, B., Wets, G., Vanhoof, K.: Traffic accident segmentation by means of latent class clustering. Accident Anal. Prevention **40**(4), 1257–1266 (2008). https://doi.org/10.1016/j.aap.2008.01.007
8. Djenouri, Y., Belhadi, A., Djenouri, D., Lin, J.C.W.: Cluster-based information retrieval using pattern mining. Appl. Intell. **51**(4), 1888–1903 (2021). https://doi.org/10.1007/s10489-020-01922-x
9. Gribel, D., Vidal, T.: Hg-means: a scalable hybrid genetic algorithm for minimum sum-of-squares clustering. Pattern Recogn. (2019). https://doi.org/10.1016/j.patcog.2018.12.022

10. Jiang, D., Tang, C., Zhang, A.: Cluster analysis for gene expression data: a survey. IEEE Trans. Knowl. Data Eng. **16**(11), 1370–1386 (2004). https://doi.org/10.1109/TKDE.2004.68
11. Karmitsa, N., Bagirov, A.M., Taheri, S.: Clustering in large data sets with the limited memory bundle method. Pattern Recogn. (2018). https://doi.org/10.1016/j.patcog.2018.05.028
12. Marowka, A.: Python accelerators for high-performance computing. J. Supercomput. **74**(4), 1449–1460 (2018)
13. Mittal, H., Pandey, A.C., Pal, R., Tripathi, A.: A new clustering method for the diagnosis of covid19 using medical images. Appl. Intell. **51**(5, SI), 2988–3011 (2021). https://doi.org/10.1007/s10489-020-02122-3
14. Mussabayev, R., Mladenovic, N., Jarboui, B., Mussabayev, R.: How to use k-means for big data clustering? Pattern Recogn. **137**, 109269 (2023)
15. Rakthanmanon, T., Keogh, E.J., Lonardi, S., Evans, S.: Mdl-based time series clustering. Knowl. Inf. Syst. **33**(2), 371–399 (2012). https://doi.org/10.1007/s10115-012-0508-7
16. de Ridder, D., de Ridder, J., Reinders, M.J.T.: Pattern recognition in bioinformatics. Briefings Bioinform. **14**(5), 633–647 (2013). https://doi.org/10.1093/bib/bbt020
17. Sculley, D.: Web-scale k-means clustering. In: Proceedings of the 19th International Conference on World Wide Web, pp. 1177–1178 (2010)
18. Tu, B., Yang, X., Li, N., Zhou, C., He, D.: Hyperspectral anomaly detection via density peak clustering. Pattern Recogn. Lett. **129**, 144–149 (2020). https://doi.org/10.1016/j.patrec.2019.11.022
19. Yeung, M., Yeo, B., Liu, B.: Segmentation of video by clustering and graph analysis. Comput. Vis. Image Understanding **71**(1), 94–109 (1998). https://doi.org/10.1006/cviu.1997.0628
20. Yin, Y., Liu, F., Zhou, X., Li, Q.: An efficient data compression model based on spatial clustering and principal component analysis in wireless sensor networks. Sensors **15**(8), 19443–19465 (2015). https://doi.org/10.3390/s150819443
21. Zhao, P., Zhang, C.Q.: A new clustering method and its application in social networks. Pattern Recogn. Lett. **32**(15), 2109–2118 (2011). https://doi.org/10.1016/j.patrec.2011.06.008
22. Zhao, W., Ma, H., He, Q.: Parallel *K*-means clustering based on MapReduce. In: Jaatun, M.G., Zhao, G., Rong, C. (eds.) CloudCom 2009. LNCS, vol. 5931, pp. 674–679. Springer, Heidelberg (2009). https://doi.org/10.1007/978-3-642-10665-1_71

Development of Optimal Feedback for Zooplankton Seasonal Diel Vertical Migration

D. Perov(✉) and O. Kuzenkov

Lobachevsky State University, Gagarin Av. 23, 603950 Nizhny Novgorod, Russia
unn@unn.ru
http://www.unn.ru

Abstract. The main objective of the research is to develop the optimal feedback for zooplankton diel vertical migrations, taking into account seasonal variations. Feedback mechanisms serve to generate zooplankton movements in response to changes in an environment. Modern approaches to modeling evolutionary stable behavior are based on the involvement of the fitness function that reflect competitive advantages of organisms. In order to model zooplankton behavior, it is crucial to formalize the fitness function and find the feedback settings that maximize this function. To achieve this, the neural network is trained using a dataset encompassing several seasons. The Survival of the Fittest algorithm (SoFa) is utilized as the training method, aiming to maximize the fitness function. The trained network generates migration strategies in different seasons, which are confirmed by field observations. The developed feedback represents an effective tool for studying the diel vertical migration of zooplankton considering seasonal variations.

Keywords: Feedback · Global optimization · Neural network · Survival of the Fittest algorithm · Zooplankton · Seasonal diel vertical migration

1 Introduction

Optimal control problems with feedback are of great importance among optimization problems. Feedback enables the system to adjust its behavior in response to external factors [19]. In the context of biological evolution, feedback ensures the organism adaptation to changing environmental conditions [11,23]. It is essential to consider that observable behavior is the result of long-term evolution, when the most competitive organisms survive. Modern approaches to modeling evolutionary stable behavior are based on the involvement of the fitness function that reflect competitive advantages [3,8]. This is consistent with Darwin idea of "Survival of the fittest". To model the fittest behavior of living organisms, it is crucial to formalize their fitness function and to find the feedback settings that maximize this function. Thus, mathematically, the problem of modeling behavior can be reduced to an optimization problem in which the fitness function is maximized by development of the feedback.

N. Olenev et al. (Eds.): OPTIMA 2023, CCIS 1913, pp. 33–43, 2024.
https://doi.org/10.1007/978-3-031-48751-4_3

This problem arises in modeling zooplankton seasonal diel vertical migrations (seasonal DVM) [4]. It is a complex phenomenon involves repetitive vertical movements and plays a significant role in the dynamics of the ocean organic matter [1,5,12]. This movements are associated with seasonal changes in environmental conditions, making it challenging to accurately predict seasonal diel vertical migration patterns [2,6,13,14,17,21,22].

The periodic nature of zooplankton seasonal diel vertical migrations makes them a convenient material to study and model the adaptive behavior of living organisms. However, most of the suggested approaches for modeling diel vertical migrations over 200 years are based on a construction of a particular migration strategy for certain environmental conditions [2]. In this work, a new approach to modeling zooplankton vertical migrations in any environmental conditions is proposed. We construct a feedback between zooplankton and their environment, using an artificial neural network as a biologically justified nervous system model of organisms. By analyzing the patterns and characteristics of zooplankton periodic movements, neural networks can generate optimal behavioral responses to environmental factors by maximizing the fitness function [14,26]. One of the approaches to training neural networks is evolutionary algorithms, which simulate the natural process of evolution [14,15,24]. In nature, biological species evolve through selection, reproduction and mutation, leading to the development of more adapted traits. In evolutionary algorithms neural networks undergo a similar process. Thus, the application of evolutionary algorithms appears to be the most natural for modeling behavior as a result of biological evolution [7].

The main objective of this research is to develop optimal feedback that reflects seasonal variations in diel vertical migrations. By this feedback, we can construct a migration strategy as a result of elementary, momentary reactions of a living organism to its current local environmental conditions. To determine the optimal feedback settings for generating the fittest migration strategies in different seasons, we apply the neural network and evolutionary algorithm Survival of the Fittest. This feedback gives us the ability to predict seasonal diel vertical migrations of zooplankton at a particular time and depth in different environmental conditions.

2 Materials and Methods

2.1 The Problem Statement

This research uses the results of field observations on zooplankton behavior, which are published in works [9,25]. These observations were conducted in Saanich Inlet, located in British Columbia, Canada, during different months, including April, June, September, and December. Additionally, various approaches were employed to model the vertical migration of zooplankton [2,6,13,14,17,21].

From the field observations, zooplankton tend to move towards the surface at night, where they can graze on phytoplankton, and migrate back down to deeper waters during the day to avoid predation. Therefore, one of the impact factors affecting vertical migration is the availability of food, denoted by F. Another important factor is mortality from predation. However, zooplankton do not directly perceive these losses, but they are associated with the light intensity throughout the day, denoted by P_t and the depth lighting P_x, which affect the visibility of zooplankton to predators. It can be observed that zooplankton upwards migration starts at sunset, and downwards migration starts some time before sunrise. Additionally, mortality, denoted by D, due to high water surface temperatures and the presence of hydrogen sulfide at depth also influences zooplankton behavior.

The impact factors of the zooplankton seasonal diel vertical migrations can be classified into two groups. The first group comprises factors that undergo insignificant changes throughout the season. These are food distribution (F), mortality due to high water surface temperatures and the presence of hydrogen sulfide at depth (D). The second group comprises factors that undergo substantial changes throughout the season, such as light intensity throughout the day (P_t) and depth lighting (P_x).

Let's consider K seasons, where k represents the season number, $k = \overline{1, K}$. Let x_k denote the vertical coordinate of zooplankton position in the k season, and t represent the time of day from 0 to 24. In this case, the factors influencing the vertical migration in the k season can be approximated as follows [13,14,20,21]:

$$F(x_k) = \frac{1}{2}(\tanh(\xi_1(x_k - c_1)) + 1), \tag{1}$$

$$P_{xk}(x_k) = \tanh(\xi_2(x_k - c_{2k})), \tag{2}$$

$$P_{tk}(t) = \cos\left(2\pi(\frac{t - t_{rk}}{t_{sk} - t_{rk}} + \frac{1}{2})\right) + 2, \tag{3}$$

$$P_k(x_k, t) = 0.5P_{xk}(x_k)P_{tk}(t), \tag{4}$$

$$D(x_k) = \frac{1}{2}(\xi_3\eta_1^{x_k - c_3} + \xi_4\eta_2^{-(x_k - c_4)}). \tag{5}$$

Here c_1, c_{2k}, c_3, c_4 (depth characteristics), t_{rk}, t_{sk} (the start time of the downward migration and the upward migration associated with the sunrise and sunset times), ξ_1, ξ_2, ξ_3, ξ_4 and η_1, η_2 are certain parameters. The function $F(x_k)$ corresponds to the distribution of phytoplankton, and $P_k(t)$ is the expression for the underwater light intensity during the day in the k season. The parameterization of mortality $D(x_k)$ is the sum of two functions used in [13].

Let $\nu_k = x_k(t)$ represent the behavioral strategy of zooplankton in the k season. It is a continuous and differentiable function on the interval $[0, 24]$, satisfying the condition $x_k(0) = x_k(24)$. Let $E(x'_k)$ denote energy costs associated with the vertical movements of zooplankton. It is directly proportional to the square of the velocity $E(x'_k) = \xi_5(\dot{x}_k)^2$.

The fitness function of the strategy ν_k can be determined as the average reproductive rate [13]. In the long term, the survival of organisms depends on their ability to implement the strategy with the highest fitness value. It directly depends on the average energy gain over a 24-hour period per individual. This energy gain is determined by the food consumption and the energy losses due to predation, adverse environmental conditions and movements:

$$J_k(\nu_k) = \int_0^{24} (\alpha F(x_k) - \beta P_k(x_k, t) - \gamma D(x_k) - \delta E(x'_k))\, dt. \quad (6)$$

Here α, β, γ and δ are the weighting coefficients that quantify the relative contribution of the corresponding factors. The selection of these coefficients values is the separate research works [16,18,20].

In the simplest case, we could find the strategy ν_k that maximizes the given fitness function J_k separately for each season using calculus of variations or optimal control methods. However, development of the feedback between the zooplankton and its environment is a more promising and natural approach. This feedback can be used to model migration strategies adapted to different seasons. In nature, there is no evolutionary selection for every change in environmental conditions. Instead, there is a gradual adjustment of feedback mechanisms that provide adaptation, characterized by the averaged fitness value within in a range of changing conditions. In other words, our goal is to find the feedback function $\sigma(F, P_k, D)$ between the state of the surrounding environment and the optimal velocity of the zooplankton in different seasons. This feedback generates optimal migration strategies $\nu_k(\sigma(F, P_k, D))$, which maximize the following average fitness function over K seasons:

$$\frac{1}{K}\sum_{k=1}^{K} J_k(\nu_k(\sigma(F, P_k, D))). \quad (7)$$

Before development of the feedback, let's define migration strategies ν_k as piecewise linear functions. To do this, we divide a day into n equal intervals of length $h = 24/n$, with the start time $t_0 = 0$ and subsequent times given by $t_i = t_0 + ih$. We assume that the depth of zooplankton immersion varies linearly within each interval. Then zooplankton movement strategies are determined by its positions at each interval, denoted by $x_{ki} = x_k(t_i)$, or by the initial positions $x_{k0} = x_k(0)$ and the displacements $\Delta x_{ki} = x_{ki} - x_{ki-1}$. We represent the displacements as $\Delta x_{ki} = v\sigma_{ki}$, where v is a fixed constant and σ_{ki} takes one of three values: -1, 0, or 1. In this case, behavioral strategies are determined by the initial positions x_{k0} and sets of coefficients σ_{ki}, $k = \overline{1, K}$, $i = \overline{1, n}$. The fitness function for the k season takes the following form $J_k(\nu_k(x_{k0}, \sigma_{k1}, \ldots, \sigma_{kn}))$. We need to find the function σ that establishes the dependence of displacements σ_{ki} on the values of surrounding factors $F(x_{ki-1})$, $P_k(x_{ki-1}, t_{i-1})$, $\Delta P_k(x_{ki-1}, t_{i-1}) \equiv P_k(x_{ki-2}, t_{i-2}) - P_k(x_{ki-1}, t_{i-1})$, $D(x_{ki-1})$:

$$\sigma_{ki} = \sigma(F(x_{ki-1}), P_k(x_{ki-1}, t_{i-1}), \Delta P_k(x_{ki-1}, t_{i-1}), D(x_{ki-1})), \quad (8)$$

in order to maximize the following functional:

$$\frac{1}{K}\sum_{k=1}^{K} J_k(\nu_k(x_{k0}, \sigma_{k1}, \ldots, \sigma_{kn})). \quad (9)$$

2.2 Developing of Optimal Feedback

To find the dependence σ, we use a neural network. We developed the four-layer neural network, illustrated in Fig. 1. The architecture was selected based on the systematic investigation and previous work [14]. All layers are fully connected. The first input layer consists of 4 neurons, while the second and third hidden layers contain 4 and 12 neurons, respectively. The fourth layer consists of 3 neurons. The weight w_i of the i connection, which amplifies the signal passing through it, is bounded by $|w_i| < w_{max}$ and b_j represents the bias corresponding to the j neuron of the second to fourth layers, with $|b_j| < b_{max}$. The activation function for each neuron of the second to fourth layer is a sigmoid function defined as $f(y) = \frac{1}{1+e^{-y}}$.

The neural network, denoted by N, is characterized by the set of weights w_i, $i = 1, 132$, and biases b_j, $j = 1, 21$, with boundaries $w_{max} = b_{max} = 100$. The input is the values of four environmental factors corresponding to the current position of the zooplankton in current environment. The output is the displacement coefficient σ_{ki}, which is determined by the following rule:

$$I(s_{1ki}, s_{2ki}, s_{3ki}) = \begin{cases} -1, \text{if} \max(s_{1ki}, s_{2ki}, s_{3ki}) = s_{1ki}; \\ 0, \text{if} \max(s_{1ki}, s_{2ki}, s_{3ki}) = s_{2ki}; \\ 1, \text{if} \max(s_{1ki}, s_{2ki}, s_{3ki}) = s_{3ki}. \end{cases} \tag{10}$$

$$\sigma_{ki} = I(s_{1ki}, s_{2ki}, s_{3ki}). \tag{11}$$

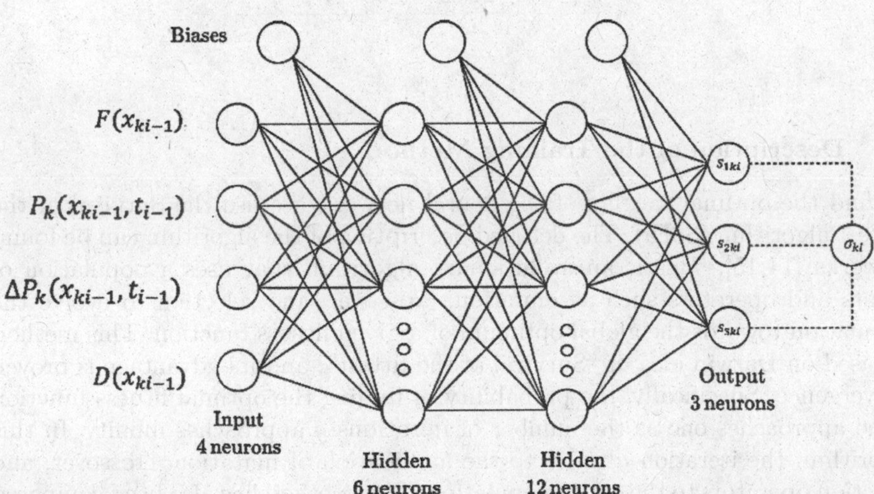

Fig. 1. The architecture of the artificial neural network.

Knowing the current zooplankton position $x_k(t_{i-1})$, we can calculate the corresponding environmental values, then pass this information through the neural network and generate the displacement σ_{ki} and new zooplankton position $x_k(t_i)$. Thus, we can obtain the zooplankton daily trajectory $\nu_k(N) = x_k(t, N)$ corresponding to a given neural network N by repeating this procedure n times. Repeating this for K seasons allows to calculate the average fitness function. Then we can train the developed neural network by selecting the values of weights and biases to maximize the average fitness function:

$$\frac{1}{K}\sum_{k=1}^{K} J_k(\nu_k(N^*)) = \max_{\substack{N \\ |w_i|\le w_{max}, \\ |b_j|\le b_{max}}} \frac{1}{K}\sum_{k=1}^{K} J_k(\nu_k(N)). \tag{12}$$

2.3 Values of Model Parameters

The following parameter values are used: $\xi_1 = 0.02$, $\xi_2 = 0.04$, $\xi_3 = \xi_4 = 0.1$, $\eta_1 = \eta_2 = 1.15$, $c_1 = -100$, $c_{21} = -80$, $c_{22} = -85$, $c_{23} = c_{24} = -92$, $c_3 = -10$, $c_4 = -100$, $\alpha = \beta = 0.35$, $\gamma = \delta = 0.02$, $n = 1440$, $v = 0.75$, $x_0 = -5.5$, $\xi_5 = 0.27 \cdot 10^{-3}$. Table 1 shows the start times of the downward migration t_{rk} and the start times of the upward migration t_{sk} in different seasons.

Table 1. The times of the downward and upward migrations (in hours).

	Apr	Jun	Sep	Dec
t_{rk}	6	3.7	5.3	7.5
t_{sk}	19	21.03	19.38	17

2.4 Description of the Training Method

To find the optimal weights of the neural network, we use the Survival of the Fittest algorithm (SoFa). The detailed description of the algorithm can be found in works [14,15]. This is an evolutionary algorithm that uses a population of agents and operators such as mutation, crossover, and selection to evolve the population towards the global optimum of a given fitness function. This method is based on Darwin idea of "Survival of the fittest", and its advantage is proven convergence. Specifically, the probability of finding the optimal fitness function value approaches one as the number of iterations g approaches infinity. In this algorithm, the iteration g refers to the application of mutation, crossover, and selection operators to the entire population. This method has also been compared to other heuristic and evolutionary algorithms, demonstrating its effectiveness [15].

To train the neural network, we consider three training environments in April, June and September ($K = 3$), each with its corresponding fitness function J_k. And for testing the developed feedback, we use the December environment with the fitness function J_4. In our work, we investigate two approaches to optimizing the average fitness function for the seasonal diel vertical migration of zooplankton. The initial approach involves calculating the average sum of the fitness functions for each season and using it as the target function for the algorithm. In contrast, the second approach aims to better mimic seasonality by dynamically changing the target function with each iteration g. Specifically, the target function takes the form of the fitness function J_k associated with the k season during each iteration g, thus simulating the seasonal variations.

3 Results

We trained the neural network using two approaches. One of them uses the target function as the average fitness function. The other uses the target function as fitness function of a particular season J_k, that changes depending on the iteration index g. In Fig. 2 showed the growth rate of average fitness function for the two training approaches of the neural network using the SoFa, averaged over 10 runs. The first graph compares the growth rate of the average sum of fitness functions in three environments between two approaches. The growth rate of the corresponding fitness function for the two approaches on the December test environment is also shown. You can observe the gradual improvement in the fitness value of living organisms as the result of mutations, competition, and selection. Additionally, it can be observed that the second approach, which simulates seasonal changes, has advantages in the early iterations but loses its effectiveness as the number of iterations increases.

We also compared the SoFa with other methods based on evolutionary strategies. This is the Evolution Strategy (ES) method proposed by the Open AI team [24] and additionally these are Simple Evolution Strategies methods based on the normal Cauchy (CES) and Gauss (GES) distributions [10]. Figure 3 illustrates the average growth rates of these methods on 10 runs. You can see that SoFA demonstrates the better performance for the training environments, as well as the more stable behavior of the average growth rate for the test environment.

Table 2. The fitness function values in the different seasons.

	Jun	Apr	Sep	Dec
J_k	6.69	5.94	6.02	6.64
J_k^*	6.7	5.97	6.03	6.69

The solid blue line in Fig. 4 represents the migration strategies generated using the neural network trained by the SoFa through the first approach. These strategies are compared to the field observations, where the dashed lines represent the upper and lower bounds of observed migration. It can be seen that the generated migration strategies are satisfy with the field observations. Table 2

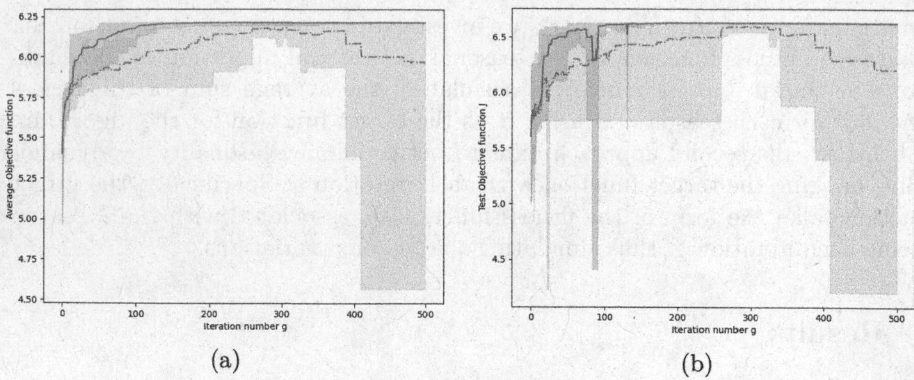

Fig. 2. The comparison of the average growth rate with its upper and lower boundary for the first approach (blue solid line) and the second (green dashed line) on (a) three training environments; (b) test environment. (Color figure online)

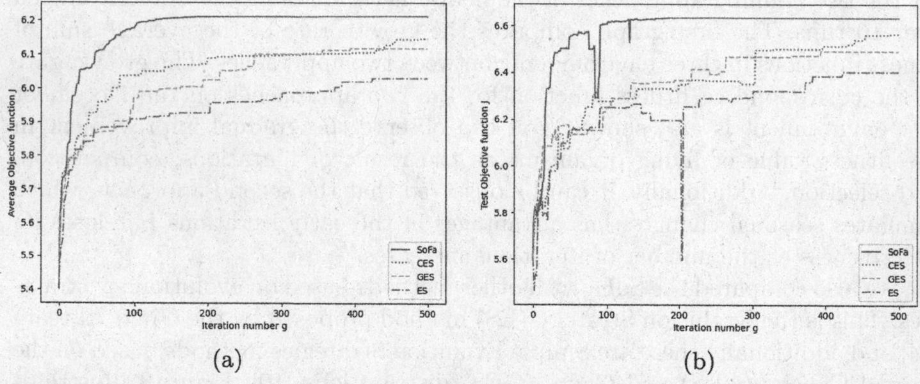

Fig. 3. The comparison of the average growth rate of SoFa, CES, GES, ES on (a) three training environments; (b) test environment.

compares the fitness values J_k of the migration strategies for each season, obtained from the neural network trained using the objective function of the average sum of fitness functions, with the best fitness values J^* obtained by retraining the neural network specifically for each environment.

4 Summary

This research presents the approach to modeling the diel vertical migration of zooplankton with seasonal variations. It is based on the feedback mechanism that links the organism behavior to the current state of the environment. This approach uses the neural network, with input values of environmental factors at a given point and time and output values of corresponding local displacement

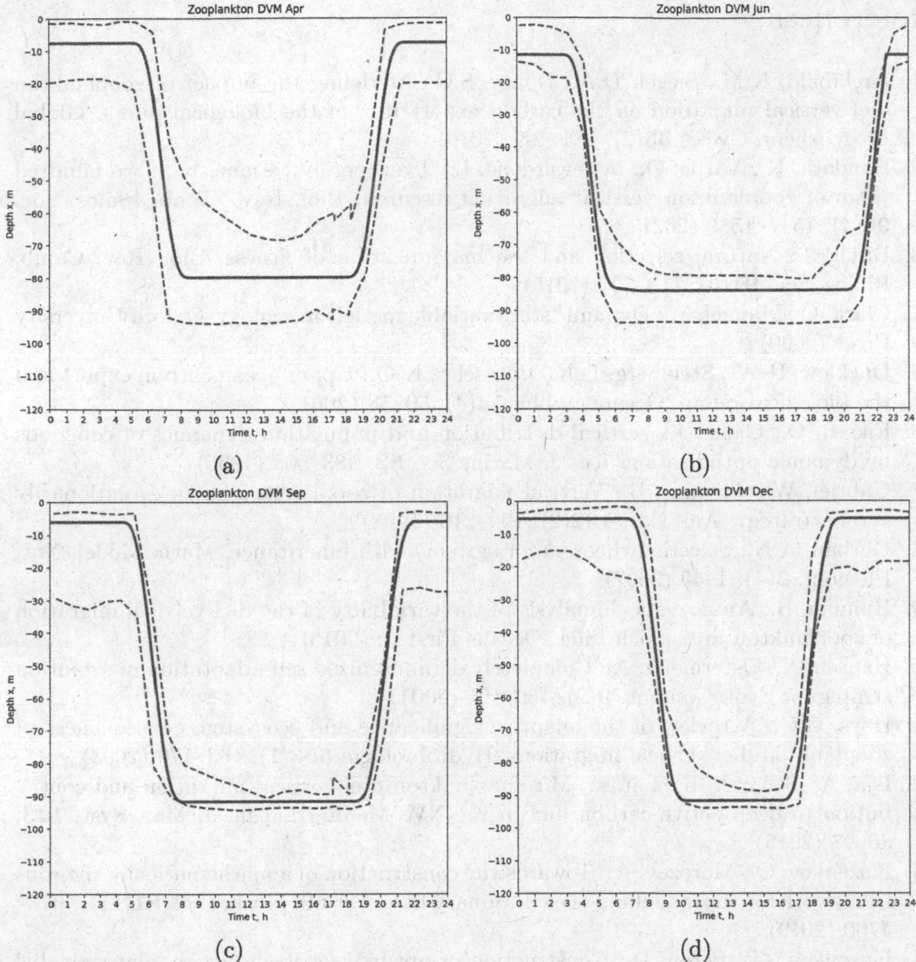

Fig. 4. The comparison of formed by the neural network seasonal DVM strategies (solid line) with field observations (dashed lines) in (a) April; (b) June; (c) September; (d) December.

of zooplankton. The net weights are optimized to maximize zooplankton fitness, using the stochastic global optimization evolutionary algorithm Survival of the Fittest (SoFa). The results demonstrate that this approach can generate optimal behavioral strategies for different environmental conditions, including seasonal changes. The trajectory obtained from this approach is the good approximation of the real pattern of seasonal DVM observed in Saanich Inlet, British Columbia, Canada. As a result, it can be noted that neural networks and evolutionary algorithms provide the powerful framework for studying the adaptation and survival strategies in response to changing environmental conditions.

References

1. Archibald, K.M., Siegel, D.A., Doney, S.C.: Modeling the impact of zooplankton diel vertical migration on the carbon export flux of the biological pump. Global Biogeochem. Cycles **33**(2), 181–199 (2019)
2. Bandara, K., Varpe, Ø., Wijewardene, L., Tverberg, V., Eiane, K.: Two hundred years of zooplankton vertical migration research. Biol. Rev. Camb. Philos. Soc. **96**(4), 1547–1589 (2021)
3. Birch, J.: Natural selection and the maximization of fitness. Biol. Rev. Camb. Philos. Soc. **91**(3), 712–727 (2015)
4. Clark, C., Mangel, M.: Dynamic state variable models in ecology. Oxford University Press (2000)
5. Ducklow, H.W., Steinberg, D.K., Buesseler, K.O.: Upper ocean carbon export and the biological pump. Oceanography **14**(4), 50–58 (2001)
6. Fiksen, O., Giske, J.: Vertical distribution and population dynamics of copepods by dynamic optimization. Ices J. Marine Sci. **52**, 483–503 (1995)
7. Gabriel, W., Thomas, B.: Vertical migration of zooplankton as an evolutionarily stable strategy. Am. Nat. **132**(2), 199–216 (1988)
8. Gorban, A.N.: Selection theorem for systems with inheritance. Math. Model. Nat. Phenom. **2**(4), 1–45 (2007)
9. Hannah, B.: An acoustical analysis of the variability of the diel vertical migration of zooplankton in Saanich Inlet. Oceans First J. (2015)
10. Hansen, N., Ostermeier, A.: Completely derandomized self-adaptation in evolution strategies. Evol. Comput. **9**(2), 159–195 (2001)
11. Hays, G.C.: A review of the adaptive significance and ecosystem consequences of zooplankton diel vertical migrations. Hydrobiologia **503**(1), 163–170 (2003)
12. Isla, A., Scharek, R., Latasa, M.: Zooplankton diel vertical migration and contribution to deep active carbon flux in the NW Mediterranean. J. Mar. Syst. **143**, 86–97 (2015)
13. Kuzenkov, O., Morozov, A.: Towards the construction of a mathematically rigorous framework for the modelling of evolutionary fitness. Bull. Math. Biol. **81**(11), 4675–4700 (2019)
14. Kuzenkov, O., Perov, D.: Construction of optimal feedback for zooplankton diel vertical migration. Commun. Comput. Inform. Sci. **1739**, 1547–1589 (2022)
15. Kuzenkov, O., Perov, D.: Global optimization method based on the survival of the fittest algorithm. Commun. Comput. Inform. Sci.**1750** (2022)
16. Kuzenkov, O., Kuzenkova, G.: Identification of the fitness function using neural networks. Proc. Comput. Sci. **169**, 692–697 (2020)
17. Kuzenkov, O., Morozov, A., Kuzenkova, G.: Recognition of patterns of optimal diel vertical migration of zooplankton using neural networks. In: 2019 International Joint Conference on Neural Networks (IJCNN), pp. 1–6 (2019)
18. Kuzenkov, O., Morozov, A., Kuzenkova, G.: Exploring evolutionary fitness in biological systems using machine learning methods. Entropy **23**(1) (2021)
19. Lee, E.B., Lawrence, M.: Foundations of optimal control theory (1967)
20. Morozov, A., Kuzenkov, O.A., Arashkevich, E.G.: Modelling optimal behavioural strategies in structured populations using a novel theoretical framework. Sci. Rep. **9**(1), 15020 (2019)
21. Morozov, A.Y., Kuzenkov, O.A.: Towards developing a general framework for modelling vertical migration in zooplankton. J. Theor. Biol. **405**, 17–28 (2016)

22. Ostrovskii, A.G., Arashkevich, E.G., Solovyev, V.A., Shvoev, D.A.: Seasonal variation of the sound-scattering zooplankton vertical distribution in the oxygen-deficient waters of the ne black sea. Ocean Sci. **17**(4), 953–974 (2021)
23. Parvinen, K., Dieckmann, U., Heino, M.: Function-valued adaptive dynamics and the calculus of variations. J. Math. Biol. **52**(1), 1–26 (2006)
24. Salimans, T., Ho, J., Chen, X., Sidor, S., Sutskever, I.: Evolution strategies as a scalable alternative to reinforcement learning (2017)
25. Sato, M., Dower, J.F., Kunze, E., Dewey, R.: Second-order seasonal variability in diel vertical migration timing of euphausiids in a coastal inlet. Mar. Ecol. Prog. Ser. **480**, 39–56 (2013)
26. Xingui, H., Shaohua, X.: Feedback process neural networks. Process Neural Netw. 128–142 (2010)

Continuous Optimization

Numerical Solution of an Inverse Problem for a General Hyperbolic Heat Equation

George Akindinov[1,2,3(✉)], Olga Krivorotko[1,3], and Vladislav Matyukhin[1]

[1] Moscow Institute of Physics and Technology, 9 Institute lane, Dolgoprudny 141701, Russia
akindinov.gd@phystech.edu
[2] IITP RAS, Bolshoy Karetny per. 19, build.1, Moscow 127051, Russia
[3] Sobolev Institute of Mathematics, 4 Acad. Koptyug avenue, Novosibirsk 630090, Russia

Abstract. In this paper we describe an algorithm of numerical solving of an inverse problem on a general case of a hyperbolic heat equation with additional second time derivative with a small parameter. The problem in this case is finding an initial distribution with given final distribution. This algorithm allows finding a solution to the problem for any admissible given precision. Algorithm allows evading difficulties analogous to the case of heat equation with inverted time. Furthermore, it allows finding an optimal grid size by learning on a relatively big grid size and small amount of iterations of a gradient method and later extrapolates to the required grid size using Richardson's method. This algorithm allows finding an adequate estimate of Lipschitz constant for the gradient of the target functional. Finally, this algorithm may easily be applied to the problems with similar structure, for example in solving equations for plasma, social processes and various biological problems. The theoretical novelty of the paper consists in the developing of an optimal procedure of finding of the required grid size using Richardson extrapolation in context of ill-posed problems, and accelerated methods were applied to this problem.

Keywords: Inverse and ill-posed problems · Hyperbolic heat equation · Inexact gradient · Richardson's method · Accelerated gradient methods · Regularization

1 Introduction

1.1 Problem Formulation

In this paper we research how the gradient methods work for the problem of finding the initial distribution from a final one for a hyperbolic heat equation. This equation is often used to describe quick processes as ones found in plasma, some social processes and other problems [4].

The research was supported by the Russian Science Foundation (project no. 23-71-10068), https://rscf.ru/en/project/23-71-10068/.

N. Olenev et al. (Eds.): OPTIMA 2023, CCIS 1913, pp. 47–60, 2024.
https://doi.org/10.1007/978-3-031-48751-4_4

The main problem(inverse problem) consists of finding the initial distribution(of temperature, concentration, population density etc.) given the final distributions as observations for the given process. The methods used to solve this problems consist of variational approaches, functional analysis methods and gradient methods. In order to find the explicit form of the gradient of the target functional we use calculus of variations. After that we face a problem of effectively solving partial differential equations, we have to find the optimal grid-size for finite differences methods, or have to come up with the correct basis of eigenfunctions.

1.2 Hyperbolic Heat Equation

Assume the problem of the form:

$$\begin{cases} \varepsilon u_{tt} + u_t = (u_x u)_x + u(1-u), & t \in (0,T),\ x \in [0,1], \\ u(0,x) = q(x), \quad \varepsilon u_t(0,x) = 0, & x \in [0,1], \\ u_x(t,0) = u_x(t,1) = 0, & t \in (0,T). \end{cases} \tag{1}$$

The first Eq. (1) is called a hyperbolic heat equation with an additional term in the Kolmogorov's form [2]. The initial distribution $q(x)$ — is bounded, is in $L_2[0,1]$. In real applications(physics of plasma processes, social, epidemiological, social processes) the initial distribution is unavailable for measuring, only the final one can be measured. The final distribution is of the form:

$$f(x) = u(T, x|q_{\text{exact}}), \quad x \in [0,1]. \tag{2}$$

The expression above means that f(x) is calculated for the exact initial distribution, and the notation $u(t,x|q)$ means the solution of the problem (1), when the initial distribution is $q(x)$. The inverse problem (1)–(2) consists of reconstruction of the initial distribution $q(x)$ by exploiting the final one($f(x)$).

In paper [4] it was shown that adding a small parameter to the equation(with $\varepsilon > 0$ improves the stability of the initial problem for parabolic equation. In this work we will research the stability of the solution of the problem (1)–(2) with respect to the parameter ε.

2 Problem Analysis

2.1 Gradient Approach to the Inverse Problem

We try to find the solution in class:

$$Q := \{q \in L_2([0,1]) : 0 \le q(x) < \infty \quad q'(0) = q'(1) = 0\}.$$

In our case $q(x)$ is a probability distribution function, we assume it to be bounded and Lebesgue integrable function.

The idea of solving this kind of problem is in applying the gradient method in this way:

$$q^{n+1}(x) = q^n(x) - h_n J'(q^n), \quad q^0(x) \in Q.$$

We assume the target quality functional to be of the form:

$$J(q) = \int_0^1 |f(x) - u(T, x|q)|^2 dx,$$

and the derivative of the aforementioned functional $J'(q)$ is considered in a sense of Fréchet derivative:

$$J(q + \delta q) - J(q) = \langle J'(q), \delta q\rangle + o(||\delta q||).$$

2.2 Sketch of the Derivation of the Target Functional Derivative

The full proof is available in the Appendix section

In order to get the gradient of the target functional w.r.t. q we will assume two admissible initial conditions from class Q: q and $q + \delta q$ which correspond to the solution of the given problem $u(t, x|q)$ and $u(t, x|q) + \delta u(t, x|q)$. After that we transform the functional $J(q)$ to the form:

$$J(q) = \min_u \{\mathcal{J}(u, q, \Psi = (p(t, x), \lambda_1(x), \lambda_2(x), \mu_1(t), \mu_2(t)\} \tag{3}$$

$$\begin{aligned}\mathcal{J}(u, q, \Psi) = \int_0^1 ||u(T, x) - f(x)||^2 dx + \int_0^1 \int_0^T p(\varepsilon u_{tt} + u_t - (uu_x)_x - u(1 - u))dtdx + \\ + \int_0^1 \lambda_1(x)(u(0, x) - q(x))dx + \int_0^1 \lambda_2(x)(u_t(0, x))dx + \\ + \int_0^T \mu_1(t)u_x(t, 0)dt + \int_0^T \mu_2(t)u_x(t, 1)dt\end{aligned}$$

In order to minimize this functional we assume variation of the (1) and use integration by parts and we are left with the partial differential equation for $p(t, x)$ of the form:

$$\begin{cases} \varepsilon p_{tt} - p_t - p_{xx}u - p(1 - 2u) = 0, & (t, x) \in (0, T) \times (0, 1), \\ 2(u(T, x|q) - f(x)) + (p(T, x) - \varepsilon p_t(T, x)) = 0, & x \in (0, 1), \\ \varepsilon p(T, x) = 0, & x \in (0, 1), \\ p_x(t, 0) = p_x(t, 1) = 0, & t \in (0, T). \end{cases} \tag{4}$$

And $\lambda_1(x) = p(0, x) - \varepsilon p_t(0, x)$. According to Danskin's theorem, the gradient of the target functional is given as $J_q(q) = -\lambda_1 = \varepsilon p_t(0, x) - p(0, x)$

In order to use the same solving device for the conjugated Eq. (4) as the (1), we will change variables as follows: $v(t,x) = p(T-t,x)$

$$\begin{cases} \varepsilon v_{tt} + v_t - v_{xx}u(T-t,x) = v(1-2u(T-t,x)), & (t,x) \in (0,T)\times(0,1), \\ 2(u(T,x|q) - f(x)) + (v(0,x) + \varepsilon v_t(0,x)) = 0, & x \in (0,1), \\ \varepsilon v(0,x) = 0, & x \in (0,1), \\ v_x(t,0) = v_x(t,1) = 0, & t \in (0,T). \end{cases} \tag{5}$$

According to this substitution the formula for gradient is defined as follows:

$$J_q = -(v(T,x) + \varepsilon v_t(T,x))$$

2.3 Evaluating the Inexact Gradient

In order to calculate the inexact gradient of the target functional we have to firstly solve the initial PDE (1), then the conjugated PDE (5). In order to calculate them we use method of finite differences: We sample the domain of the PDE using step sizes Δt for time and Δx for space. We will denote the function value as $U_j^i = u(t_i, x_j)$. We approximate derivatives as follow:

$$u_t = \frac{U_j^{i+1} - U_j^{i-1}}{2\Delta t}, \quad u_{tt} = \frac{U_j^{i+1} - 2U_j^i + U_j^{i-1}}{\Delta t^2}$$

$$u_x = \frac{U_{j+1}^i - U_{j-1}^i}{2\Delta x}, \quad u_{xx} = \frac{U_{j+1}^i - 2U_j^i + U_{j-1}^i}{\Delta x^2}$$

Therefore the equations are written as follows:

$$U_j^{i+1} = a^T U + U^T A U,$$

where $U = [U_j^i, U_j^{i-1}, U_{j+1}^i, U_{j-1}^i]^T$, $a^T = \left[\frac{2\varepsilon+\Delta t^2}{2\varepsilon+\Delta t}, \frac{-2\varepsilon+\Delta t}{2\varepsilon+\Delta t}, 0, 0\right]$,

$$A = \begin{bmatrix} -4(\Delta x^2+2) & 0 & 2 & 2 \\ 0 & 0 & 0 & 0 \\ 2 & 0 & 1 & -1 \\ 2 & 0 & -1 & 1 \end{bmatrix} \cdot \frac{\Delta t^2}{2(2\varepsilon+\Delta t)\Delta x^2}.$$

Auxiliary conditions in case of $\varepsilon \neq 0$ will be as follows:

$$U_j^0 = q_j,\ U_j^1 = U_j^0,\ U_0^i = U_1^i,\ U_n^i = U_{n-1}^i.$$

Noticing, that U^TAU can significantly affect the solution because of its quadratic nature, we should minimize the matrix norm(assuming the spectral norm of the matrix), therefore we can find the optimal ratio between time and space step sizes. In real problems it is required for this matrix to have norm significantly less than 1, we have:

$$||A||_2 < 1 \Rightarrow \frac{\sqrt{(\Delta x^2+2)^2 + (\Delta x^2+2)\sqrt{4+\Delta x^2} + 1}}{2} \cdot \frac{2\Delta t^2}{(2\varepsilon+\Delta t)\Delta x^2} < 1$$

$$\sqrt{(\Delta x^2+2)^2+(\Delta x^2+2)\sqrt{4+\Delta x^2}+1} \leq 4.1$$

$$\frac{4.1\Delta t^2}{(2\varepsilon+\Delta t)\Delta x^2}=c<1$$

We can define a decent step size ratio according to the aforementioned formula.

The conjugated PDE is solved similarly, and its finite differences scheme is as follows:

$$V_j^{i+1}=b^T V+V^T BV,$$

where $V=[V_j^i, V_j^{i-1}, V_{j+1}^i, V_{j-1}^i, U_j^{n-i}]^T$, $\quad b^T=\left[\frac{2\varepsilon+\Delta t^2}{\Delta t^2}, \frac{2\varepsilon-\Delta t}{2\Delta t^2}, 0, 0, 0\right]\cdot\frac{2\Delta t^2}{2\varepsilon+\Delta t}$

$$B=\begin{bmatrix} 0\,0\,0\,0\,s \\ 0\,0\,0\,0\,0 \\ 0\,0\,0\,0\,1 \\ 0\,0\,0\,0\,1 \\ s\,0\,1\,1\,0 \end{bmatrix}\cdot\frac{\Delta t^2}{(2\varepsilon+\Delta t)\Delta x^2}, \quad s=-2(\Delta x^2+1).$$

Matrix norm inequality, analogous to the one for the initial PDE: Co

$$||B||_2=\sqrt{s^2+1}\cdot\frac{\Delta t^2}{(2\varepsilon+\Delta t)\Delta x^2}<\frac{4.2\Delta t^2}{(2\varepsilon+\Delta t)\Delta x^2}=c_1<1.$$

Optimal step size ratio difference between the initial PDE and the conjugated PDE is negligible, therefore we can establish the same step size ratio for both of them.

Aforementioned procedure worked well for different initial conditions, the target functional converged to zero sub-linearly, $\int_0^1 ||\hat{q}(x)-q_n(x)||^2 dx$ was converging similarly to the target functional. q_n represents the n-th approximation of the solution using the gradient method.

However, there is one downside to this approach: often times the explicit finite difference methods are numerically unstable, so we can use the same results for implicit finite difference schemes due to their being more numerically stable than the explicit ones. In case of implicit finite difference schemes the analog to the Courant's condition:

$$\frac{\Delta t^2}{(2\varepsilon+\Delta t)\Delta x^2}\leq\frac{1}{4}. \tag{6}$$

2.4 Finding Optimal Step Size

We have discovered the optimal ratio between time and space step sizes, but we still have to get some ideas about getting the optimal step size for space(for example). Figuring out Δx will be easier in our case because our target functional is defined in terms of the integral over space. We can assume that our target functional is of the form:

$$J[q(x|\Delta x)]=C\Delta x^p+o(\Delta x^{p+1})$$

In this paper we used the largest space step size(in terms of needing to have a certain number of points on our sampled grid) and after that, for example, we use a half of our step size and we are supposed to get a better estimate. This idea is usually referred to as **Richardson's extrapolation method**, where the connection between $J[q(x|\Delta x)]$ and Δx is being studied.

$$A(h) - A^* = Ch^p + O(h^{p+1}) \Rightarrow A_1(h) = \frac{s^p A_0(h/s) - A_0(h)}{s^p - 1} = A^* + O(h^{p+1}).$$

According to the aforementioned formula, the recurrent relation is established as follows:

$$A_{i+1}(h) = \frac{s^{p_i} A_i(h/s) - A_i(h)}{s^{p_i} - 1} = A^* + O(h^{p_{i+1}}).$$

Therefore Richardson's extrapolation formula proves to be a powerful tool allowing us to estimate the power p and the constant C by running gradient method(for a small number of iterations) on different step sizes and by interpolating this relation using MSE we can find the optimal step size for our problem and after that use our method with this optimal step size with the desired number of iterations. If the desired precision is γ, the optimal grid size in terms of Δx is:

$$\Delta x \approx \left(\frac{\gamma}{C}\right)^{\frac{1}{p}} \tag{7}$$

The time step size will be calculated according to (6)

3 Analysis of the Gradient Method

3.1 Lipschitz Constant for the Gradient of the Target Functional

Assume the variation of the gradient of the target functional:

$$\delta J'(q) = -\delta v(T, x) - \varepsilon \delta v_t(T, x),$$

where $\frac{\delta v(t,x)}{\delta q}$ can be substituted as the solution of (w):

$$\begin{cases} \varepsilon w_{tt} + w_t = -g(T-t,x)(v_{xx} + 2v) + u(T-t,x)w_{xx} + (1 - 2u(T-t,x))w, & t \in (0,T), x \in [0,1], \\ w(0,x) + \varepsilon w_t(0,x) = -2g(T,x), & x \in [0,1], \\ w(0,x) = 0, & x \in [0,1], \\ w_x(t,0) = w_x(t,1) = 0, & t \in (0,T), \end{cases}$$

where $g \equiv \frac{\delta u}{\delta q}$, is defined from the system:

$$\begin{cases} \varepsilon g_{tt} + g_t = u g_{xx} + u_{xx} g + 2u_x g_x + g - 2ug, & t \in (0,T), x \in [0,1], \\ g(0,x) = 1, & x \in [0,1], \\ \varepsilon g_t(0,x) = 0, & x \in [0,1], \\ g_x(t,0) = g_x(t,1) = 0 & t \in (0,T). \end{cases}$$

Therefore by solving these systems we can define the Hessian matrix for our target functional and while the gradient method is running we can estimate the Lipschitz constant for the gradient of the target functional as max $\frac{||\delta J'(q^i)||}{||\delta q^i||}$. Therefore we can use it to estimate the optimal step size **for the gradient method**. These two systems guarantee that we indeed have well-defined continuous solutions, it means that we indeed have the smoothness condition, therefore the gradient method(in the assumption that the gradient we calculated is exact) converges to the stationary point. Therefore we can preserve the values of $v^n(T,x)$ and $v^n_t(T,x)$(where n stands for the n-th approximation using the gradient method) and we can calculate the variational derivative of $J'(q)$ numerically.

As far as the variation of the initial distribution is concerned, it may be estimated as follows: $q^i - q^{i-1}$. After having done several iterations we can use the calculated estimate for the Lipschitz constant.

After running several experiments, we discovered that the norm of the second derivative of the final distribution is a decent initial estimate for the Lipschitz constant while running Richardson extrapolation procedure. The exact Lipschitz constant may be found using the linear search.

3.2 Optimization of the Small Parameter

We can determine the magnitude of the small parameter by analyzing the dimensions of the left hand side of the (1), therefore $\varepsilon < \Delta t$. According to the the experiments, the quality functional(time elapsed, number of operations) is a quasiconvex function with respect to the small parameter ε, therefore we study the small parameter's behaviour while running Richardson's extrapolation and determine its optimal value using linear search.

3.3 Accelerated Methods

The problem is not convex, therefore we cannot guarantee in theory that the acceleration will have a significant effect on the convergence of the method, but we studied the accelerated method that is guaranteed to converge at least to something. In this section we studied primal-dual accelerated gradient methods with small-dimensional relaxation oracle as shown in [5]. This method worked well for our problem, however it was not significantly faster in our synthetic experiments to make up for the additional needed computational power. However, the approach that proved to be more efficient in our case, is that we used the aforementioned accelerated method in Richardson's extrapolation with a small amount of iterations, and we got a more accurate representation of the connection between $\log J(f[q(x|\Delta x)])$ and Δx. As far as the actual process of computing q^* is concerned, the method didn't prove to be much better than the general gradient descent in terms of trade-off between precision and the computational cost of one iteration of accelerated gradient method. We might have not picked the best examples for this method, and these methods might work better for some real data, we are planning to further research this later.

4 Experiments

In order to check how our approaches to this problem work, we assumed to know the initial distribution q_{exact}, after that we computed the final distribution function by solving (1) for the case $\varepsilon = 0$, and after that we worked with $f(x) = u(T, x|q_{exact})$. After that we use Richardson's extrapolation with small number of iterations for $J(f)$ in order to calculate the optimal grid size for x(using accelerated methods). And after getting to know the optimal grid size with respect to the desired precision, we start our gradient method. The procedure of the gradient method is as follows:

a) If given q_{exact}:

0. Calculate $f(x) = U(T, x|q_{exact})$ and reach the desired precision using accelerated Richardson's extrapolation. Erase q_{exact} from the memory, initializing with another $q = q_0 \in Q$.
   - i.1. Solve for $u(t, x|q_i)$ using finite differences method.
   - i.2. Solve for $v(t, x|q_i)$ using finite differences method.
   - i.3. Calculate the gradient $J'(q_i) = -(v(T, x)+\varepsilon v_t(T, x))$ and make a gradient method iteration $q_{i+1} = q_i - hJ'(q_i)$

b) If given $f(x) = u(T, x|q_{exact})$

0) initializing with some $q_0 \in Q$ from some general assumptions
   - i.1. Solve for $u(t, x|q_i)$ using finite differences method.
   - i.2. Solve for $v(t, x|q_i)$ using finite differences method.
   - i.3. Calculate the gradient $J'(q_i) = -(v(T, x)+\varepsilon v_t(T, x))$ and make a gradient method iteration $q_{i+1} = q_i - hJ'(q_i)$

In reality we can estimate $J(h) = Ch^p + o(h^p)$. We can improve this estimate by using Richardson's extrapolation. We ran the gradient method with 20–40 iterations with a large step size on the x axis ($\Delta x = 0.1$ for $T = 0.01$), and accordingly set $\Delta t \approx \frac{\Delta x^2}{4}$, after that reduce the Δx by half and get J_2, and after that assume $J^* = J_1 + \frac{J_1 - J}{2^p - 1}$, which will better represent the reality. This procedure can be applied several times to get a better estimate for the relation $J(\Delta x) = C\Delta x^p + o(\Delta x^p)$, which is being studied for the value of C and the power value p.

After that we explore the relationship between $\log J(q)$ and $\log \Delta x$ and get an estimate of $\ln(C)$ and p using the MSE. Using these numbers we can get the optimal Δ in terms of the required precision according to (7). For all of the studied functions we got a good linear relationship between $\log J(q)$ and $\log \Delta x$ and numbers' C and p values were reasonable.

This connection was studied for $T = 0.01$ for functions:
$q_{exact}(x) \in \{\sin^2(\pi x), \sin^2(\pi x/2), x\}$ and
$f(x) = u(T, x|q_{exact}) \in \{\sin^2(\pi x), \sin^2(\pi x/2), x\}$.
see Figs. 1, 2, 3, 4, 5 and 6 for the results of the aforementioned experiments

It is impossible to reach any desired precision while working with this algorithm due to the inequality $\Delta t \leq T/3$, and to the inequality $\Delta x \geq \sqrt{T}$. Therefore the best precision $\hat{\gamma}$ can be denoted in terms of p, C, T as follows:

$$\hat{\gamma} \sim CT^{\frac{p}{2}}$$

5 Summary

In this paper the connection between the grid size for time and space was explored, method of finding the optimal grid size for space was developed, and the approach to finding the optimal small parameter ε was suggested, and an adequate estimate for Lipschitz constant for the gradient of the target functional was suggested. All of the methods were used for synthetic examples and proved to be quite effective. Moreover, the approaches described in this paper may be applied to a wide range of problems of similar structure.

At the end of this paper we want to express our gratitude towards Gasnikov Alexander for an enormous impact on this paper and seemingly limitless number of ideas for developing and researching this kind of problems.

6 Appendix

Theorem 1. *In a problem of the kind* (3) *the gradient of the target functional can be computed as* $-\lambda_1$ *from the Lagrangian formalism and* $J'(q) = \varepsilon p_t(0,x) - p(0,x)$

Proof. Let's establish a full formal Lagrangian formula:

$$\mathcal{J} = \int_0^1 |u(T,x|q) - f(x)|^2 dx + \int_0^1 \int_0^T p(\varepsilon u_{tt} + u_t - (u_x u)_x - G(t,x,u))dtdx +$$

$$+ \int_0^1 \lambda_1(u(0,x) - q(x))dx + \int_0^1 \lambda_2 u_t(0,x)dx + \int_0^T \mu_1 u_x(t,0)dt + \int_0^T \mu_2 u_x(t,1)dt$$

Let's assume two initial distributions q and $q + \delta q$. Both of them should satisfy the auxiliary conditions for the initial problem. Subtracting one from another, we get:

$$\begin{cases} \delta u(0,x) = \delta q, \\ \delta u_t(0,x) = 0, \\ \delta u_x(t,0) = \delta u_x(t,1) = 0. \end{cases} \tag{8}$$

Assuming the analogous variation of the $\mathcal{J}$ term by term:

$$(A) = \delta_q \int_0^1 |u(T,x|q) - f(x)|^2 dx = \int_0^1 2(u(T,x) - f(x))\delta u(T,x)dx$$

$$(B) = \delta_q \int_0^1 \int_0^T p(\varepsilon u_{tt} + u_t)dtdx = \int_0^T \int_0^1 p(\varepsilon \delta u_{tt} + \delta u_t))dxdt =$$

$$= \int_0^1 [p(\varepsilon\delta u_t + \delta u)]\Big|_{t=0}^{t=T} dx - \int_0^1\int_0^T p_t(\varepsilon\delta u_t + \delta u)dtdx =$$

$$= \int_0^1\int_0^T (\varepsilon p_{tt} - p_t)\delta u dt dx + \int_0^1 [p - \varepsilon p_t)\delta u + \varepsilon p\delta u_t]\Big|_{t=0}^{t=T} dx$$

$$(C) = \delta_q \int_0^1 \delta_0^T - p(u_x u)_x dxdt = \int_0^T\int_0^1 -p(u_x\delta u + u\delta u_x)_x dxdt =$$

$$= \int_0^T -[p(u_x\delta u + u\delta u_x]\Big|_{x=0}^{x=1} dt + \int_0^T\int_0^1 p_x u_x \delta u dxdt +$$

$$+ \int_0^T [p_x u\delta u]\Big|_{x=0}^{x=1} dt - \int_0^T\int_0^1 (p_x u)_x \delta u dxdt =$$

$$= \int_0^T\int_0^1 -p_{xx} u\delta u dxdt + \int_0^T [(p_x u - u_x p)\delta u - pu\delta u_x]\Big|_{x=0}^{x=1} dt$$

$$(D) = \delta_q \int_0^T\int_0^1 -pG dxdt = \int_0^1\int_0^T -\frac{\partial G}{\partial u} p\delta u dxdt$$

$$(E) = \delta_q\Big\{\int_0^1 \lambda_1(u((0,x) - q(x))dx + \int_0^1 \lambda_2 u_t(0,x)dx + \int_0^T \mu_1 u_x(t,0)dt + \int_0^T \mu_2 u_x(t,1)dt\Big\} =$$

$$= \int_0^1 \lambda_1\delta u(0,x)dx + \int_0^1 \lambda_2\varepsilon\delta u_t(0,x)dx + \int_0^T \mu_1\delta u_x(t,0)dt + \int_0^T \mu_2\delta u_x(t,1)dt$$

Combining δu terms:

$$(a) = \int_0^1\int_0^T \left(\varepsilon p_{tt} - p_t - p_{xx}u - p\frac{\partial G}{\partial u}\right)\delta u dt dx$$

Combining $\delta u(0,x)$ terms:

$$(b) = \int_0^1 (\varepsilon p_t(0,x) - p(0,x) + \lambda_1)\delta u(0,x)dx$$

Combining $\delta u_t(0,x)$ terms:

$$(c) = \int_0^1 (-\varepsilon p_t(0,x) + \lambda_2)\delta u_t(0,x))dx$$

Combining $\delta u(T,x)$ terms:

$$(d) = \int_0^1 \{2(u(T,x) - f(x)) + (p(T,x) - \varepsilon p_t(T,x))\}\delta u(T,x)dx$$

Combining $\delta u_t(T,x)$ terms:

$$(e) = \int_0^1 \varepsilon p(T,x)\delta u_t(T,x)dx$$

Combining $\delta u(t,0)$ terms:

$$(f) = \int_0^T p_x(t,0)u(t,0)\delta u(t,0)dt$$

Combining $\delta u(t,1)$ terms:

$$(g) = \int_0^T p_x(t,1)u(t,1)\delta u(t,1)dt$$

Combining $\delta u_x(t,0)$ terms:

$$(h) = \int_0^T (\mu_1 + p(t,0)u(t,0))\delta u_x(t,0)dt$$

Combining $\delta u_x(t,1)$ terms:

$$(i) = \int_0^T (\mu_2 - p(t,1)u(t,1))\delta u_x(t,1)dt$$

Keeping in mind (8) and applying the fundamental lemma of the calculus of variations(and using Danskin's theorem for the gradient of the target functional), we get the following PDE for $p(t,x)$:

$$\begin{cases} \varepsilon p_{tt} - p_t - p_{xx}u - p\frac{\partial G}{\partial u} = 0, & (t,x) \in (0,T) \times (0,1), \\ 2(u(T,x|q) - f(x)) + (p(T,x) - \varepsilon p_t(T,x)) = 0, & x \in (0,1), \\ \varepsilon p(T,x) = 0, & x \in (0,1), \\ p_x(t,0) = p_x(t,1) = 0, & t \in (0,T), \\ J'(q) = -\lambda_1 = \varepsilon p_t(0,x) - p(0,x). \end{cases}$$

Q.E.D.

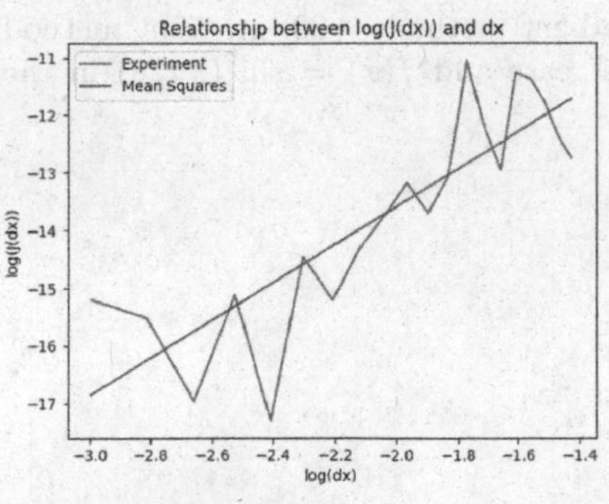

case with given q_{exact}

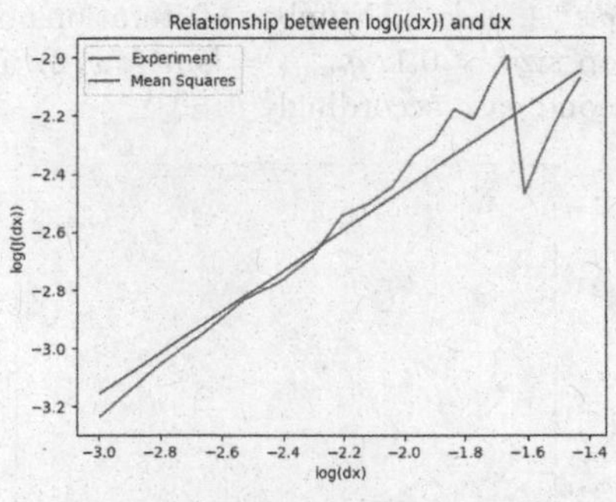

case with unknown q_{exact}

Fig. 1. Graph for log of the target functional and values of C and p for $J(dx) = Cdx^p + o(dx^p)$. Number of iterations of the gradient method = 20, gradient method step size = 0.1. $q_{exact} = \sin^2(\pi x)$ in the first case and $f(x) = \sin^2(\pi x)$ in the second case accordingly

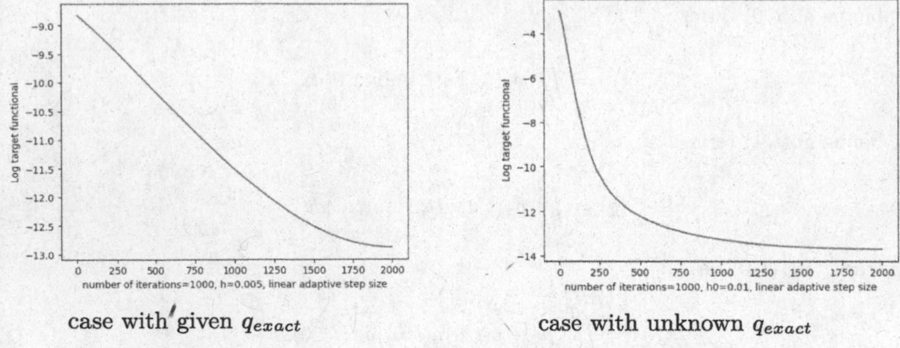

case with given q_{exact} case with unknown q_{exact}

Fig. 2. Graph for the natural log of the target functional with an optimal grid size calculated with respect to the C and p calculated in the previous figure. $q_{exact} = \sin^2(\pi x)$ in the first case and $f(x) = \sin^2(\pi x)$ in the second case accordingly

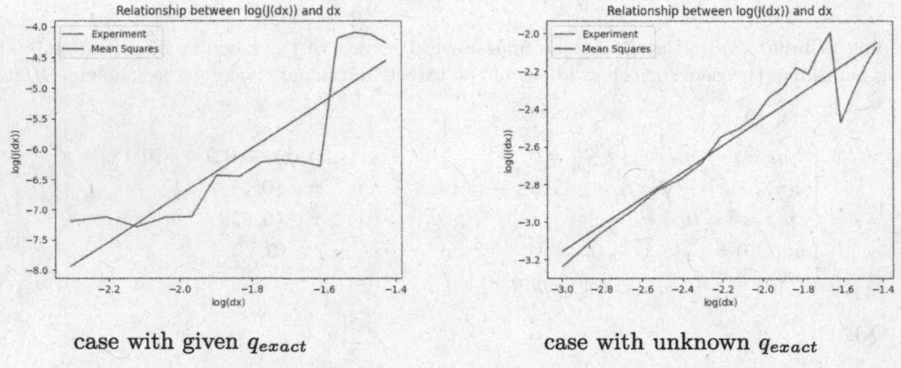

case with given q_{exact} case with unknown q_{exact}

Fig. 3. Graph for log of the target functional and values of C and p for $J(dx) = Cdx^p + o(dx^p)$. Number of iterations of the gradient method = 20, gradient method step size = 0.1. $q_{exact} = \sin^2(\pi x/2)$ in the first case and $f(x) = \sin^2(\pi x/2)$ in the second case accordingly

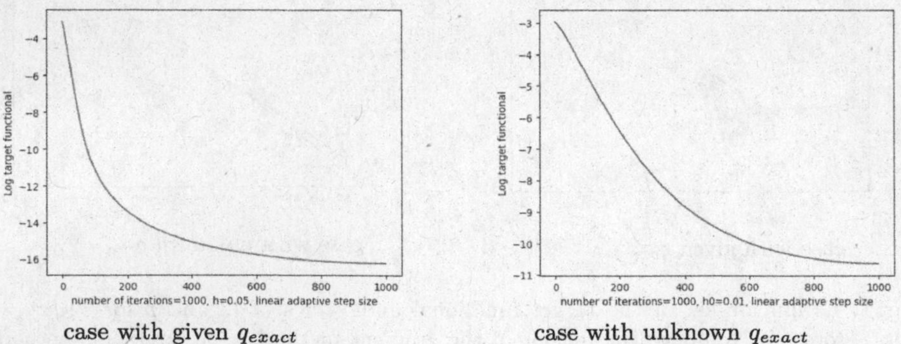

case with given q_{exact} case with unknown q_{exact}

Fig. 4. Graph for the natural log of the target functional with an optimal grid size calculated with respect to the C and p calculated in the previous figure. $q_{exact} = \sin^2(\pi x/2)$ in the first case and $f(x) = \sin^2(\pi x/2)$

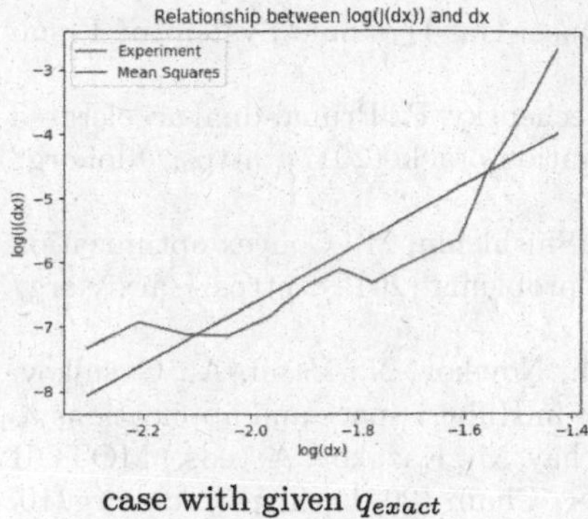

case with given q_{exact}

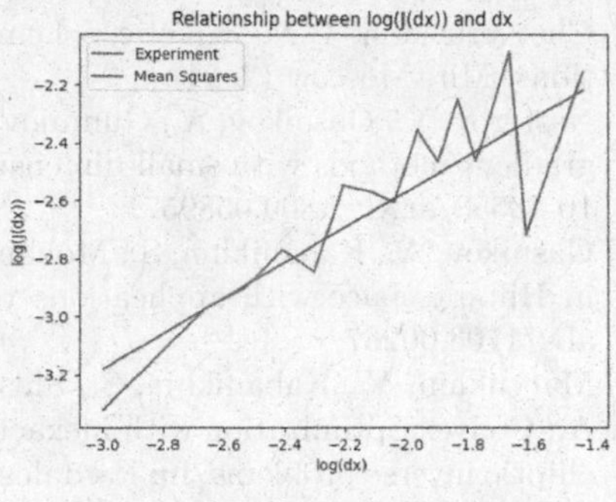

case with unknown q_{exact}

Fig. 5. Graph for log of the target functional and values of C and p for $J(dx) = Cdx^p + o(dx^p)$. Number of iterations of the gradient method = 20, gradient method step size = 0.1. $q_{exact} = x$ in the first case and $f(x) = x$ in the second case accordingly. One can notice that $q(x) = x$ is inconsistent with the boundary conditions, however the algorithm still works.

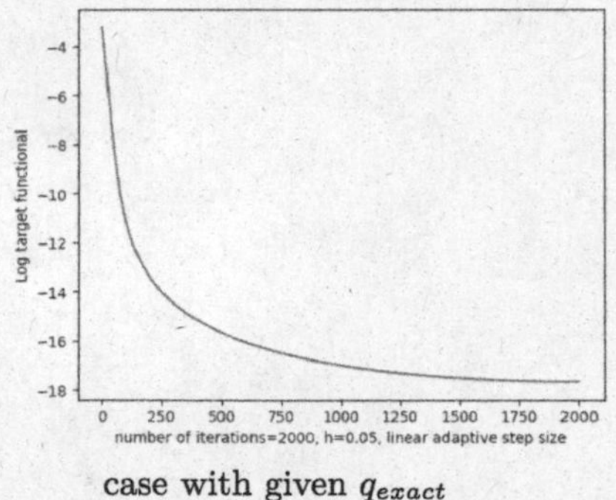

case with given q_{exact}

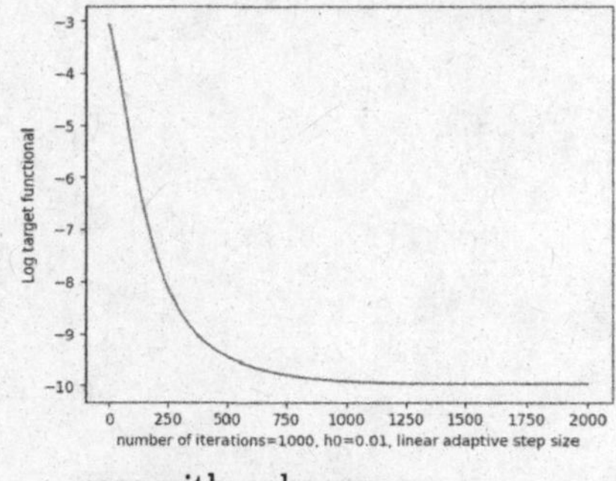

case with unknown q_{exact}

Fig. 6. Graph for the natural log of the target functional with an optimal grid size calculated with respect to the C and p calculated in the previous figure. $q_{exact} = x$ in the first case and $f(x) = x$ in the second case accordingly

References

1. Kabanihin, S.I.: Inverse and Ill-posed Problems, vol. 4. FUE "Publishing House SB RAS", Novosibirsk (2018)
2. Kolmogorov, A., Petrovskii, A., Piscunov N.: A study of the equation of diffusion with increase in the quantity of matter, and its application to a biological problem. Byul. Moskovskogo Gos. Univ. 1(6), 1–25 (1937): Reprinted. In: V.M. Tikhomirov (ed.) Selected Works of A. N. Kolmogorov, vol. 1, pp. 242–270. Kluwer, Dordrecht (1991). Also in: I. G. Petrowsky Selected Works, Part II, pp. 106–132. Gordon and Breach, Amsterdam (1996)
3. Moiseev, T.E., Myshetskaya, E.E., Tishkin, V.F.: On the proximity of solutions of unperturbed and hyperbolized heat equations for discontinuous initial data. Keldysh Inst. Prepr. **86**, 15 (2017)

4. Chetverushkin, B.N.: Kinetic Schemes and Quasi-Gas-Dynamic System of Equations. Mir, Moscow (2004)
5. Nesterov, Y., Gasnikov, A., Guminov, S., Dvurechensky, P.: Primal-dual accelerated gradient methods with small-dimensional relaxation oracle (2017). https://doi.org/10.48550/arXiv.1809.058957
6. Gasnikov, A., Kabanikhin, S., Mohammed, A., Shishlenin, M.: Convex optimization in Hilbert space with applications to inverse problems (2019). https://arxiv.org/abs/1703.00267
7. Matyukhin, V., Kabanikhin, S., Shishlenin, M., Novikov, N., Vasin, A., Gasnikov, A.: Convex optimization with inexact gradients in Hilbert space and applications to elliptic inverse problems. In: Pardalos, P., Khachay, M., Kazakov, A. (eds.) MOTOR 2021. LNCS, vol. 12755, pp. 159–175. Springer, Cham (2021). https://doi.org/10.1007/978-3-030-77876-7_11
8. Krivorotko, O., Kabanikhin, S.I., Zhang, S., Kashtanova, V.: Global and local optimization in identification of parabolic systems. J. Inverse Ill-posed Probl. **28**(6), 899–913 (2020)
9. Polyak, B.T.: Introduction to Optimization. Optimization Software, New York (1987)

Linear Optimization by Conical Projection

Evgeni Nurminski[1] and Roman Tarasov[2(✉)]

[1] Far Eastern Centre for Research and Education in Mathematics, Far Eastern Federal University, Vladivostok, Russia
nurminskiy.ea@dvfu.ru

[2] Skolkovo Institute of Science and Technology, Moscow, Russia
troman85235@gmail.com

Abstract. This article focuses on numerical efficiency of projection algorithms for solving linear optimization problems. The theoretical foundation for this approach is provided by the basic result that bounded finite dimensional linear optimization problem can be solved by single projection operation on the feasible polyhedron. The further simplification transforms this problem into projection of a special point onto a convex polyhedral cone generated basically by inequalities of the original linear optimization problem.

Keywords: Linear optimization · Orthogonal projection · Polyhedral cone

1 Introduction

Linear optimization remains a workhorse of many practical applications and modern general-purpose industrial quality simplex-based algorithms and interior point methods demonstrated remarkable success in the area. Nevertheless there is an intellectual challenge to develop new approaches which may find their applications in these or that situations. In this article, we consider a projection-based algorithm for solving linear optimization problems in the standard form

$$\min_{Ax \leq b} cx \;=\; cx^\star = \max_{uA=c,\; u \leq 0} ub \;=\; u^\star b \tag{1}$$

which is written here with its dual.

Here in more or less standard notation x, c and $x^\star$ are vectors of finite (n) dimensional euclidean space E with the inner product xy and norm $\|x\|^2 = xx$. The right-hand side b of the constraints in (1) belongs to m-dimensional space E' and A is a linear operator ($m \times n$ matrix) from E to E' — $A : E \to E'$. The dual part of this problem has m-dimensional vector u of dual variables and its optimal solution is denoted as $u^\star$.

Research was supported by the subsidy of the Ministry of Science and High Education of Russian Federation 075-02-2023-946 from February 16, 2023.

N. Olenev et al. (Eds.): OPTIMA 2023, CCIS 1913, pp. 61–71, 2024.
https://doi.org/10.1007/978-3-031-48751-4_5

Solution of (1) can be reduced to solving primal-dual system of linear inequalities or convex and even polyhedral feasibility problem (PFP):

$$cx \geq bu, \ Ax \leq b, \ uA = c, \ u \leq 0. \tag{2}$$

The PFP and Convex Feasibility Problem (CFP) for the general convex sets were in gun sights of many mathematicians since the middle of 20 century and projection methods are amongst the most popular for solving it since pioneering works [1–3] (see the extensive review of H. Bauschke and J.M. Borwein [4]). However in the area of linear optimization projection methods were not very successful in practical sense, mainly for slow convergence and computational difficulties of solving multiple high dimensional projection problems for polyhedrons of the general type [10].

We hope nevertheless that our experiments inspire new interest in projection agenda and algorithms. These hopes are based on certain intrinsic properties of projection operation which can operate on reduced basis sets and on additional decomposition possibilities, see f.i. [11].

2 Notations and Preliminaries

As it defined in Introduction let E be a finite-dimensional vector space of the primal variables with the standard inner product xy and the norm $\|x\|^2 = xx$. This space is then self-conjugate with the duality relation induced by the inner product. The dimensionality of this space, if needed, is determined as $\dim(E)$ and the space of dimensionality n when necessary is denoted as E^n. The non-negative part of a space E will be denoted as E_+.

Among the others special vectors and sets we mention the null vector $\mathbf{0}$, vector of ones $\mathbf{1} = (1, 1, \dots, 1)$, and the standard simplex $\Delta = E_+ \cap \{x : \mathbf{1}x = 1\}$. Linear envelope, convex and conical hull of a set X are denoted as $\text{lin}(X)$, $\text{co}(X)$ and $\text{Co}(X)$ respectively.

We define linear operators, acting from E into E' with $\dim(E') = m$ as collections of vectors $\mathcal{A} = \{a^1, a^2, \dots, a^m\}$ with $a^i \in E$ which produce vector $y = (y_1, y_2, \dots, y_m) \in E^m$ according to following relations $y_i = a^i x, i = 1, 2, \dots, m$. In the classical matrix-vector notation vectors $\mathcal{A}$ form the rows of the matrix A and $y = Ax$. At the same time we will consider the row subspace E' as the linear envelope of $\mathcal{A}$:

$$E' = \text{lin}(\mathcal{A}) = \{x = \sum_{i=1}^{m} a^i z_i = A^T z, z \in E^m\} \subset E.$$

The projection operator of a point p onto a closed convex set X in E is defined as

$$p \downarrow X = \text{argmin}_{x \in X} \|p - x\|,$$

that is $\min_{x \in X} \|p - x\| = \|p - p \downarrow X\|$. For closed convex X, this operator is well-defined and Lipschitz-continuous with the Lipschitz constant less or equal

1. The point-to-set projection operation is naturally generalized for sets: $X \downarrow A = \{z = x \downarrow A, x \in X\}$.

We will also notice that this operator is idempotent: $(p \downarrow X) \downarrow X = p \downarrow X$ and linear for projection on linear subspace L of E: $\alpha p \downarrow L = \alpha(p \downarrow L)$ for $\alpha \in \mathbb{R}$ and $(p+q) \downarrow L = p \downarrow L + q \downarrow L$. Of course $p = p \downarrow L + p \downarrow L^{\perp}$.

For a closed convex set X denote as $(X)_z$ its support function

$$(X)_z = \min_{x \in X} xz. \tag{3}$$

In this notation the standard linear optimization problem

$$\min_{Ax \le b} cx = \min_{x \in X} cx \tag{4}$$

becomes just $(X)_c$.

Basically the same holds and for nonlinear problems

$$\min_{x \in X} f(x) = \min_{\bar{x} \in \bar{X}} \bar{c}\bar{x} = (\bar{X})_{\bar{c}} \tag{5}$$

for $\bar{x} = (x, \xi)$, $\bar{X} = \{\bar{x} : x \in X, f(x) \le \xi$, $\bar{c} = (\mathbf{0}, 1)$.

There is a general result which connect support functions with projection [5].

Theorem 1. *Let X — closed bounded subset of E and $c \in E$. Then for any x^0*

$$(X)_c = \lim_{\tau \to \infty} c((x^0 + \tau c) \downarrow X). \tag{6}$$

For the formal correctness of application of the Theorem 1 to the set $\bar{X}$ it is necessary to ensure boundness of $\bar{X}$. This, generally speaking, formal requirement can be easily satisfied by adding an arbitrary upper bound $\bar{f} \ge \inf_{x \in X} f(x)$ for ξ. Toward this purpose any $x^0 \in X$ will provide trivial upper bound $\bar{f} = f(x^0)$.

For linear optimization problems (4) where X is a bounded polyhedron, exact equivalence can be proved [6]:

Theorem 2. *If (4) has a unique solution $x^\star$, then for any x^0 there existsm $\theta_c > 0$ such that*

$$(x^0 - \theta c) \downarrow X = x^\star \tag{7}$$

for any $\theta \ge \theta_c$.

In more details the problem (7) can be written down as:

$$\min_{x \in X} \|x - x^0 + \theta c\|^2 = \min_{y \in X_{\theta,c}} \|y\|^2, \tag{8}$$

where $X_{\theta,c} = X - x^0 + \theta c$ is the original feasible set X, shifted by $x^c = \theta c - x^0$.

If the polyhedron X is described by a system of linear inequalities

$$X = \{x : Ax \leq b\} \tag{9}$$

then

$$X_{\theta,c} = \{x : Ax \leq b^c\}, \tag{10}$$

$b^c = b - A(x^0 - \theta c)$.

The latter problem (8) does not look as something essentially different however it can be transformed into the conical projection problem,

$$\min_{\bar{x} \in \text{Co}(\bar{A})} \|\bar{x} - p\|^2 \tag{11}$$

where $\bar{x} = (x, \xi)$ — is the vector from extended space $\bar{E} = E \times \mathbb{R}$, and $\bar{A} = |A, -b^c|$. The rows of this matrix can be considered as vector of $\bar{E}$. Then $\text{Co}(\bar{A})$ is the conical envelope of these vectors which can be represented as

$$\text{Co}(\bar{A}) = \{\bar{x} = \bar{A}^T z, z \in E^r_+\},$$

where E^r_+ — non-negative orthant of the correspondent dimensionality.

Finally the algorithm for solution of (4) can be represented by the algorithmic scheme 1.

Data: The dataset $(A, b, c$ of the original problem, and scaling constant $\theta > 0$
Result: The solution $x^\star$ of the linear optimization problem (4).
Step 1. Data preparation for projection problem;

$$\begin{aligned} x^c = x - 0c;\ b^c = b - Ax^c; \\ \bar{A} = [A, -b^c];\ \bar{p} = [_n, 1] \end{aligned} \tag{12}$$

Step 2. Solution of the projection problem;

$$\min_{\bar{x} \in \text{Co}(\bar{A})} \|\bar{x} - \bar{p}\|^2 = \|\bar{p} \downarrow \text{Co}(\bar{A}) - \bar{p}\|^2 \tag{13}$$

Step 3. Getting back to (4);
By representing solution of the problem (13) as $\bar{p} \downarrow \text{Co}(\bar{A}) = (y^c, \xi)$, where $y^c \in E$, a $\xi \in \mathbb{R}$ compute

$$x^\star = y^c/\xi + \theta c.$$

Algorithm 1: Solving (4) by projection.

3 Inside Out

The subject of this section is the least norm problem $\min_{x \in X} \|x\|^2$ in an n-dimensional euclidean space E for a bounded closed convex polyhedron X. Here we do not make a great distinction between row and column vectors which are assumed of any type depending on context. Polyhedron X most commonly described as the intersection of half-spaces

$$X = \{x : a^i x \leq \beta_i, i = 1, 2, \ldots, m\} = \{x : Ax \leq b\} \tag{14}$$

where vectors $a^i, i = 1, 2, \ldots, m$ of the dimensionality n can be considered as rows of the matrix A, and the m-vector $b = (\beta_1, \beta_2, \ldots, \beta_m)$ is the corresponding right-hand side vector. It can be considered as the "outer" description of X in contrast with the "inner" description

$$X = \mathrm{co}(\hat{x}^j, \hat{x}^j \in \mathrm{Ext}(X), j = 1, 2, \ldots, J) \tag{15}$$

as the convex hull of the set $\mathrm{Ext}(X)$ of extreme points of the same set X. The later is often considered as "polytope" description. These are equivalent descriptions for this class of polyhedrons/polytopes, but direct conversion between them is complicated as any of them may be exponentially long even for the polynomially long in n, m counterparts.

The polyhedron description is more common so the vast majority of computational algorithms is developed namely for this description of X. The notable exceptions are possibly game problems with probability simplexes and nondifferentiable optimization algorithms in which subdifferentials are approximated by convex hulls of known subgradients. However convex hull-like description has its own computational advantages, for instance as linear optimization problem over convex hulls it has low nm complexity for the trivial direct algorithm and can be reduced to logarithmic complexity if parallel computations allowed. In [7] we considered the transformation of the least norm problem with the polyhedral description (14) into the close relative of (15) with practically the same datasize as (14). The original version of this transformation was rather convoluted and here we present its alternative derivation which uses basically only standard duality arguments.

To begin with we expand our basic space E with one additional variable into $\bar{E} = E \times \mathbb{R}$ and transform the initial least norm problem into something which is almost homogeneous:

$$\min_{Ax \leq b} \frac{1}{2}(\|x\|^2 + 1) = \min_{\substack{\bar{A}\bar{x} \leq 0 \\ \bar{e}\bar{x} = 1}} \frac{1}{2}\|\bar{x}\|^2 \tag{16}$$

with $\bar{A} = |A, -b|$, $\bar{x} = (x, \xi)$, and vector $\bar{e} = (0, 0, \ldots, 0, 1) \in \bar{E}$. The saddle point reformulation of this problem goes as follows:

$$\min_{\substack{\bar{A}\bar{x} \leq 0 \\ \bar{e}\bar{x} = 1}} \tfrac{1}{2}\|\bar{x}\|^2 = \tfrac{1}{2}\|\bar{x}^\star\|^2 = \max_{u \geq 0,\ \theta} \min_{\bar{x}} \{\tfrac{1}{2}\|\bar{x}\|^2 + \theta(1 - \bar{e}\bar{x}) + u\bar{A}\bar{x}\} =$$
$$\max_{u \geq 0,\ \theta} \{\theta + \min_{\bar{x}} \{\tfrac{1}{2}\|\bar{x}\|^2 + (u\bar{A} - \theta\bar{e})\bar{x}\} = \max_{u \geq 0,\ \theta} \{\theta - \tfrac{1}{2}\|u\bar{A} - \theta\bar{e}\|^2\}$$

Introducing the cone $\mathcal{K} = \{z : z = u\bar{A}, u \geq 0\}$ we can rewrite the last problem as

$$-\min_{\theta} \{\frac{1}{2} \min_{z \in \mathcal{K}} \|z - \theta\bar{e}\|^2 - \theta\} = -\min_{\theta} \{\frac{1}{2}\theta^2 \min_{z \in \mathcal{K}} \|z - \bar{e}\|^2 - \theta\} = -\min_{\theta} \{\frac{1}{2}\gamma^2\theta^2 - \theta\} = 1/2\gamma^2,$$

where we made use of $\alpha\mathcal{K} = \mathcal{K}$ for any $\alpha > 0$ and denoted $\gamma^2 = \min_{z \in \mathcal{K}} \|z - \bar{e}\|^2$. The solution of the last minimum is attained for $\theta^\star = 1/\gamma^2$. As solution of $\min_{z \in \mathcal{K}} \|z - \bar{e}\|^2 = \|z^\star - \bar{e}\|^2$ is unique we obtain $\bar{x}^\star = \theta^\star(z^\star - \bar{e})$.

4 Numerical Experiments

For numerical experiments we implemented the projection algorithm in the `OCTAVE` open-source MATLAB-like system [8]. For comparison, we used the `GLPK` linear programming kit implemented in C programming language and also built into `OCTAVE` as internal function. This function however has a limited functionality in comparison with stand-along `GLPK` solver [9], but still allows for some basic comparison of projection algorithm and contemporary symplex method.

We considered the following 2 types of linear optimization problems. The first one consists of linear optimization problems of different dimensions with the following structure

$$\min_{\substack{Ax \leq \mathbf{0} \\ -g \leq x \leq f}} cx = \min_{\bar{A}x \leq \bar{b}} cx \tag{17}$$

where $\bar{A} = [A;\ I;\ -I]$, $\bar{b} = [\mathbf{0};\ f;\ g]$. Here I is the identity matrix, the elements of the matrix A and the vectors f, g are randomly generated independently and uniformly from the segment $[0, 1]$, and the elements of the vector c are generated from the segment $[-5, 5]$. Also for matrices A, B) of matching dimensions $[A; B]$ denotes (by following `MATLAB`/`OCTAVE` convention) a stacked up A, B, that is $[A^T B^T]^T$. This form of problems on one hand ensures the feasibility ($\mathbf{0}$ is a feasible point) and boundness (due to simple two-sided bounds on variables) but intersection of the cone and the box produces sufficiently demanding feasible polyhedron.

However, this form provides one-sided advantages to `GLPK` as it immediately makes use of built-in presolver which takes into account sparsity and specific structure of box constraints. Of course our experimental implementation of projection algorithm is missing such advanced features, so our second type of test problems was a simple modification of (17) which consisted in replacing x with the new variable z, such that $x = Qz$ with 100%-dense random unitary matrix Q. Then the test problems become

$$\min_{\bar{A}_Q z \leq \bar{b}} c_Q z \tag{18}$$

where $\bar{A}_Q = \bar{A}Q$, $c_Q = cQ$. After such changes the problem constraints became fully dense and GLPK presolver does not interfere with optimization.

Figure 1 shows that presolver keeps the GLPK solver noticeably faster, but projection algorithm demonstrates at least the similar dynamics when problem dimensions increase.

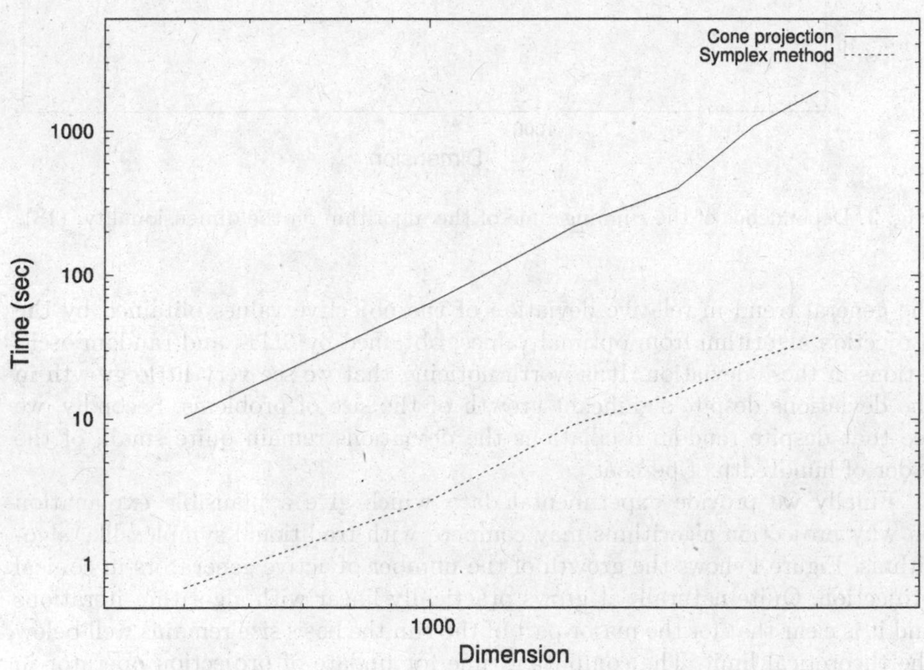

Fig. 1. Dependence of the running time of the algorithm on the dimensionality: (17).

For dense problems (18) projection algorithm demonstrates however some speed-up (Fig. 2) which slightly increases when problem size grows. It was a pleasant surprise that despite the very different levels of implementation the projection algorithm was faster than GLPK.

The important characteristic of the quality of solution is its relative accuracy which the projection algorithm manages to attain. The Fig. 3 demonstrates

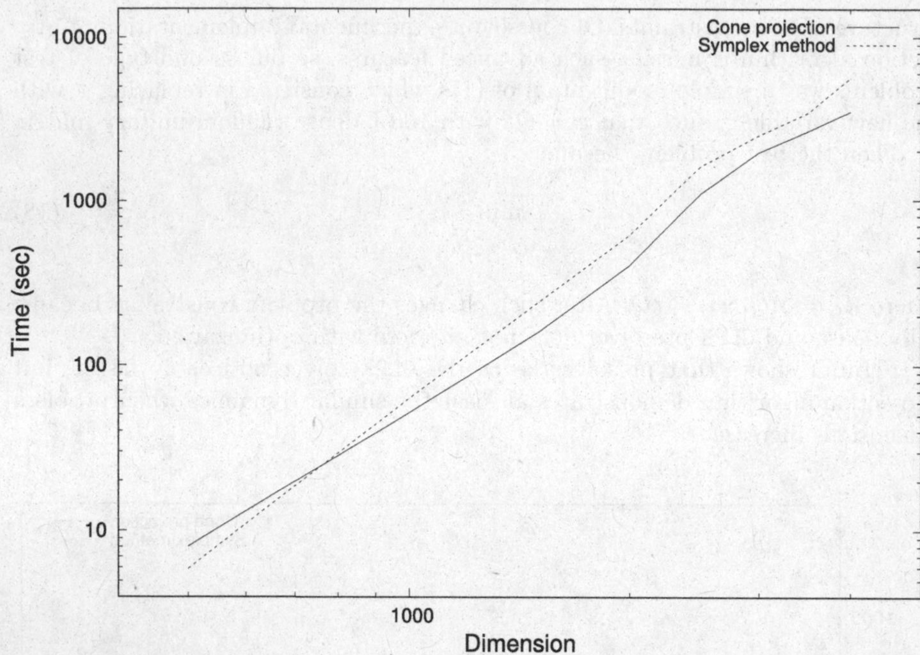

Fig. 2. Dependence of the running time of the algorithm on the dimensionality: (18).

the general trend in relative deviation of the objective values obtained by the projection algorithm from optimal values, obtained by GLPK, and random oscillations in these deviation. It is worth noticing that we see very little growth in the deviations despite significant growth of the size of problems. Secondly, we see that despite random oscillations the deviations remain quite small, of the order of hundredth of percent.

Finally we provide experimental data which give a plausible explanation on why projection algorithms may compete with traditional symplex-like algorithms. Figure 4 shows the growth of the number of active generators in conical projection. Quite naturally it grows practically linear with algorithm iterations and it is clear that for the major part of the run the basis size remains well below the theoretical limit. The computing time for update of projection operator in projection algorithm also follows the growth of the basis size (Fig. 5). It implies that the best part of it essentially smaller then the maximal run time.

Figure 1 and 2 demonstrate generally polynomial growth of computing time both for GLPK and projection algorithm as a function of problems size. We provide the additional Fig. 6 to demonstrate more explicitly the difference in computing time between GLPK and projection algorithm. It can be seen from Fig. 6 that despite rather large fluctuations in the ratio between computing times for GLPK and projection algorithm the general tendency is in favor of projection when problems become larger.

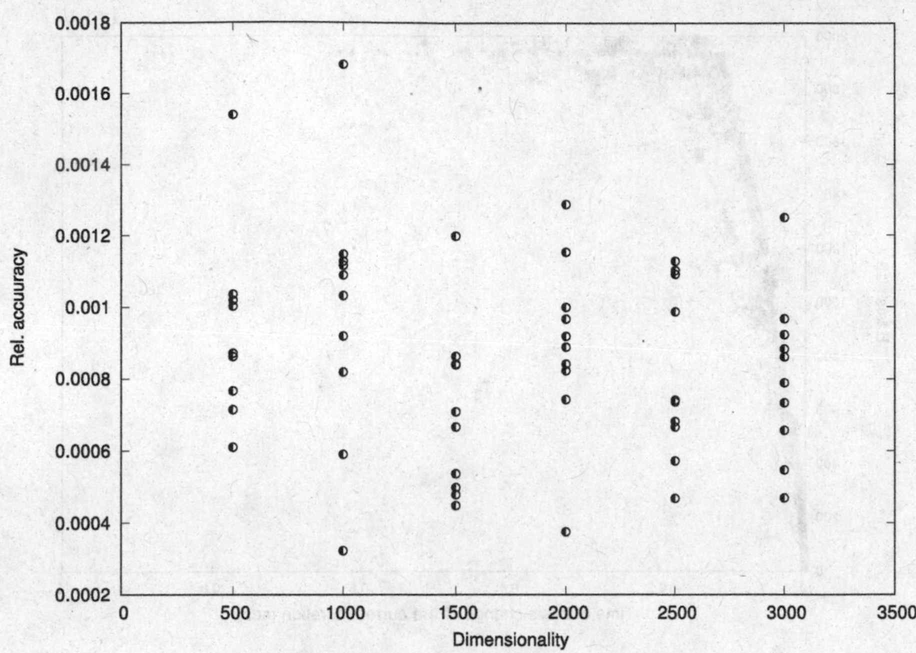

Fig. 3. Deviations from the optimums

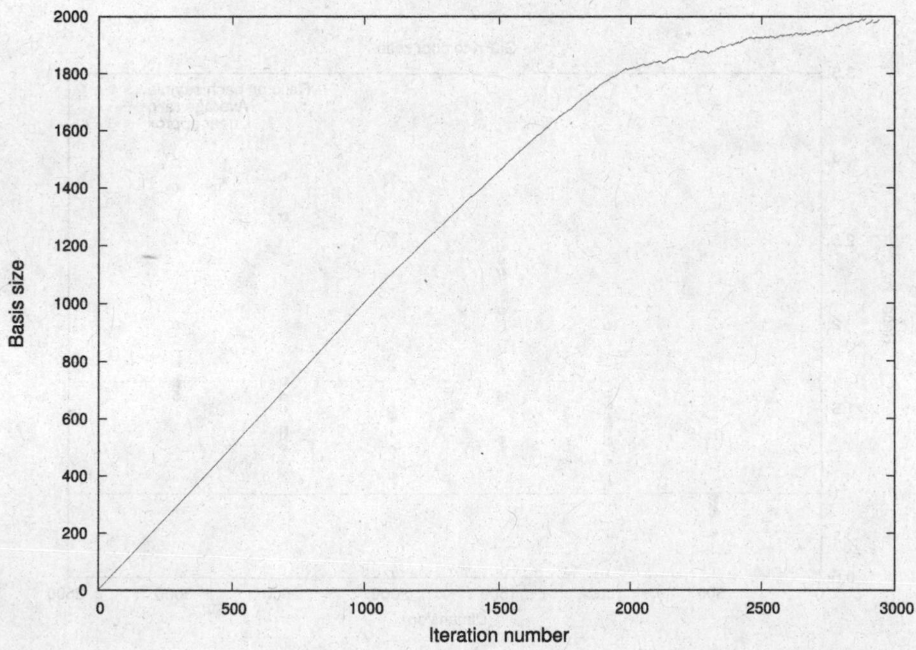

Fig. 4. Analysis of the projection algorithm: Basis size over iterations.

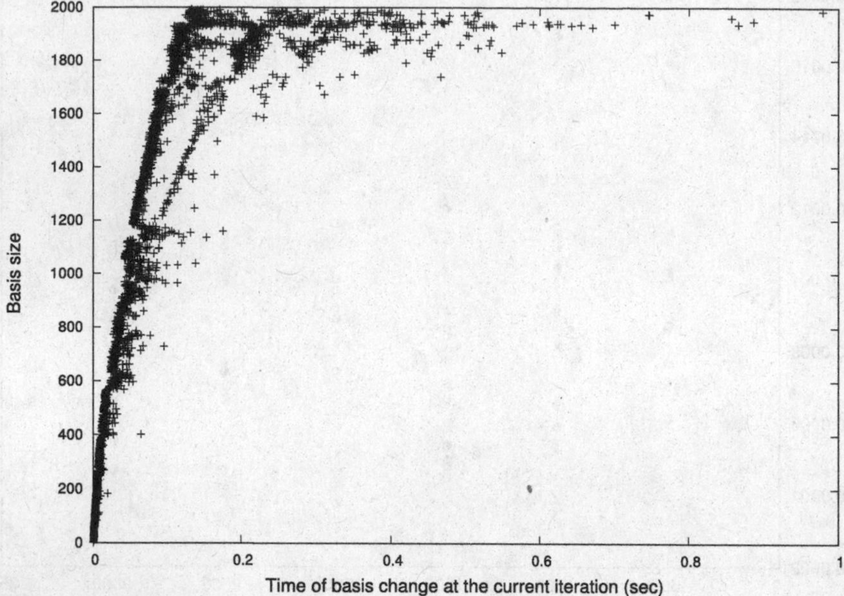

Fig. 5. Analysis of the projection algorithm: The time it takes to recalculate the basis depending on its size.

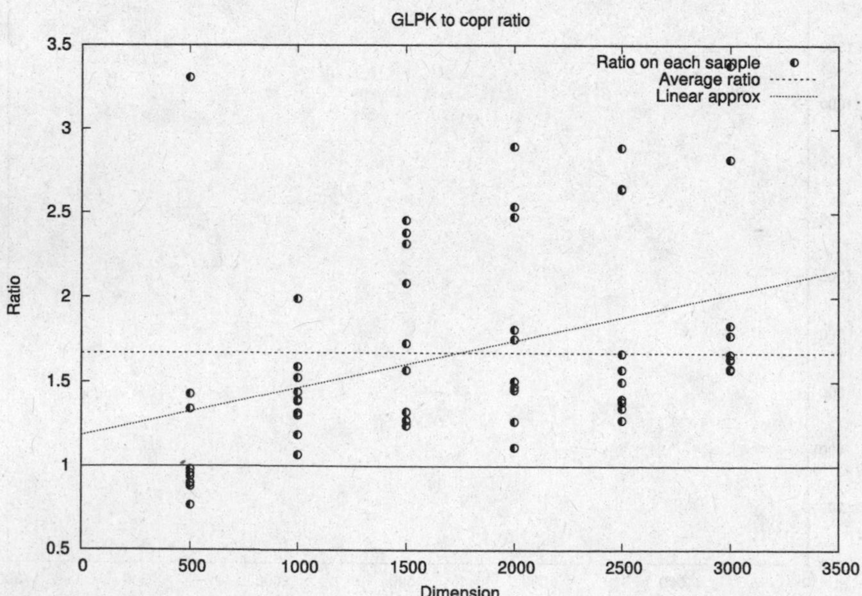

Fig. 6. The ratio of the running times of GLPK to copr on each sample.

Acknowledgements. This research was initiated and supported at Sirius supplementary educational program «Actual methods of information theory and optimization», November, 2022.

References

1. Kaczmarz, S.: Angenäherte Auflösung von Systemen Linearer Gleichungen. Bull. Int. de l'Académie Polonaise des Sciences et des Lettres **35**, 355–357 (1937)
2. Cimmino, G.: Calcolo approssimate per le soluzioni dei sistemi di equazioni lineari. La Ricerca scientifica ed il Progresso tecnico nell' Economia nazionale **9**, 326–333 (1938)
3. von Neumann, J.: Functional operators, Vol. II. The geometry of orthogonal spaces. Ann. Math. Stud. **22** (1950)
4. Bauschke, H., Borwein, J.: On projection algorithms for solving convex feasibility problems. SIAM Rev. **38**(3), 367–426 (1996)
5. Nurminski, E.: Equivalencies in convex optimization. Disc. Anal. Oper. Res. **30**(2) (2023)
6. Nurminski, E.: Single-projection procedure for linear optimization. J. Global Optim. **66**(1), 95–110 (2016)
7. Nurminski, E.: Projection onto polyhedra in outer representation. Comput. Math. Math. Phys. **48**(3), 367–375 (2008)
8. GNU Octave. Scientific Programming Language https://octave.org/
9. GLPK (GNU Linear Programming Kit) https://www.gnu.org/software/glpk/
10. Gould, N.: How good are projection methods for convex feasibility problems? Comput. Optim. Appl. **40**, 1–12 (2008)
11. Censor, Y., Chen, W., Combettes, P., et al.: On the effectiveness of projection methods for convex feasibility problems with linear inequality constraints. Comput. Optim. Appl. **51**, 1065–1088 (2012)

Discrete and Combinatorial Optimization

LIP Model in Solving RCPSP at the Flow Type Production

V. A. Rasskazova(✉)

Moscow Aviation Institute, National Research University, Moscow, Russia
varvara.rasskazova@mail.ru

Abstract. The paper is dedicated to investigating the problem of optimal planning at the steelmaking converter shop-floor. An integer linear program is proposed, which takes into account all major technological constraints. The objective is to minimize the total cost of daily task processing, leading to a significant reduction in overall production costs. Numerical experiments were conducted using real-world data, demonstrating the strength and effectiveness of the proposed approach.

Keywords: Combinatorial Optimization · Linear Integer Programming (LIP) · Scheduling Theory

Introduction

Linear integer programming (LIP) is widely used in various fields, including transportation and industrial planning. In [1,2], LIP models were developed to address applied problems related to transportation planning in railway systems. Reference [3] explores the use of LIP methods to solve problems involving network flows. Scheduling theory problems, discussed in [4,5], were effectively transformed into associated integer linear programs. In [6], LIP models were proposed for solving industrial planning problems. The main challenges when reducing applied problems to associated integer linear programs lie in the complex system of constraints inherent in any real-world problem. In this paper, we present a LIP model that tackles the problem of optimal planning at the steelmaking converter shop-floor, fully capturing the true technological constraints of the system.

References [7–10] provide an extensive overview of LIP applications, as well as modern methods for solving them. In [7,8], classical statements and methods for solving linear and integer programming problems, including Boolean programming, are examined in detail. References [9,10] place special emphasis on the development of LIP models for solving various applied problems related to management, planning, and decision-making. In this paper, we propose a scalable technique that can be extended to address related optimization problems in the field of metallurgical production. Examples of such related problems include

Supported by Russian Science Foundation. Project No. 23-21-00293.

N. Olenev et al. (Eds.): OPTIMA 2023, CCIS 1913, pp. 75–88, 2024.
https://doi.org/10.1007/978-3-031-48751-4_6

transportation logistics at the shop-floor and out-of-furnace production, among others.

Given that LIP is a well-known $\mathcal{NP}$-complete problem, there are ongoing efforts to actively develop and improve methods for solving it. An extensive overview of modern techniques for solving LIP problems is provided in [11,12]. However, in this paper, our focus is not on analyzing methods for solving LIP problems. The contribution lies in the development of a LIP model that enables the effective and high-quality solution of the specific applied problem at hand.

References [13–15] discuss various applied problems related to optimal planning in metallurgical production. In [13], a methodology is proposed for solving complex decision-making problems in management within the context of metallurgical production. References [16–18] introduce robust optimization approaches to address related problems in steelmaking production. Algorithms aiming to improve the final product quality at the rolling mill are developed in [14,15]. However, there is relatively limited coverage of the more specific class of problems concerning dispatching at the metallurgical shop-floor in current publications. This paper focuses on investigating the problem of optimal planning for the fundamental technological route from the releasing machine to the casting machine. It is worth noting that the effective execution of this stage is crucial for subsequent decisions that contribute to the overall profitability of the production process.

It is widely recognized that various instances of Machine Scheduling and Shop Scheduling Problems can be effectively reduced to the Resource Constrained Project Scheduling Problem (RCPSP). The problem investigated in this paper shares similar characteristics with RCPSP. In this case, a schedule needs to be constructed for processing melts, taking into account the given resources of converters and a set of technological constraints. Thus, the set of melts corresponds to a set of tasks, denoted as j with $j = 1, 2, ..., n$, and the set of converters represents the set of resources, denoted as k with $k = 1, 2, ..., K$. Additionally, each task j has a specified processing interval of $[r_j, d_j]$. Consequently, the problem at hand can be classified as an RCPSP variant with fixed durations for processing at each resource and constraints on the start and finish times for each task. For a comprehensive overview of various RCPSP cases, readers can refer to [19].

The RCPSP remains an $\mathcal{NP}$-complete problem even in its simplest formulation. The best exact algorithm, proposed in [20], is capable of solving instances with up to $n = 60$ tasks. However, it is clear that such a limitation is insufficient when dealing with real-world applied problems. Even in the case of $K = 1$ (a single resource), there are no polynomial-time algorithms that guarantee a solution with a known error bound. Popular approaches involve constructing upper and lower bounds for optimal solutions of RCPSP. The most efficient bounds utilize a linear programming framework, as introduced in [21,22]. List Scheduling (LS) is a commonly used polynomial-time algorithm for solving RCPSP. However, it does not provide an optimal solution for even small instances. Furthermore, it fails to account for significant technological constraints related to the number of tasks processed within a given time interval. The same limitations apply to various modifications of LS, including Ant Colony and other metaheuristics. Therefore, it is necessary to propose a new model that incorporates constraints related to both task duration and sequential processing, as well as other constraints mentioned earlier. This paper presents a Linear Integer Programming (LIP) model that addresses these challenges. By using LIP, one can solve large instances within a reasonable time frame, while incorporating all relevant technological constraints.

The paper is organized as follows. Section 1 (Introduction) provides an overview of the problem and its constraints. Section 2 (Model Domain) delves into the specific domain of the model being developed. Section 3 (Model Constraints and Objectives) discusses the constraints and objectives of the model in detail. Finally, Sect. 4 (Numerical Experiment and Results) presents a description of the numerical experiment conducted and provides the corresponding results.

1 Problem Statement

The problem at hand involves the dispatching of the converter shop-floor in the context of metallurgical production. Several key concepts are essential to understanding the problem:

- Melt of liquid metal: This is the material that moves through the shop-floor using ladles and cranes. It undergoes chemical and thermal processing to achieve the desired target chemical composition, as specified by the user based on technological requirements.
- Converter (CV): A converter is a machine used for purging melts, aiming to attain the initial chemical composition necessary for further processing.
- Out-of-furnace processing machine (PM): This machine is responsible for various tasks such as heating, vacuuming, degassing, desulfurization, and other technological processes. Its purpose is to maintain the required temperature and control the chemical composition of the melts.
- Continuous steel casting machine (CCM): The CCM forms metal briquettes from the melts, adhering to predetermined standards for chemical composition and fulfilling technological requirements for the final product. This machine plays a crucial role in the production process, as the metal briquettes become the input units for further processing at the rolling shop-floor, eventually transforming into metal sheets, which serve as the final product.

These concepts form the foundation of the problem and will guide the development of the model.

The initial stage of dispatching the converter shop-floor involves creating a projected schedule for the movement of melts from the converter (CV) to the continuous steel casting machine (CCM). This allows for the planning and coordination of the overall production process. The subsequent stage entails constructing an operational traffic schedule, which goes into further detail by specifying the route for each assignment, such as CV-PM1-PM2-...-CCM. However, the problem of developing the operational traffic schedule is beyond the scope of this paper.

Let us now focus on the problem of constructing a forecast schedule for the movement of melts at the converter shop-floor, which is the first stage mentioned earlier. For this purpose, we consider a planning period of 24 h, starting at 15:30 on the first day and ending at 15:30 on the second day. The given input data includes:

1) A set of casting series denoted as:

$$S = \{s_1, s_2, ..., s_n\},$$

where each series s_i is represented by parameters:

$$(\alpha_i, \sigma_i, \mu_i, \hat{\mu}_i, \eta_i, \omega_i),$$

Here, α_i represents the number of the continuous casting machine assigned to the series, and σ_i denotes the start time of casting for the corresponding series s_i. The parameters μ_i and $\hat{\mu}_i$ indicate the minimum and maximum allowed treatment time for each melt within the series s_i, respectively. Additionally, η_i represents the number of melts in the series, and ω_i denotes the casting cycle duration for each melt.

The minimum and maximum treatment times determine the required minimum and the allowable maximum time for the melt's entire lifecycle, starting from the CV and ending at the CCM at the start of casting. It is considered unacceptable to assign a CV-CCM pair if the time from the CV departure until arrival at the CCM is less than the minimum treatment time. Similarly, the assignment is considered unacceptable if the time from CV departure until arrival at the CCM exceeds the maximum treatment time. Other important parameters associated with the casting series, such as steel assortment (chemical composition of the steel), recommended technological routes, and the relative sequence number of melting within the series, are not considered in this paper as they do not significantly affect the problem under investigation.

2) The set of technological inspections and repairs of CVs (TI) is denoted as:

$$R = \{\rho_1, \rho_2, ..., \rho_m\},$$

Each ρ_i is represented by parameters:

$$(\kappa_i, \tau_i, \hat{\tau}_i),$$

where κ_i represents the converter number, and τ_i and $\hat{\tau}_i$ indicate the start and finish times, respectively, for the corresponding technological inspection ρ_i.

3) The shift plan for CVs is represented as:

$$P = \{p_1, p_2, p_3\},$$

where $p_i, i \in [1, 3]$, denotes the number of melts to be released from all converters during the corresponding shift of the planning period. In the context of the applied problem, shifts 1, 2, and 3 correspond to the time periods from 15:30 to 23:30 on the first day of the planning period, from 23:30 on the first day to 07:30 on the second day, and from 07:30 to 15:30 on the second day. The shift plan for converters is designed to maintain regularity in production and is specified in terms of forming daily casting tasks.

Additionally, it is important to note that:

$$p_j \neq \frac{1}{3} \cdot \sum_{i=1}^{n} \eta_i, j \in \{1, 3\}.$$

This condition is due to the fact that casting series from the set S may fall outside the planning period. This is most common for series that fall outside shift 3 (the last shift), but it can also occur for those outside shift 1 (the first shift).

4) The operating conditions of CVs are represented as:

$$K = \{k_1, k_2, ..., k_l\},$$

where each element k_i in the set K characterizes the operating conditions of CV number i. In this case, the actual operating conditions k_i are determined by the following parameters:

$$(\beta_i, \hat{\beta}_i).$$

Here, β_i represents the purge cycle time, which is the duration of melt processing on CV number i. This purge cycle time is necessary and sufficient for preparing the melt for further out-of-furnace processing. The parameter $\hat{\beta}_i$ indicates the inter-melt delay time, which is the time required for the preparation of CV number i when purging two sequential melts one after another. There is another parameter related to inter-shift delay, but in terms of implementing corresponding constraints, it does not significantly differ from the inter-melt delay. Therefore, the parameter of inter-shift delay will not be considered further in this paper.

Let us shift our focus back to the set of casting series S. Each element $s_i \in S$ is connected with a subset of melts:

$$\Pi(i) = \{\pi_{1i}, \pi_{2i}, ..., \pi_{ni}\},$$

where $n = \eta_i$, and each π_{ji} is characterized by parameters:

$$(\alpha_{ji} = \alpha_i, \sigma_{ji}, \mu_{ji} = \mu_i, \hat{\mu}_{ji} = \hat{\mu}_i).$$

In other words, each melt in the series needs to be cast at the same CCM, following the same minimum and maximum allowed treatment times. However, the casting start time for each melt in the series needs to be determined directly, taking into account the given number of melts in the series and the casting cycle.

This can be achieved as follows:

$$\sigma_{1i} = \sigma_i$$
$$\sigma_{2i} = \sigma_i + \omega_i$$
$$\ldots$$
$$\sigma_{ji} = \sigma_i + (j-1) \cdot \omega_i,$$

where $i \in \{1, n\}$ and $j \in \{1, \eta_i\}$. Thus, each casting series means a separate set of melts, and the goal of the proposed investigation is to address the following problem.

Problem (Construction of the Forecasting Schedule for Melts Movement at the Converter Shop-floor): Given the resources of CVs, the set of casting series $S = \{s_1, s_2, ..., s_n\}$ with connected sets of melts $\Pi(i) = \{\pi_{ji}\}$ for all $i = \{1, n\}$ and $j = \{1, \eta_i\}$, as well as the shift plan for CVs $P = \{p_1, p_2, p_3\}$, the set of TIs $R = \{\rho_1, \rho_2, ..., \rho_m\}$, and the operating conditions of CVs $K = \{k_1, k_2, ..., k_l\}$, the objective is to construct a mapping of the form:

$$f \colon \bigcup_{i \in \{1,n\}} \bigcup_{j \in \{1,\eta_i\}} \{\pi_{ji}\} \longrightarrow \{(t_{ji}, k)\}, \tag{1}$$

where t_{ji} corresponds to the departure time from CV with number k. The constructed mapping should satisfy the following conditions for each assignment in (1):

1. $\mid \sigma_{ji} - t_{ji} \mid \in [\mu_{ji}, \hat{\mu}_{ji}]$: The time from the departure of a melt from CV to its casting start at CCM should be within the minimum and maximum treatment times for the corresponding melt π_{ji}.
2. $t_{ji} \notin [\tau_l, \hat{\tau}_l], l \in \{1, m\}$: No assignment should use a CV during a planned TI on that CV.
3. $\mid t_{j_1 i_1} - t_{j_2 i_2} \mid \geq \beta_k + \hat{\beta}_k$: For any sequence of melts assigned to the same CV, the operating conditions (purge cycle and inter-melt delay) should be satisfied.
4. $t_{j_1 i_1} \neq t_{j_2 i_2}$: No two different melts should be simultaneously released from the same CV.
5. $\mid \bigcup \{(t_{ji}, k) \colon t_{ji} - \beta_k \geq T_s^{st}, t_{ji} - \beta_k < T_{s+1}^{st}\} \mid = p_s$: The total number of assignments, where the departure from the CV falls within shift s (where $s \in \{1, 3\}$), should correspond to the given shift plan. Here, T_s^{st} corresponds to the start time of shift s, specified as T_1^{st} for 15:30, T_2^{st} for 23:30 on the first day, T_3^{st} for 07:30, and T_4^{st} for 15:30 on the second day of the planning period.

The goal is to construct a forecast schedule that meets these conditions for melts movement at the converter shop-floor.

Throughout the subsequent discussion, the condition in the form of (1) will be referred to as "constraints regarding arrival at CCM". Hence, when solving the problem at hand, each melt from each series must be assigned a departure time from a CV, taking into account the constraints outlined in items 1)-5) above. The constraints in item 2) will be referred to as "constraints regarding TI", while the constraints in item 3) will be referred to as "constraints regarding operation of CVs". The constraints in item 4) will be referred to as "constraints regarding the number of melts output from CVs", and the constraints in item 5) will be referred to as "constraints regarding the shift plan". To solve the problem, a LIP model is proposed.

2 Model Domain

Based on the problem statement, it is evident that each variable in the model being developed can be represented as a set of parameters in the form (π_{ji}, t_{ji}, k). For clarity and further explanation, we can present this set in a broader form as $(i, j, k, t_{ji}, \alpha_{ji}, \sigma_{ji})$, which represents the series number, melt number, CV number, departure time from the CV, CCM, and the casting start time, respectively. It is important to note that the broader form is a direct expansion of the first form, as parameters i and j, as well as α_{ji} and σ_{ji}, can be uniquely determined from the parameter π_{ji}.

The domain of the function defined by (1) is the set of pairs (t_{ji}, k) representing the departure time from the CV and the corresponding CV. At the outset, the departure time from the CV is considered to be arbitrary. However, considering the constraints outlined in item 1), the departure time becomes dependent on the parameters μ_i and $\hat{\mu}_i$ for each series s_i, with $i \in \{1, n\}$.

We introduce a parameter h_k for all CVs, denoted by $k \in \{1, l\}$. This parameter represents the periodicity at which pairs (t_{ji}, k) can be generated for potential assignments of the form CV-CCM. For the purposes of this paper, we will consider a fixed value of $h = 10$ minutes for all CVs. However, in practice, this value can be dynamic and dependent on the range of the corresponding interval $[\mu_i, \hat{\mu}_i]$ for which the subset of variables is generated.

Without loss of generality, let us consider a subset of variables for a specific casting series. Let i be a fixed casting series index, where $i \in \{1, n\}$, and j be a fixed melt index for that series, where $j \in \{1, \eta_i\}$. As mentioned before, each element π_{ji} is associated with a set of parameters α_{ji}, which represent the specific CCM that needs to be delivered for casting. Additionally, σ_{ji} represents the casting start time calculated using the starting time of the series with respect to the values of η_i and ω_i. Furthermore, μ_{ji} and $\hat{\mu}_{ji}$ denote the minimum and maximum treatment times for the corresponding melt. Now, for each CV k, a subset of variables corresponding to the j-th melt from the i-th series can be represented as $\{x_{ijkl}\}$, where the parameters of the variables are given by

$$(i, j, k, t_{ji} = \sigma_{ji} - \hat{\mu}_{ji} + h \cdot (l - 1), \alpha_i, \sigma_{ji}), \tag{2}$$

Here, $l \in \left\{1, \left[\frac{1}{h} \cdot (\hat{\mu}_i - \mu_i) + 1\right]\right\}$ represents the number of variables that correspond to the departure time from CV and fall within the permissible interval of treatment time for the corresponding melt.

We define $x_{ijkl} = 1$ if the melt π_{ji} of series i is assigned to depart from CV k at the time moment t_{ji}, which is determined by the casting start time σ_{ji}. Similarly, $x_{ijkl} = 0$ otherwise.

3 Model Constraints and Objectives

In Sect. 1, we introduced several types of constraints that are considered in this paper. In this section, we will provide more details on each of these constraints.

3.1 Constraints Related to Arrival at CCM

Let us denote the set of variables in the model as $X = \{x^r\}$, where $r \in \{1, N\}$. Each variable x^r is represented by the parameters described in Eq. (2), i.e.,

$$x_{ijkl} = x^r \Rightarrow \begin{cases} i(r) = i, \\ j(r) = j, \\ k(r) = k, \\ t(r) = t_{ji}, \\ \alpha(r) = \alpha_i, \\ \sigma(r) = \sigma_{ji}. \end{cases}$$

It can be derived from equation (1) that for each melt in each series, a specific CV with an exact departure time needs to be assigned. Therefore, we have the constraint:

$$\sum_{r:\ \alpha(r)=\alpha, \sigma(r)=\sigma} x^r = 1, \tag{3}$$

for all α and σ, which represent the CCM number at the shop-floor and the casting start time, respectively.

3.2 Technological Inspections (TI) Constraints

The constraints related to TI mean that no variable can take the value of 1 if there exists $\rho_i \in R$ that satisfies the following conditions:

$$\begin{cases} k(r) = \kappa_i, \\ t(r) \in [\tau_i, \hat{\tau}_i], \\ t(r) - \beta_{k(r)} \in [\tau_i, \hat{\tau}_i]. \end{cases}$$

In other words, no variable can be included in the solution if a CV is used during the planned TI. Therefore, we have the constraint:

$$\sum_{r:\ k(r)=\kappa_i, t(r)\in[\tau_i+\beta_{k(r)}; \hat{\tau}_i]} x^r = 0, \tag{4}$$

for all TIs enumerated by $i \in \{1, m\}$.

It is clear that if some variables are equal to 0, the dimensionality of the problem can be significantly reduced. Therefore, the constraints in Eq. (4) can serve as a filter for the constraints in Eq. (3). Specifically, from the set of constraints in Eq. (3), we can remove all terms that are included in the constraints of Eq. (4).

3.3 Constraints Related to CV Operation

The constraints regarding the operation of CVs prohibit processing multiple melts on the same CV with a delay that is less than the blow cycle. For a fixed $r \in \{1, N\}$, we define the following subset:

$$X'(r) = \{x^r\} \cup \{x^s \colon s \neq r, k(s) = k(r), |t(s) - t(r)| \leq \beta_{k(r)} + \hat{\beta}_{k(r)}\},$$

where $s \in \{1, N\}$, and $k(\cdot)$ corresponds to the CV number at the shop-floor. The constraints regarding the operation of CVs can then be expressed as:

$$\sum_{x^s \in X'(r)} x^s \leq 1, \tag{5}$$

for all $r \in \{1, N\}$.

3.4 Constraints Related to the Number of Melts Output from CVs

Constraints regarding the number of melts output from CVs are similar to the constraints in Eq. (5), and are also characterized by inequalities. These constraints are designed to ensure the uniqueness of the melt output from a specific CV at a given time moment.

Due to the irregularity of the set of variables, where no casting series starts at the same time and with the same cycle, these constraints may degenerate into an empty set. In other words, if no pair of variables is equal in terms of the parameters $k(\cdot)$ and $t(\cdot)$, then the conditions of the constraints regarding the number of melts output from CVs cannot be satisfied. However, if these conditions are satisfied, the constraints in Eq. (5) may or may not include the additional conditions.

For a fixed $r \in \{1, N\}$, we define the following:

$$X''(r) = \{x^r\} \cup \{x^s \colon s \neq r, k(s) = k(r), t(s) = t(r)\},$$

where $s \in \{1, N\}$, and $k(\cdot)$ corresponds to the CV number at the shop-floor. The constraints regarding the number of melts output from CVs can then be expressed as:

$$\sum_{x^s \in X''(r)} x^s \leq 1, \tag{6}$$

for all $r \in \{1, N\}$.

It is important to note that the subsets of the form $X'(r)$ and $X''(r)$ may overlap. Therefore, any well-known methods can be used to extract a subset of maximal inequalities of the form in Eqs. (5) and (6).

3.5 Constraints Related to the Shift Plan

Constraints related to the shift plan are characterized by equality. These constraints require the construction of a schedule for the movement of melts such that the total number of melts output from all CVs during each shift matches a given value. By satisfying this equality, the condition regarding the regular operation of CVs is also fulfilled. Specifically, we have the constraint:

$$\sum_{r:\, T_s^{st} \le t(r) - \beta_{k(r)} < T_{s+1}^{st}} x^r = p_s, \tag{7}$$

for all $s \in \{1, 3\}$, corresponding to each shift of the planning period.

As mentioned before, the constraints in Eq. (4) can serve as a filter for the constraints in Eqs. (5)–(7). This means that if a variable is included in the constraints in Eq. (4), it can be excluded from the constraints in Eqs. (5)–(7) because it equals 0.

3.6 Objective Function

Given the constraints described in Eqs. (3)–(7), the problem of constructing a forecast schedule for melts movement at the shop-floor can be formulated as an integer linear program. The objective is to minimize the total processing time for all melts. Let us define the following:

$$c(r) = \sigma(r) - t(r), \quad r \in \{1, N\},$$

where $c(r)$ corresponds to smaller values of coefficients for variables that model longer processing time.

Thus, the objective is to minimize the following:

$$\sum_{r=1}^{N} x^r \cdot c(r) \longrightarrow \min, \tag{8}$$

subject to the constraints:

$$\begin{cases} x^r + \sum_{q, q \ne r} \{x^q \colon \alpha(q) = \alpha(r), \sigma(q) = \sigma(r)\} = 1, \\ x^r + \sum_{q, q \ne r} \left\{x^q \colon k(q) = k(r), |t(q) - t(r)| \le \beta_{k(r)} + \hat{\beta}_{k(r)}\right\} \le 1, \\ \sum_r \{x^r \colon T_s^{st} \le t(r) - \beta_{k(r)} < T_{s+1}^{st}\} = p_s, \\ \sum_r \{x^r \colon k(r) = \kappa_i, t(r) \in [\tau_i + \beta_{k(r)}; \hat{\tau}_i]\} = 0, \\ x^r \in \{0, 1\}, \end{cases} \tag{9}$$

where:

- index $q \in \{1, N\}$ is the same as index $r \in \{1, N\}$,
- symbol $\alpha(\cdot)$ corresponds to the CCM number at the shop-floor,
- the value of $\sigma(\cdot)$ represents the casting start time for the corresponding melt,

- symbol $k(\cdot)$ corresponds to the CV number at the shop-floor, with given parameters $\beta_{k(\cdot)}$ for the purge cycle and $\hat{\beta}_{k(\cdot)}$ for the required delay when multiple melts are output sequentially from the corresponding CV,
- the value of $t(\cdot)$ represents the departure time from the CV,
- index $s \in \{1, 3\}$ corresponds to the shift number in a planning period of 24 h, where T_1^{st} and T_2^{st} correspond to 15:30 and 23:30 of the first day, and T_3^{st} and T_4^{st} correspond to 07:30 and 15:30 of the second day, respectively.
- index $i \in [1, m]$ corresponds to a specific TI with given parameters κ_i for the CV number, as well as τ_i and $\hat{\tau}_i$ for the start and finish times of maintenance, respectively.

To solve problem (8), (9), any software with an integrated solver can be used. The next section presents numerical results using the PuLP library for Python.

4 Numerical Experiment and Results

The proposed LIP model was implemented using Python 3.8.5 and the PuLP software for solving. Real-world data regarding processes at the converter shop-floor of the Novolipetsk Metallurgical Enterprise was used as instances. The planning period under consideration is from May 2020 to April 2021. The experiments were conducted on an Intel Core m3 1.2 GHz processor with 8GB of 1867 MHz LPDDR3 RAM, running macOS 10.13.6.

For each daily task in the production of casting series, an optimized production scenario was calculated using the proposed LIP model. The objective function value obtained from the model was compared with the actual cost. Tables 1 and 2 present partial computational results for the months of May 2020 and June 2020. The columns in the tables contain the following data:

- The *Date* column corresponds to the calendar day for which a daily task for the production of casting series was defined, and a comparison of costs between the actual scenario and the optimized one was conducted.
- The value of F_1 represents the actual cost value, which is the total cost of producing a given amount of steel, calculated in hundreds of monetary units. The total cost of production is directly related to the duration of production for each melt in the daily task. Minimizing the total processing time for the given set of casting series is equivalent to minimizing the total processing cost.
- The value of F_2 represents the optimized cost obtained using the proposed model. It is the minimum possible cost for producing the same amount of steel while satisfying all technological restrictions. The value of F_2 is also calculated in hundreds of monetary units.
- The parameter $\Delta, \%$ shows the difference between the values of F_1 and F_2, expressed as a percentage. If the value of $\Delta, \%$ is negative, it means that the obtained cost is higher than the actual cost. This can occur due to the possibility of conflicts between some model restrictions in the actual scenario (e.g., some melts in the actual scenario may violate the minimum allowed processing time).

- The parameter t, sec represents the CPU time required to solve the corresponding instance, measured in seconds.

Table 1. Computational results (2020, May)

Date	F_1	F_2	Δ, %	t, sec	Date	F_1	F_2	Δ, %	t, sec	Date	F_1	F_2	Δ, %	t, sec
01	1961,6	1852,1	5,58	7,11	12	1426,8	1101,8	22,78	0,55	22	1820,3	1723,2	5,33	16,07
02	1859,8	1759	5,42	5,20	13	2921,7	1131,3	61,28	0,74	24	1990,7	2065,5	-3,76	33,47
03	1730,9	1690,2	2,35	3,22	14	1632,0	931,5	42,92	0,44	25	1787,1	1632,2	8,67	34,85
05	1953,4	1843,8	5,61	6,64	15	1364,2	1080,5	20,80	0,88	26	1887,2	1925,6	-2,03	10,34
06	1816,3	1627,7	10,38	3,92	16	1434,8	1086,9	24,25	0,50	27	1905,0	1770,6	7,06	3,94
07	1894,8	1835	3,16	4,64	17	1488,4	1133,2	23,86	0,42	28	2136,1	1982,8	7,18	3,71
08	1893,3	1832,8	3,20	9,25	18	1527,8	1283,7	15,98	0,74	29	1887,9	1990,8	-5,45	25,17
09	1801,9	1717,2	4,70	8,69	19	1641,3	1233,4	24,85	0,51	30	2017,1	1891,7	6,22	26,67
10	2162,4	1823,8	15,66	60,75	20	1781,4	1383	22,36	0,52	31	2052,3	1935,3	5,70	15,39
11	1673,3	1251,6	25,20	1,08	21	1877,2	1704,7	9,19	2,65					

Table 2. Computational results (2020, June)

Date	F_1	F_2	Δ, %	t, sec	Date	F_1	F_2	Δ, %	t, sec	Date	F_1	F_2	Δ, %	t, sec
01	2297,0	2205,9	3,97	23,05	11	2228,7	1903,2	14,60	32,12	21	1956,8	1566,8	19,93	3,37
02	1952,1	1833,3	6,09	16,67	12	1723,8	1507,4	12,55	2,69	22	2305,3	2126,6	7,75	23,25
03	2291,1	1815,4	20,76	7,52	13	1483,7	1228	17,23	0,66	23	1986,4	1567,8	21,07	24,78
04	2045,3	1659,5	18,86	10,29	14	1554,5	984,9	36,64	0,46	24	1839,5	1807,7	1,73	41,31
05	1997,5	2005,2	-0,39	24,68	15	1195,7	1030,9	13,78	0,67	25	1732,3	1556	10,18	7,64
06	1915,3	1880,5	1,82	4,63	16	1104,1	873,9	20,85	0,71	26	1721,3	1742,4	-1,23	22,03
07	1942,8	2044,4	-5,23	6,93	17	1461,5	1123,6	23,12	0,55	27	1921,4	1691,6	11,96	4,98
08	1599,5	1321,3	17,39	5,67	18	918,9	748,1	18,59	0,67	28	1679,7	1424,8	15,18	13,14
09	1648,8	1299,8	21,17	3,79	19	1692,3	1439,8	14,92	1,71	29	1831,0	1462,7	20,11	3,86
10	1786,3	1535,1	14,06	5,11	20	1959,7	1790,9	8,61	20,54	30	1963,9	1503,8	23,43	4,13

The same calculations were conducted for each month during the planning period. Table 3 presents the aggregated results for the parameter Δ, % averaged across all months:

- The column labeled *Month* indicates the month within the considered period.
- The value *Avg*, % represents the average improvement of the objective function for all days within the respective month. It is calculated as the arithmetic mean of the Δ, % values for all days in the given month.

One can therefore conclude on the effectiveness of the proposed model in solving real-world optimization problems, as it achieves an average improvement of the objective function up to 19.16% (the maximum value of *Avg*, % in Table 3).

Table 3. Average Improvement

Month	Avg, %	Month	Avg, %	Month	Avg, %
2020, May	13,05	2020, August	8,81	2020, November	9,64
2021, February	13,58	2020, June	13,65	2020, September	11,10
2020, December	19,16	2021, March	10,59	2020, July	9,76
2020, October	18,09	2021, January	9,39	2021, April	18,15

This improvement is achieved by selecting the optimal departure time from the converters, which in turn reduces the processing time for most melts. In other words, the optimized scenario results in longer processing times for the same melts compared to the actual scenario. Consequently, the total production cost for the daily task is lower in the optimized scenario.

5 Conclusion

The paper presents a LIP model for optimizing the construction of a forecast schedule for melts movement on the converter shop-floor. It describes the domain, constraints, and introduces an objective function related to the total processing time.

Numerical experiments were conducted using real-world data from the converter shop-floor processes, demonstrating an average improvement of the objective function ranging from 8.81% to 19.16% (the minimum and maximum values of $Avg, \%$ in Table 3, respectively). The proposed approach has been implemented in the digital service "HEPHAESTUS: Shop-Floor Logistics and Dispatching at Steelmaking Production (Novolipetsk Metallurgical Enterprise)" [23], which was the winner of the RusBase Digital Awards 2021 in the "Production" category [24].

Future research will focus on the development of operative schedules for melts movement on the converter shop-floor, as well as planning routes for out-of-furnace processing and transportation logistics.

References

1. Lazarev, A.A., Musatova, E.G.: Integer statements of the problem of forming railway trains and schedules of their movement. Manag. Large Syst. **38**, 161–169 (2012)
2. Gainanov, D.N., Ignatov, A.N., Naumov, A.V., Rasskazova, V.A.: On track procession assignment problem at the railway network sections. Autom. Remote. Control. **81**(6), 967–977 (2020)
3. Hu, T.: Integer programming and threads in networks. Mir, Moscow (1974)
4. Ryan, D.M., Foster, B.A.: An integer programming approach to scheduling. In: Wren, A. (ed.) Computer Scheduling of Public Transport Urban Passenger Vehicle and Crew Scheduling, pp. 269–280. North-Holland, Amsterdam (1981)

5. Wagner, H.M.: An integer linear-programming model for machine scheduling. Nav. Res. Logist. Quart. **6**(2), 131–140 (1959)
6. Pochet, Y., Wolsey, L.A.: Production planning by mixed integer programming. In: Mikosh, T.V., Resnick, S.I., Robinson, M, (eds.) Springer Series in Operations Research & Financial Engineering (2006)
7. Shevchenko, V.N., Zolotykh, N.Y.: Linear and integer linear programming. Nizhny Novgorod State University named after N. I. Lobachevsky, Nizhny Novgorod (2004)
8. Schraver, A.: Theory of linear and integer programming. Mir, Moscow (1991)
9. Segal, I.K., Ivanova, A.P.: Introduction to applied discrete programming: models and computational algorithms. FIZMATLIT, Moscow (2007)
10. Appa, G.M., Pitsoulis, L.S., Paul, W.H. (eds.): Handbook on Modeling for Discrete Optimization. Springer Series, International Series in Operations Research & Management Science, vol. 88, XXII (2006). https://doi.org/10.1007/0-387-32942-0
11. Wolsey, L.A.: Integer programming. John Wiley & Sons, NJ (2020)
12. Hu, T.C., Kahng, A.B.: Linear and Integer Programming Made Easy. Springer, Cham (2016). https://doi.org/10.1007/978-3-319-24001-5
13. Kabulova, E.G.: Intelligent management of multi-stage systems of metallurgical production. Modeling Optimiz. Inform. Technol. **71**(24), 341–351 (2018)
14. Gitman, M.B., Trusov, P.V., Fedoseev, S.A.: On optimization of metal forming with adaptable characteristics. J. Appl. Math. Comput. **7**(2), 387–396 (2020)
15. Gainanov, D.N., Berenov, D.A.: Algorithm for predicting the quality of the product of metallurgical production. In: CEUR Workshop Proceedings, vol. 1987, pp. 194–200 (2017)
16. Qiu, Y., Wang, L., Xu, X., Fang, X., Pardalos, P.M.: Scheduling a realistic hybrid flow shop with stage skipping and adjustable processing time in steel plants. Appli. Soft Comput. **64**, 536–549 (2018)
17. Kong, M., Pei, J., Xu, J., Liu, X., Pardalos, P.M.: A Robust Optimization Approach for Integrated Steel Production and Batch Delivery Scheduling with Uncertain Rolling Times and Deterioration Effect. Int. J. Prod. Res. **58**(17), 5132–5154 (2020). https://doi.org/10.1080/00207543.2019.1693659
18. Long, J., Sun, Z., Pardalos, P.M., Bai, Y., Zhang, S., Li, C.: A robust dynamic scheduling approach based on release time series forecasting for the steelmaking continuous casting production. Appli. Soft Comput. **92**, 106271 (2020). https://www.sciencedirect.com/science/article/pii/S1568494620302118
19. Lazarev, A.A., Gafarov, E.R.: Scheduling Theory. Nauka, Problems and Algorithms. M. (2011)
20. Brucker, P., Knust, S.: Complex scheduling. Springer-Verlag, Berlin (2006). https://doi.org/10.1007/978-3-642-23929-8
21. Burkov, V.N.: Problems of optimum distribution of recources. Control Cibernetics **11**(2), 27–41 (1972)
22. Mingozzi, A., Maniezzo, V., Ricciardelli, S., Bianco, L.: An exact alforithm for project scheduling with recource constraints based on new mathematical formulation. Manage. Sci. **44**, 714–729 (1998)
23. Digital service of HEPHAESTUS. https://habr.com/ru/users/AlexanderSkorniakov/posts/
24. RusBase Homepage. https://rb.ru/awards/

Optimal Control

Optimal Control of Sources with Feedback with Optimization of the Placement of Measurement Points Along Given Trajectories

Kamil R. Aida-zade[1,2(✉)] and Vugar A. Hashimov[1]

[1] Institute of Control Systems of the Ministry of Science and Education of Republic of Azerbaijan, Baku, Azerbaijan
kamil_aydazade@rambler.ru

[2] Azerbaijan University of Architecture and Construction, Baku, Azerbaijan
https://www.azmiu.edu.az, https://www.isi.az

Abstract. In this paper, we study the problem of synthesizing the optimal control of the power of point-wise heat sources for heating a two-dimensional plate. The current power values are determined depending on the plate temperature values measured by the measuring devices. At the same time, these devices move along predefined individual trajectories. The movement velocities of the devices depend on the results of the current measurements. Necessary optimality conditions are obtained for the feedback parameters determined by the current values of the source powers and the velocities of the measuring devices. The conditions contain formulas for the gradient components of the objective functional with respect to the optimized feedback parameters. The results of computer experiments obtained with the use of first-order numerical optimization methods are presented.

Keywords: Plate heating · Point-wise source · Feedback control · Moving measuring devices · Feedback parameters

1 Introduction

The work is devoted to the problem of feedback control of the power of point-wise sources of plate heating. The most important specific feature of the formulation of the problem under consideration is that the devices for measuring temperature at the points of the plate are movable and move along the plate by given individual trajectories. The velocities of movement of these devices are controllable and depend on the current measured temperature values and the temperature distribution of the plate, which is desirable to obtain at the end of the heating process.

The problem under consideration belongs to the class of problems of optimal synthesis of control systems with distributed parameters [1–4]. Note that these

N. Olenev et al. (Eds.): OPTIMA 2023, CCIS 1913, pp. 91–104, 2024.
https://doi.org/10.1007/978-3-031-48751-4_7

problems are difficult compared to the problems of control synthesis for systems with lumped parameters [5–8]. The difficulty is due both to the complexity of the study and solution of initial-boundary value problems with respect to partial differential equations, and the technical feasibility of the obtained controls in the corresponding automatic control systems for objects [1–3].

In the article, an approach is proposed for constructing the dependence of control actions on the measured information. The finite-dimensional vector of parameters (coefficients) involved in these dependencies is optimizable and determines the feedback parameters.

In this paper, the optimality conditions for the feedback parameters are obtained, which are used to construct first-order numerical optimization methods. The results of computer experiments on the test problem are presented.

2 Formulation of the Problem

The problem of controlling the heating process of a thin two-dimensional plate is considered, which is described by a parabolic partial differential equation [9]:

$$u_t(x,t) = a^2 \text{div}\left(\text{grad}\, u(x,t)\right) - \lambda_0 \left[u(x,t) - \theta\right] + \sum_{i=1}^{N_c} q_i\left(t\right) \delta\left(x - \eta^i\right), \qquad (1)$$

$$x \in \Omega, \quad t \in \left(t_0, t_f\right],$$

with boundary condition

$$\frac{\partial u(x,t)}{\partial \mathrm{n}} = \lambda \left[u(x,t) - \theta\right], \quad x \in \Gamma, \quad t \in \left(t_0, t_f\right], \qquad (2)$$

Here: $u(x,t)$ is the temperature of the plate at point $x \in \Omega$ at time t; $\Omega \in \mathrm{R}^2$ is the area occupied by the plate, with the boundary Γ; n is internal normal to Γ; R^m – m-dimensional Euclidean space; $a^2 > 0$, λ_0, λ are given parameters of the heating process; θ is ambient temperature; $\eta = (\eta^1, \eta^2, \ldots, \eta^{N_c})$, $\eta^i = (\eta_1^i, \eta_2^i) \in \Omega$, $i = 1, 2, \ldots, N_c$ are coordinates of the placement points of lumped point-wise heat sources on the plate; $\delta(\cdot)$ is a two-dimensional Dirac function, for which, for an arbitrary function $f(x)$ continuous in Ω and $\tilde{x} \in \Omega$, the following holds [9]:

$$\int_{\Omega} f(x) \delta\left(x - \tilde{x}\right) dx = f(\tilde{x}).$$

The temperature of the plate at the initial time t_0 is not set, but the set of its possible values is given, determined by a parametric class of functions depending on the s-dimensional vector of parameters b

$$u(x, t_0) = \phi(x; b), \quad x \in \Omega, \quad b \in \mathrm{B} \subset \mathrm{R}^s. \qquad (3)$$

Here B is a given set of parameter values of the initial function $\varphi(x;b)$, and the distribution density function $\rho_{\mathrm{B}}(b)$ is known such that

$$\rho_{\mathrm{B}}(b) \geq 0, \quad b \in \mathrm{B}, \quad \int\limits_{\mathrm{B}} \rho_{\mathrm{B}}(b)db = 1.$$

θ is the time-constant ambient temperature, the exact value of which is not given, but the set Θ of possible values θ is known and the distribution density function $\rho_{\Theta}(\theta)$ is such that

$$\rho_{\Theta}(\theta) \geq 0, \quad \theta \in \Theta, \quad \int\limits_{\Theta} \rho_{\Theta}(\theta)d\theta = 1.$$

The function $u(x,t)$, which determines the value of the plate temperature at point $x \in \Omega$ at time t, is a generalized unique solution to the (1)–(3) problem, [10], namely: for an arbitrary function $\psi(x,t) \in H^{2,1}\left(\Omega \times [t_0,t_f]\right)$ such that $\psi(x,t_f) = 0$, $x \in \Omega$, $\dfrac{\partial\psi(x,t)}{\partial \mathrm{n}} - \lambda\psi(x,t) = 0$, $x \in \Gamma$, $t \in [t_0,t_f]$, we have

$$\int\limits_{t_0}^{t_f}\int\limits_{\Omega} u(x,t)\left[-\psi_t(x,t) - a^2 \mathrm{div}\left(\mathrm{grad}\,\psi(x,t)\right) + \lambda_0\psi(x,t)\right] dxdt$$

$$= \sum_{i=1}^{N_c}\int\limits_{t_0}^{t_f} q_i\left(t\right)\psi(\eta^i,t)dt + \int\limits_{\Omega}\psi(x,t_0)\phi(x;b)dx$$

$$+ \int\limits_{t_0}^{t_f}\int\limits_{\Omega}\lambda_0\theta\psi(x,t)dxdt - a^2\int\limits_{t_0}^{t_f}\int\limits_{\Gamma}\lambda\theta\psi(x,t)dxdt.$$

The plate is heated by N_c point-wise sources with controlled power $q_i\left(t\right)$, defined by piece-wise continuous functions in t, such that

$$q_i\left(t\right) \in Q^i = \left[\underline{q_i}, \overline{q_i}\right], \quad i = 1,2,\ldots,N_c \quad t \in \left[t_0,t_f\right], \tag{4}$$

where $\underline{q_i}$, $\overline{q_i}$, $i = 1,2,\ldots,N_c$ are given.

The problem of optimal control of the plate heating process is to find the admissible optimal values $q = q\left(t\right) = \left(q_1\left(t\right), q_2\left(t\right), \ldots, q_{N_c}\left(t\right)\right) \in Q$, $t \in [t_0,t_f]$ control powers of sources delivering, on average over all possible parameters, the values of the initial states $b \in \mathrm{B}$ and the ambient temperature $\theta \in \Theta$, the minimum value to the following objective functional:

$$J(q) = \int\limits_{\mathrm{B}}\int\limits_{\Theta} I(q;b,\theta)\rho_{\Theta}(\theta)\rho_{\mathrm{B}}(b)d\theta db, \tag{5}$$

$$I(q;b,\theta) = \int\limits_{\Omega}\mu(x)[u(x,t_f;q,b,\theta) - U(x)]^2dx + \varepsilon\left\|q\left(t\right) - \hat{q}\right\|^2_{L_2^{N_c}[t_0,t_f]}. \tag{6}$$

Here: $u(x,t) = u(x,t;q,b,\theta)$ is the solution of the initial-boundary value problem (1)–(3) with initial condition $u(x,t_0) = \phi(x,b)$, ambient temperature θ at admissible source powers $q\,(t)$; $U(x)$ is a given continuous differentiable function that determines the temperature distribution of the plate, which is required to be achieved at the end of the heating process; $\mu(x) \geq 0$, $x \in \Omega$, is given weight function; ε, $\hat{q}$ are regularization parameters of the functional of problem [10].

Let the along the given trajectories $\xi^j\,(t) = (\xi_1^j\,(t)\,,\xi_2^j\,(t)) \in \Omega$, $j = 1,2,\ldots,N_o$, by the plate moves measuring devices that continuously measure the current temperature at the points of the trajectory:

$$\hat{u}_j(t) = u\left(\xi^j\,(t)\,,t\right), \quad t \in [t_0,t_f]\,, \quad \xi^j\,(t) \in \Omega, \quad j = 1,2,\ldots,N_o. \tag{7}$$

Without loss of generality, we will assume that these trajectories are defined by explicit functions:

$$\xi_2^j\,(t) = F^j(\xi_1^j\,(t)), \quad t \in [t_0,t_f]\,, \quad j = 1,2,\ldots,N_o.$$

with given initial values:

$$\xi_1^j\,(t_0) = \breve{\xi}_1^j,\quad \xi_2^j\,(t_0) = \breve{\xi}_2^j, \quad j = 1,2,\ldots,N_o, \tag{8}$$

Let the velocities along the first coordinate of the measurement points be controllable and described by the following differential equations:

$$\dot{\xi}_1^j(t) = a^j\,(t)\,\xi_1^j\,(t) + \vartheta^j\,(t)\,, \quad j = 1,2,\ldots,N_o. \tag{9}$$

where $\xi^j\,(t) \in \mathrm{R}^2$ are the coordinates of the location of the jth measurement points on the plate at time t; $\breve{\xi}^j \in \Omega$, $j = 1,2,\ldots,N_o$ are given; $a^j\,(t)$, $j = 1,2,\ldots,N_o$ are given piece-wise continuous functions. Piece-wise continuous functions $\vartheta^j\,(t)$, $j = 1,2,\ldots,N_o$, which are control actions, determine the rules of movement of measurement points along given trajectories and satisfy technological constraints:

$$\vartheta^j\,(t) \in V^j = \left[\underline{\vartheta^j}, \overline{\vartheta^j}\right], \quad j = 1,2,\ldots,N_o \quad t \in [t_0,t_f]\,, \tag{10}$$

where $\underline{\vartheta^j}$, $\overline{\vartheta^j}$, $j = 1,2,\ldots,N_o$ are given.

Comment. Similarly, we can consider the case when measurements can be carried out by measuring devices only at discrete moments in time.

It is proposed to use the results of continuous measurements to form the current values of control actions: the powers for each of the sources $q_i\,(t)$, $i = 1,2,\ldots,N_c$ and the velocities of movement of the measuring points $\vartheta^j\,(t)$, $j = 1,2,\ldots,N_o$. To do this, consider the following linear feedback of control actions from continuously measured temperature values $u\left(\xi^j\,(t)\,,t\right)$ and the desired values of functions $U(\xi^j\,(t))$:

$$q_i\,(t) = \sum_{j=1}^{N_o} \alpha_i^j \|\xi^j\,(t) - \eta^i\|^2 \left[u\left(\xi^j\,(t)\,,t\right) - U(\xi^j\,(t))\right], \tag{11}$$

$$t \in [t_0, t_f], \quad i = 1, 2, \ldots, N_c.$$

$$\vartheta^j(t) = \sum_{s=1}^{N_o} \beta_j^s \|\xi^s(t) - \xi^j(t)\|^2 \left[u(\xi^s(t), t) - U(\xi^s(t))\right], \tag{12}$$

$$t \in [t_0, t_f], \quad j = 1, 2, \ldots, N_o.$$

Here the constants α_i^j, β_j^s, $i = 1, 2, \ldots, N_c$, $j = 1, 2, \ldots, N_o$, $s = 1, 2, \ldots, N_o$ are synthesized feedback parameters. The parameters α_i^j, β_j^s, by analogy with synthesis problems for objects with lumped parameters, will be called gain factors.

Combine the parameters $\alpha = ((\alpha_i^j))$, $\beta = ((\beta_j^s))$, $i = 1, 2, \ldots, N_c$, $j = 1, 2, \ldots, N_o$, $s = 1, 2, \ldots, N_o$ into one $\mathcal{N} = N_o(N_c + N_o)$ dimensional synthesizable vector of feedback parameters $y = (\alpha, \beta) \in \mathrm{R}^{\mathcal{N}}$.

The objective functional in this case can be written as follows:

$$J(y) = \int_{\mathrm{B}} \int_{\Theta} I(y; b, \theta) \rho_\Theta(\theta) \rho_{\mathrm{B}}(b) d\theta db, \tag{13}$$

$$I(y; b, \theta) = \int_{\Omega} \mu(x) [u(x, t_f; y, b, \theta) - U(x)]^2 dx + \varepsilon \|y - \hat{y}\|^2_{\mathrm{R}^{\mathcal{N}}}. \tag{14}$$

Substituting the expressions for control actions with continuous feedback (11), (12) into the differential equations (1) and (9), respectively, we obtain the following equations:

$$u_t(x, t) = a^2 \mathrm{div}\,(\mathrm{grad}\, u(x, t)) - \lambda_0 [u(x, t) - \theta] \tag{15}$$

$$+ \sum_{j=1}^{N_o} \alpha_i^j \|\xi^j(t) - \eta^i\|^2 \left[u(\xi^j(t), t) - U(\xi^j(t))\right], \quad x \in \Omega, \quad t \in (t_0, t_f],$$

$$\dot{\xi}^j(t) = a^j(t) \xi^j(t) + \sum_{s=1}^{N_o} \beta_j^s \|\xi^s(t) - \xi^j(t)\|^2 \left[u(\xi^s(t), t) - U(\xi^s(t))\right], \tag{16}$$

$$t \in (t_0, t_f], \quad j = 1, 2, \ldots, N_o.$$

The specificity of the Eqs. (15), (16) is, firstly, that they are point loaded with respect to the spatial variable. Secondly, the initial-boundary value problems with respect to the differential Eqs. (15), (16) with respect to the time variable must be solved simultaneously. Note that loaded differentiable equations with ordinary and partial derivatives were studied in such works as [11–14].

Thus, the original problem of controlling the power of sources and moving measuring points (1)–(10) with feedback (11), (12) is reduced to the parametric problem of optimal control (13)–(16) [6].

Let us note the following features of the obtained problem.

First, the process under study is described by a system consisting of both loaded partial and ordinary differential equations.

Secondly, the problem is specific due to the objective functional (5), (6), which evaluates the behavior of not a single trajectory, but a bunch of phase trajectories with values of initial conditions and ambient temperature from given sets.

In general, the resulting problem can be attributed to the class of finite-dimensional optimization problems with respect to the vector $y \in \mathrm{R}^{\mathcal{N}}$. In this problem, to calculate the objective functional for admissible values of the feedback parameters, it is required to solve initial-boundary value problems with respect to loaded differential equations with partial and ordinary derivatives.

3 Necessary Conditions for the Optimality of the Solution. Formulas for the Gradient of the Problem Functional with Respect to the Feedback Parameters

For the numerical solution of the obtained parametric optimal control problem, it is proposed to use first-order iterative optimization methods.

It is easy to verify that the functional of the original optimal control problem (1)–(6) is convex in $q(t)$. The functional of the problem with feedback (11), (12), as can be seen from (6), is generally not convex in the optimized feedback parameters y. This follows from the generally obvious non-linear dependence of the solution of the boundary value problem $u(x, t; y, b, \theta)$ on the parameters $y = (\alpha, \beta)$ [15]. Therefore, the optimality conditions formulated in this section are only local. The formulas involved in these conditions for the components of the functional gradient can be used in the numerical solution of the problem to determine the locally optimal values of the feedback parameters or to refine these values locally. To determine the optimal values of the feedback parameters, global optimization methods can be used together with methods for local improvement of the parameter values by first-order optimization methods using the following functional gradient formulas.

Restrictions (4), (10) on control actions will be written using (11), (12) in the following form

$$\underline{q_i} \le \sum_{j=1}^{N_o} \alpha_i^j \|\xi^j(t) - \eta^i\|^2 \left[u\left(\xi^j(t), t\right) - U(\xi^j(t))\right] \le \overline{q_i}, \tag{17}$$

$$t \in [t_0, t_f], \quad j = 1, 2, \ldots, N_o.$$

$$\underline{\vartheta^j} \le \sum_{s=1}^{N_o} \beta_j^s \|\xi^s(t) - \xi^j(t)\|^2 \left[u\left(\xi^s(t), t\right) - U(\xi^s(t))\right] \le \overline{\vartheta^j}, \tag{18}$$

$$t \in [t_0, t_f], \quad j = 1, 2, \ldots, N_o.$$

Let us introduce the notation

$$g^i(t; y) = |g_0^i(t; y)| - \frac{\overline{q_i} - \underline{q_i}}{2}, \quad t \in [t_0, t_f], \quad i = 1, 2, \ldots, N_c,$$

$$g_0^i(t;y) = \sum_{j=1}^{N_o} \alpha_i^j \|\xi^j(t) - \eta^i\|^2 \left[u\left(\xi^j(t), t\right) - U(\xi^j(t))\right] - \frac{\overline{q_i} + \underline{q_i}}{2}.$$

$$w^j(t;y) = |w_0^j(t;y)| - \frac{\overline{\vartheta^j} - \underline{\vartheta^j}}{2}, \quad t \in [t_0, t_f], \quad i = 1, 2, \ldots, N_c,$$

$$w_0^j(t;y) = \sum_{s=1}^{N_o} \beta_j^s \|\xi^s(t) - \xi^j(t)\|^2 \left[u\left(\xi^s(t), t\right) - U(\xi^s(t))\right] - \frac{\overline{\vartheta^j} + \underline{\vartheta^j}}{2}.$$

Then the constraints (17) and (18) can be written as follows:

$$g^i(t;y) \le 0, \quad t \in [t_0, t_f], \quad i = 1, 2, \ldots, N_c, \tag{19}$$

$$w^j(t;y) \le 0, \quad t \in [t_0, t_f], \quad j = 1, 2, \ldots, N_o. \tag{20}$$

To take into account the constraints (19) and (20) in the problem of synthesizing the y parameters, we use the external penalty method, adding to the functional (13), (14) a penalty term for violation of the constraints (19) and (20):

$$J_{\mathcal{R}}(y) = \int\limits_{\mathrm{B}} \int\limits_{\Theta} I_{\mathcal{R}}(y; b, \theta) \rho_\Theta(\theta) \rho_{\mathrm{B}}(b) d\theta db, \tag{21}$$

$$I_{\mathcal{R}}(y; b, \theta) = \int\limits_{\Omega} \mu(x)[u(x, t_f; y, b, \theta) - U(x)]^2 dx + \varepsilon \|y - \hat{y}\|^2_{\mathrm{R}^{\mathcal{N}}} \tag{22}$$

$$+\mathcal{R}_1 \sum_{i=1}^{N_c} \int\limits_{t_0}^{t_f} \left[g_+^i(t;y)\right]^2 dt + \mathcal{R}_2 \sum_{j=1}^{N_o} \int\limits_{t_0}^{t_f} \left[w_+^j(t;y)\right]^2 dt.$$

The functional (21) is repeatedly minimized taking into account the fact that, the penalty coefficients $\mathcal{R} = (\mathcal{R}_1, \mathcal{R}_2)$ tend to $+\infty$. The notation $g_+^i(t;y)$ and $w_+^j(t;y)$ means that $g_+^i(t;y) = g^i(t;y)$ and $w_+^j(t;y) = w^j(t;y)$ if $g^i(t;y) > 0$, $w^j(t;y) > 0$ and $g_+^i(t;y) = 0$, $w_+^j(t;y) = 0$ if $g^i(t;y) \le 0$, $w^j(t;y) \le 0$, $i = 1, 2, \ldots, N_c$, $j = 1, 2, \ldots, N_o$.

Further, the function $\operatorname{sgn}(\cdot)$ will be used, equal to -1 for negative values of the argument, zero for the argument equal to 0 and 1 for positive values of the argument.

In general, to solve the problem of optimal synthesis of the feedback parameters y, it is proposed to use iterative minimization methods with a combination of penalty function methods.

In particular, the gradient method with respect to the penalty function can be used, which we write in the form

$$y^{k+1} = y^k - \nu^k \operatorname{grad}_y J_{\mathcal{R}}(y^k), \tag{23}$$

$$\nu^k = \arg\min_{\nu \ge 0} J_{\mathcal{R}}\left(y^k - \nu \operatorname{grad}_y J_{\mathcal{R}}(y^k)\right), \quad k = 0, 1, \ldots. \tag{24}$$

or other more efficient first order optimization methods.

Here $\text{grad}_y J_{\mathcal{R}}(y^k)$ is the gradient of the penalty functional (21), (22), ν^k is step size in the direction of the anti-gradient of the functional $J_{\mathcal{R}}(y)$ [10].

It is clear that the main element needed to implement the procedure (21) is the gradient of the $\text{grad}_y J_{\mathcal{R}}(y)$ functional. The components of the gradient of the penalty functional with respect to the parameters of continuous feedback are determined from the following theorem.

Theorem 1. *Objective functional $J_{\mathcal{R}}(y)$ of the problem (15), (2), (3), (16), (8), (21), (22) is differentiable with respect to the synthesized parameters $y = (\alpha, \beta)$ with continuous feedback* (7). *The functional gradient components are determined by the formulas:*

$$\frac{\partial J_{\mathcal{R}}(y)}{\partial \alpha_i^j} = \int_{\mathrm{B}} \int_{\Theta} \Bigg\{ - \int_{t_0}^{t_f} \left[\psi(\eta^i, t) + 2\mathcal{R}_1 g_0^i(t; y) g_+^i(t; y)\right] \|\xi^j(t) - \eta^i\|^2 \tag{25}$$

$$\times \left[u\left(\xi^j(t), t\right) - U(\xi^j(t))\right] dt + 2\varepsilon \left(\alpha_i^j - \hat{\alpha}_i^j\right) \Bigg\} \rho_\Theta(\theta) \rho_{\mathrm{B}}(b) d\theta db,$$

$$\frac{\partial J_{\mathcal{R}}(y)}{\partial \beta_j^s} = \int_{\mathrm{B}} \int_{\Theta} \Bigg\{ - \int_{t_0}^{t_f} \left[\varphi^j(t) + 2\mathcal{R}_2 w_0^j(t; y) w_+^j(t; y)\right] \|\xi^s(t) - \xi^j(t)\|^2 \tag{26}$$

$$\times \left[u\left(\xi^s(t), t\right) - U(\xi^s(t))\right] dt + 2\varepsilon \left(\beta_j^s - \hat{\beta}_j^s\right) \Bigg\} \rho_\Theta(\theta) \rho_{\mathrm{B}}(b) d\theta db,$$

where $i = 1, 2, \ldots, N_c$, $j = 1, 2, \ldots, N_o$, $s = 1, 2, \ldots, N_o$, $\psi(x,t) = \psi(x,t;y,b,\theta)$ and $\varphi^i(t;y) = \varphi^i(t;y,b,\theta)$ with the current vector of parameters y allowed by the initial condition $b \in \mathrm{B}$ and ambient temperature $\theta \in \Theta$ are solutions of the following conjugate-boundary value problem and the conjugate Cauchy problem, respectively:

$$\psi_t(x,t) = -a^2 div\left(grad\, \psi(x,t)\right) + \lambda_0 \psi(x,t) \tag{27}$$

$$-\sum_{j=1}^{N_o} \delta\left(x - \xi^j(t)\right) \sum_{i=1}^{N_c} \alpha_i^j \psi(\eta^i, t) \|\xi^j(t) - \eta^i\|^2$$

$$-\sum_{s=1}^{N_o} \delta\left(x - \xi^s(t)\right) \sum_{j=1}^{N_o} \beta_j^s \varphi^j(t) \|\xi^s(t) - \xi^j(t)\|^2,$$

$$x \in \Omega, \quad t \in [t_0, t_f),$$

$$\psi(x, t_f) = -2\mu(x) \left(u(x, t_f) - U(x)\right), \quad x \in \Omega, \tag{28}$$

$$\frac{\partial \psi(x,t)}{\partial n} = \lambda \psi(x,t), \quad x \in \Gamma, \quad t \in [t_0, t_f), \tag{29}$$

$$\dot{\varphi}^j(t) = -a_i(t) \varphi^j(t) \tag{30}$$

$$-\sum_{i=1}^{N_c}\left\{2\alpha_i^j\psi(\eta^i,t)(\xi_1^j(t)-\eta_1^i)\right\}\left[u\left(\xi^j(t),t\right)-U(\xi^j(t))\right]$$

$$-\sum_{i=1}^{N_c}\left\{2\alpha_i^j\psi(\eta^i,t)(\xi_2^j(t)-\eta_2^i)\left.\frac{dF^j(\gamma)}{d\gamma}\right|_{\gamma=\xi_1^j(t)}\right\}\left[u\left(\xi^j(t),t\right)-U(\xi^j(t))\right]$$

$$-\sum_{i=1}^{N_c}\left\{\alpha_i^j\psi(\eta^i,t)\|\xi^j(t)-\eta^i\|^2\right\}\left.\left[\frac{\partial u(x,t)}{\partial x_1}-\frac{\partial U(x)}{\partial x_1}\right]\right|_{x=\xi^j(t)}$$

$$-\sum_{i=1}^{N_c}\left\{\alpha_i^j\psi(\eta^i,t)\|\xi^j(t)-\eta^i\|^2\right\}\left.\left[\frac{\partial u(x,t)}{\partial x_2}-\frac{\partial U(x)}{\partial x_2}\right]\right|_{x=\xi^j(t)}\left.\frac{dF^j(\gamma)}{d\gamma}\right|_{\gamma=\xi_1^j(t)}$$

$$-\varphi^j(t)\sum_{s=1}^{N_o}\left\{2\beta_j^s(\xi_1^s(t)-\xi_1^j(t))\left[u\left(\xi^s(t),t\right)-U(\xi^s(t))\right]\right\}$$

$$-\varphi^j(t)\sum_{s=1}^{N_o}\left\{2\beta_j^s(\xi_2^s(t)-\xi_2^j(t))\left.\frac{dF^j(\gamma)}{d\gamma}\right|_{\gamma=\xi_1^s(t)}\left[u\left(\xi^s(t),t\right)-U(\xi^s(t))\right]\right\}$$

$$+\varphi^j(t)\sum_{s=1}^{N_o}\left\{2\beta_j^s(\xi_1^s(t)-\xi_1^j(t))\left[u\left(\xi^s(t),t\right)-U(\xi^s(t))\right]\right\}$$

$$+\varphi^j(t)\sum_{s=1}^{N_o}\left\{2\beta_j^s(\xi_2^s(t)-\xi_2^j(t))\left[u\left(\xi^s(t),t\right)-U(\xi^s(t))\right]\right\}\left.\frac{dF^j(\gamma)}{d\gamma}\right|_{\gamma=\xi_1^j(t)}$$

$$-\varphi^j(t)\sum_{s=1}^{N_o}\left\{\beta_j^s\|\xi^s(t)-\xi^j(t)\|^2\left.\left[\frac{\partial u(x,t)}{\partial x_1}-\frac{\partial U(x)}{\partial x_1}\right]\right|_{x=\xi^s(t)}\right\}$$

$$-\varphi^j(t)\sum_{s=1}^{N_o}\left\{\beta_j^s\|\xi^s(t)-\xi^j(t)\|^2\left.\left[\frac{\partial u(x,t)}{\partial x_2}-\frac{\partial U(x)}{\partial x_2}\right]\right|_{x=\xi^s(t)}\left.\frac{dF^j(\gamma)}{d\gamma}\right|_{\gamma=\xi_1^s(t)}\right\},$$

$$\varphi^j(t_f)=0,\quad j=1,2,\ldots,N_o. \tag{31}$$

The theorem is proven using calculation of the variation of the functional obtained by increasing the feedback parameters [10].

Using the variational formula, we can formulate the necessary condition for the optimality of the feedback parameters with respect to the penalty functional.

Theorem 2. *Let y^* be a solution to the problem (15), (2), (3), (16), (8), (21), (22) for an arbitrary $\mathcal{R}>0$. Then the following inequality holds:*

$$\langle grad_y J_{\mathcal{R}}(y^*), y^*-y\rangle \le 0$$

for all admissible y satisfying the conditions (19), (20).

4 Results of Numerical Experiments

Let us present the results of solving the test problem of controlling the plate heating process with continuous feedback. The following parameter values were used in the problem:

$$a^2 = 1, \quad \lambda_0 = 0.01, \quad \lambda_1 = 0.001, \quad \Omega = [0;1] \times [0;1], \quad t_0 = 0, \quad t_f = 1,$$

$$\mu(x) \equiv 1, \quad U(x) = 30, \quad x \in [0;\ 1], \quad \varepsilon = 0.1, \quad N_c = 4, \quad N_o = 2,$$

$$a^j(t) = 0, \quad \xi^j(t) \in [0.05;\ 0.95], \quad t \in [0;\ 1], \quad j = 1, 2, \ldots, 4,$$

$$\mathrm{B} = [4.8; 5.2], \quad \rho_{\mathrm{B}}(b) = 2.5\,(1 + \cos(5(x-5)\,\pi)),$$

$$\Theta = [4.75; 5.25], \quad \rho_\Theta(\theta) = 2\,(1 + \cos(4(x-5)\pi)),$$

$$0 \le q_1(t) \le 80, \quad 0 \le q_2(t) \le 65, \quad t \in [0;\ 1],$$

$$-3.5 \le \vartheta_1(t) \le 3.5, \quad -3.5 \le \vartheta_2(t) \le 3.5, \quad t \in [0;\ 1],$$

$$F^1(\xi_2^1(t)) = 0.5 + 0.4\sin(2\pi\xi_1^1(t)), \quad F^2(\xi_2^2(t)) = 0.5 + 0.4\cos(2\pi\xi_1^2(t)), \quad t \in [0,1],$$

$$a^1(t) \equiv a^2(t) \equiv 0, \quad t \in (0,1], \quad \xi_1^1(0) = 0.9, \quad \xi_1^2(0) = 0.5,$$

$$\mathcal{R}_k = 10\mathcal{R}_{k-1}, \quad \mathcal{R}_0 = 5.$$

Let us give a general description of the algorithm for solving the problem of synthesizing a parameter vector y with a selected penalty coefficient $\mathcal{R}$ and regularization parameters ε, $\hat{y}$. When implementing the procedure (23) at each iteration for a given value of $\mathcal{R}$ with the current values of the optimized parameters y^k for all possible values of $b \in \mathrm{B}$ and $etz \in \Theta$ the following steps are performed:

1) the direct initial-boundary value problem (15), (2), (3) is solved and the Cauchy problems (16), (8) are solved;
2) the corresponding conjugate-boundary value problem (27)–(29) and conjugate Cauchy problems (30), (31) are solved;
3) the components (25), (26) of the penalty functional gradient (21), (22) are calculated;
4) one-dimensional minimization along in the direction of the obtained anti-gradient of the functional is performed according to the procedure (24).

These steps are repeated until the stopping criterion for the functional or for the argument y is met. At the next iteration, the penalty coefficient $\mathcal{R}$ increases by 10 times and steps 1–4 are repeated.

To solve two-dimensional direct (15), (2), (3) and conjugate (27)–(29) loaded boundary value problems, a modified scheme of the alternating direction method with steps in space variables $h_{x_1} = h_{x_2} = 0.01$ and in time variable $h_t = 0.01$.

To approximate the two-dimensional Dirac $\delta(\cdot)$-function, we used the following trigonometric everywhere smooth (differentiable) function [16,17]:

$$\delta_\sigma(x;\eta) = \begin{cases} 0, & |x_1 - \eta_1| > \sigma_1 \text{ or } |x_2 - \eta_2| > \sigma_2, \\ \prod_{i=1}^{2} \frac{1}{2\sigma_i}\left[1 + \cos\left(\frac{x_i - \eta_i}{\sigma_i}\pi\right)\right], & |x_1 - \eta_1| \le \sigma_1 \text{ and } |x_2 - \eta_2| \le \sigma_2. \end{cases}$$

It is easy to check that in this case, for an arbitrary value $\sigma_i > 0$, $i = 1, 2$, the following equality holds:

$$\int_{\eta_1-\sigma_1}^{\eta_1+\sigma_1}\int_{\eta_2-\sigma_2}^{\eta_2+\sigma_2} \delta_\sigma(x;\eta)dx = 1.$$

In test experiments, the values of the parameters $\sigma_i > 0$, $i = 1, 2$ of the function $\delta_\sigma(x;\eta)$ were respectively set equal to $3h_{x_1}$, $3h_{x_2}$, where h_{x_1}, h_{x_2} steps of grid approximation of the domain $x \in \Omega$. Such a choice of the type of approximation of the Dirac $\delta(\cdot)$-function ensures sufficient smoothness of the $J_{\mathcal{R}}(y)$ functional with respect to the optimized current positions of measurement points $\xi(t)$ and coordinates of point heat sources η.

Tables 1 and 2 show the results of solving the problem obtained at different iterations for two different initial points y_1^0 and y_2^0.

Table 1. The results of iterations of the solution of the problem obtained with the initial vector y_1^0.

k	$y_1 = (\alpha, \beta)$								$J_{\mathcal{R}}(y)$
0	0.01835	0.05531	0.04851	−0.28695	−0.03567	−0.15711	−0.06500	0.02128	2.1949
	0.43687	0.28709	0.78072	0.13950					
1	−0.13114	−0.14900	−0.17940	−0.15568	−0.19126	−0.42137	−0.30511	−0.13718	1.7658
	0.27089	0.46873	0.80772	0.95000					
2	0.06331	0.08869	0.02837	−0.03721	−0.26277	−0.31114	−0.21000	0.63007	0.3256
	0.14827	0.40810	0.61449	0.71282					
3	−0.14555	−0.09502	−0.16813	−0.19548	−0.12544	−0.12519	−0.07036	−0.11151	0.0014
	0.19417	0.40926	0.38373	0.18266					
4	−0.05158	−0.01977	−0.27408	−0.06466	−0.01923	−0.07352	0.15347	−0.05349	0.0001
	0.58265	0.71245	0.06539	0.48854					

On Fig. 1 shows the plots of the power of heat sources for the synthesized optimal vector of feedback parameters y^*.

On Fig. 2 shows the plots of the function

$$J_{\mathcal{R}}(t;y^*) = \int_{\mathrm{B}}\int_{\Theta}\left\{\int_{\Omega}\mu(x)[u(x,t;y^*,b,\theta) - U(x)]^2dx\right\}\rho_\Theta(\theta)\rho_{\mathrm{B}}(b)d\theta db,$$

where $u(x,t;y^*,b,\theta)$ is the solution to the initial-boundary value problem (15), (2), (3) with the optimal vector of feedback parameters y^* obtained from the numerical solution of the problem.

From the plots in Fig. 2, one can see how the temperature distribution on the plate approached in time the desired distribution.

Table 2. The results of iterations of the solution for the problem obtained with the initial vector y_2^0.

k	$y_2 = (\alpha, \beta)$								$J_{\mathcal{R}}(y)$
0	0.08351	0.01854	0.01535	−0.03567	−0.29865	−0.15117	−0.02128	0.00065	1.9852
	0.82709	0.34687	0.28077	0.47343					
1	−0.10049	−0.24113	−0.51568	−0.94017	−0.34271	−0.26119	−0.17813	−0.05131	0.6758
	0.27089	0.46873	0.80772	0.95000					
2	0.01336	0.86908	0.03872	−0.30721	−0.62727	−0.14113	−0.00210	0.06307	0.3562
	0.27184	0.30108	0.24329	0.62712					
3	−0.15455	−0.13681	−0.02950	−0.41254	−0.18954	−0.00367	−0.25119	−0.11511	0.0075
	0.42046	0.47116	0.24583	0.18266					
4	−0.22197	−0.07408	−0.18021	−0.01923	−0.03456	−0.12347	0.05352	−0.05345	0.0001
	0.65265	0.27451	0.39065	0.28085					

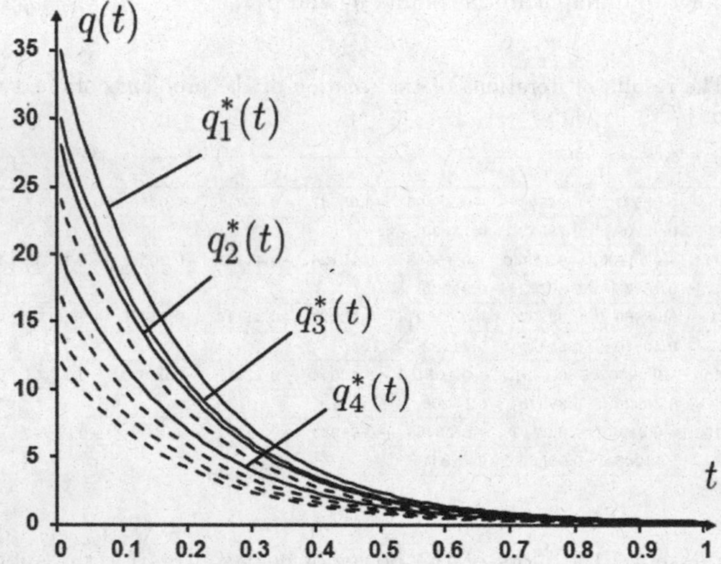

Fig. 1. Plots of point-wise sources powers for initial vector y_1^0 (- - -) and synthesized optimal vector y_1^* (——).

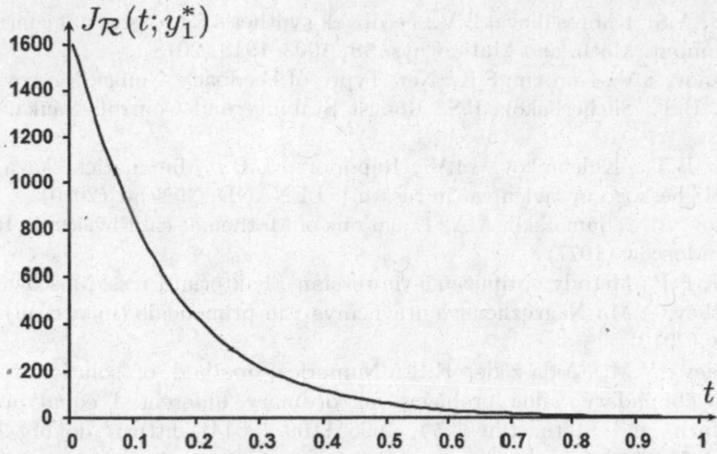

Fig. 2. Plots of the function $J_{\mathcal{R}}(t; y^*)$ values in $t \in [0; 1]$ for the feedback parameters y_1^*.

5 Conclusion

An approach to feedback control of the power of lumped point-wise heat sources and the movement of measurement points along given trajectories is proposed. The power of the sources and the velocity of movement of the measurement points are determined in the form of the proposed dependencies on the results of the measurements. The differentiability of the functional with respect to the feedback parameters is shown, formulas for the gradient of the functional with respect to the synthesized parameters are obtained. The formulas make it possible to solve the problem of lumped source control synthesis using efficient first-order numerical optimization methods and available standard software packages.

Note that the proposed approach to synthesis leads to the problem of parametric optimal control of a process described by loaded differential equations with ordinary and partial derivatives.

The results of numerical experiments on a test problem are presented.

The proposed approach to the control of lumped sources with feedback can be used in systems for automatic control and regulation of lumped sources for many other technological processes and technical objects.

References

1. Ray, W.H.: Advanced Process Control. McGraw-Hill, New York (1980)
2. Butkovskii, A.G., Pustyl'nikov, L.M.: Teoriya podvizhnogo upravleniya sistemami s raspredelennymi parametrami (in russian). Nauka, Moscow (1980)
3. Sirazetdinov, T.K.: Optimizatsiya sistem s raspredelennymi parametrami (in russian). Nauka, Moscow (1977)
4. Egorov, A.I.: Osnovy teorii upravleniya (in russian). Fizmatlit, Moscow (2004)

5. Antipin, A.S., Khoroshilova, E.V.: Feedback synthesis for a terminal control problem. Comput. Math. and Math. Phys. **58**, 1903–1918 (2018)
6. Emelyanov, S.V., Korovin, S.K.: New Types of Feedback. Nauka, Moscow (1997)
7. Polyak, B.T., Shcherbakov, P.S.: Robust Stability and Control. Nauka, Moscow (2002)
8. Polyak, B.T., Khlebnikov, M.V., Rapoport, L.B.: Matematicheskaya teoriya avtomaticheskogo upravleniya (in russian). LENAND, Moscow (2019)
9. Tikhonov, A.N., Samarskii, A.A.: Equations of Mathematical Physics (in Russian). Nauka, Moscow (1977)
10. Vasil'ev, F.P.: Metody optimizatsii (in russian). Faktorial Press, Moscow (2002)
11. Nakhushev, A.M.: Nagruzhennye uravneniya i ih primenenie (in russian). Nauka, Moscow (2012)
12. Abdullaev, V.M., Aida-zade, K.R.: Numerical method of solution to loaded nonlocal boundary value problems for ordinary differential equations. Comput. Math. and Math. Phys. **54**, 1096–1109 (2014). https://doi.org/10.1134/S0965542514070021
13. Abdullayev, V.M., Aida-zade, K.R.: Approach to the numerical solution of optimal control problems for loaded differential equations with nonlocal conditions. Comput. Math. and Math. Phys **59**, 696–707 (2019). https://doi.org/10.1134/S0965542519050026
14. Alikhanov, A.A., Berezgov, A.M., Shkhanukov-Lafishev, M.X.: Boundary value problems for certain classes of loaded differential equations and solving them by finite difference methods. Comput. Math. and Math. Phys. **48**, 1581–1590 (2008). https://doi.org/10.1134/S096554250809008X
15. Guliyev, S.Z.: Synthesis of zonal controls for a problem of heating with delay under nonseparated boundary conditions. Cybern. Syst. Anal. **54**, 110–121 (2018). https://doi.org/10.1007/s10559-018-0012-5
16. Aida-zade, K.R., Hashimov, V.A., Bagirov, A.H.: On a problem of synthesis of control of power of the moving sources on heating of a rod. Proc. Inst. Math. Mech. Natl. Acad. Sci. Azerb. **47**(1), 183–196 (2021). https://doi.org/10.30546/2409-4994.47.1/183
17. Aida-zade, K.R., Abdullayev, V.M.: Optimizing placement of the control points at synthesis of the heating process control. Autom. Remote. Control. **78**, 1585–1599 (2017). https://doi.org/10.1134/S0005117917090041

Game Theory and Mathematical Economics

Unimodality of Equilibrium Welfare in International Trade Under Monopolistic Competition

Igor Bykadorov(✉)

Sobolev Institute of Mathematics, 4 Koptyug Avenue, 630090 Novosibirsk, Russia
bykad@math.nsc.ru

Abstract. We study the classical monopolistic competition trade model. The utility of consumers is additively separable with pro-competitive sub-utility function, the production costs are linear, and the transport cots are "of iceberg type". We examine the local comparative statics of market equilibrium with respect to transport costs. Earlier we found the following results: (1) (intuitive!) an increase w.r.t. transport costs near free trade leads to a decrease in social welfare; (2) (counter intuitive!) an increase w.r.t. transport costs near autarky leads to an increase in social welfare. Thus, social welfare is non-monotonic. In this regard, a natural question arises about the unimodality of social welfare. We show the unimodality of social welfare for an important class of elementary utility function, namely, for the Behrens–Murata function.

Keywords: Monopolistic Competition · International Trade · Market Equilibrium · Comparative Statics · Free Trade · Autarky · Unimodality of Welfare

1 Introduction

A market situation where there are many independent sellers and many independent buyers, and competition is imperfect due to product differentiation and other factors, is called "monopolistic competition". At the moment, monopolistic competition, perhaps, most adequately simulates real economic processes.

It is generally accepted that the basis of this theory was laid almost simultaneously in 1933 by the American economist E. H. Chamberlin (see [1], it was here that the fundamental concept of "product differentiation" was introduced) and, a few months later, the British economist J. Robinson [2]. The "imperfection" of competition is confirmed by many empirical observations. However, for a long time, the development of the theory was hampered by its insufficient formalization, which did not allow the use of mathematical tools.

The work was carried out within the framework of the state contract of the Sobolev Institute of Mathematics of the Siberian Branch of the Russian Academy of Sciences, project FWNF-2022-0019.

N. Olenev et al. (Eds.): OPTIMA 2023, CCIS 1913, pp. 107–118, 2024.
https://doi.org/10.1007/978-3-031-48751-4_8

A real breakthrough should be considered the work of A. K. Dixit and J. E. Stiglitz [3], where a model for the additive-separable utility function is built and investigated. In the model of A. K. Dixit and J. E. Stiglitz, the sub-utility function is of CES type, i.e. "Constant Elasticity of Substitution". Hereinafter, P. R. Krugman generalizes this model to the case of international trade [4] (see also [5, 6]). Further, M. J. Melitz generalizes the model to the heterogeneous case [7] (see also [8]).

Unfortunately, the CES type of sub-utility function generates some inadequacy of the results obtained. For example, in equilibrium, an increase in market size is accompanied by an increase in competition (an increase in mass of firms), but it can also be accompanied by a constant prices.

An important example of a model with a non-CES type of sub-utility function is the work of K. Behrens and Y. Murata [9].

However, it is always of interest to develop a technique that allows to analyze general models, and not containing specific types of functions. In other words, it is interesting to abandon a specific functional type of functions (utility functions, cost functions, etc.).

In 2012 E. Zhelobodko, S. Kokovin, M. Parenti and J.-F. Thisse publish the work [10], where the mathematical apparatus is finally formalized, which makes it possible to study models of monopolistic competition of a sufficient general kind.

In 2012, C. Arkolakis, A. Costinot and A. Rodríguez-Clare published the work [11] (see also [12]), where doubts are expressed about the universal benefits of international trade (or rather, about the universal harm of the "fading" of international trade). This work prompts researchers to try to explain this phenomenon.

In [13] we compare the magnitude of social welfare at the point of free trade and at the point of autarky and show that, generally speaking, any situation is possible: both the benefits of free trade and the benefits of autarky.

Moreover, in [14] we get the following "counter intuitive" result: with the additive-separable utility function, linear production costs and transport (trade) costs of the "iceberg type", if autarky arises, then with an increase in transport costs "near" autarky, social welfare does not decrease, but, on the contrary, increases.

Thus, social welfare is non-monotonic. In this regard, a natural question arises about the unimodality of social welfare.

Remark 1. Let us note that Unimodality of social welfare function is something widely challenged in social welfare and social choice literature. But the result of [14] is that "with an increase in **transport costs** near autarky, social welfare increases" is "counter intuitive" because of usually an increase in transport costs means a deterioration in the terms of trade, and therefore a decrease in social welfare.

We show the unimodality of social welfare for an important class of sub-utility function, namely, for the Behrens–Murata function [9].

The article is organized as follows. Section 2 contains a description of the model, the definitions of symmetric equilibrium. In Sect. 3 wee study the case linear production costs. Here we consider the sub-utility function a'la Behrens–Murata [9], formulate the main result, see Corollary 2. Section 4 concludes.

2 The Basic Model of Open Economy

In this section we set the basic monopolistic competition model for open economy (international trade case). Let be two countries. For simplicity, let us assume that the countries are the same in terms of population (consumers).

2.1 Main Assumptions of Monopolistic Competition

As it is usual in monopolistic competition, we assume that (cf. [1,3,4])

- consumers are identical, each endowed with one unit of labor;
- labor is the only production factor; consumption, output, prices etc. are measured in labor;
- firms are identical, but produce "varieties" ("almost the same") of good;
- each firm produces one variety as a price-maker, but its demand is influenced by other varieties;
- each variety is produced by one firm that produces a single variety;
- each demand function results from additive utility function;
- number (mass) of firms is big enough to ignore firm's influence on the whole industry/economy;
- free entry drives all profits to zero;
- labor supply/demand in each country is balance.

2.2 Consumers

Let in each country,

- L be the number of consumers,
- N be the number (mass) of firms.

Note that L is parameter (the known constant) while N is the variable determined endogenously. Moreover, let us recall that, in monopolistic competition models, number of firms is big enough. Therefore, instead of standard "number of firms is N)" we consider the interval $[0, N]$ with uniformly distributed firms.

Now we introduce two kinds of the *individual* consumption and prices. Let [1]

- X_i be the amount of the variety produced in a country by firm $i \in [0, N]$ and consumed by a consumer in the same country ("domestic individual consumption"),

[1] Hereinafter, due to the tradition of monopolistic competition, we use the notation X_i for the function $X(i)$, etc.

- Z_i be the amount of the variety produced in a country by firm $i \in [0, N]$ and consumed by a consumer in another country ("import individual consumption"),
- p_i^X be the price of the unit of the variety produced in a country by firm $i \in [0, N]$ and consumed by a consumer in the same country,
- p_i^Z be the price of the unit of the variety produced in a country by firm $i \in [0, N]$ and consumed by a consumer in another country.

Let $u(\cdot)$ be a sub-utility function. As usual, we assume that $u(\cdot)$ is twice differentiable and satisfies the conditions

$$u(0) = 0,\ u'(\xi) > 0,\ u''(\xi) < 0,$$

i.e., it is strictly increasing and strictly concave.

Further, let w be the wage rate in each country normalizing to one, $w = 1$. Thus, the problem of representative consumer in each country is

$$\int_0^N u(X_i)\, di + \int_0^N u(Z_i)\, di \to \max$$

s.t.

$$\int_0^N p_i^X X_i di + \int_0^N p_i^Z Z_i di \le 1.$$

Using First Order Conditions, we get the inverse demand functions

$$p_i^X(X_i, \Lambda) = \frac{u'(X_i)}{\Lambda}, \tag{1}$$

$$p_i^X(Z_i, \Lambda) = \frac{u'(Z_i)}{\Lambda}, \tag{2}$$

where Λ is Lagrange multiplier.

Note that due to (1) and (2) one has

$$\Lambda > 0, \tag{3}$$

2.3 Producers

To introduce the production amount of the firms (the "size" of the firm), let us introduce the parameter $\tau \ge 1$ as transport costs of "iceberg type".[2] Each firm in each country produces for consumers in each country. Thus,

$$Q_i = LX_i + \tau LZ_i, i \in [0, N], \tag{4}$$

is the size of firm $i \in [0, N]$ in each country.

[2] To sell in another country y units of the goods, the firm must produce $\tau \cdot y$ units. "During transportation, the product melts like an iceberg ...".

Let the production costs be determined for each firm in each country by the *increasing* twice differentiable function V. Then the profits $\Pi_i, i \in [0, N]$, of firm i in each country are

$$\Pi_i = Lp_i^X X_i + Lp_i^Z Z_i - 1 \cdot V(Q_i), i \in [0, N]. \tag{5}$$

Of course, the firms choose inverse demand functions (1) and (2) as the prices. Let us introduce so-called "normalized revenue"

$$R(\xi) := u'(\xi) \cdot \xi. \tag{6}$$

Let us substitute (1) and (2) in (5). Using (6), we get

$$\Pi_i = L \cdot \frac{R(X_i)}{\Lambda} + L \cdot \frac{R(Z_i)}{\Lambda} - V(Q_i). \tag{7}$$

The Labor Balance in each country is

$$\int_0^N V(Q_i)\, di = L. \tag{8}$$

2.4 Symmetric Case

Let us recall that the consumers are assume identical, the producer are assumed identical. Thus, as usual, we consider the symmetric case. More precisely, we omit index i in consumption, inverse demand functions (1) and (2), sizes of the firms (4), profits (7) and Labor Balance (8). This way (1), (2), (4), (7), (8) are

$$p(X, \Lambda) = \frac{u'(X)}{\Lambda}, \tag{9}$$

$$p(Z, \Lambda) = \frac{u'(Z)}{\Lambda}, \tag{10}$$

$$Q = L \cdot (X + \tau \cdot Z); \tag{11}$$

$$\Pi = \frac{L}{\Lambda} \cdot (R(X) + R(Z)) - V(Q), \tag{12}$$

$$N \cdot V(Q) = L. \tag{13}$$

As to the Social Welfare (total utility) in each country, it is

$$W = L \cdot (N \cdot u(X) + N \cdot u(Z)) = L \cdot N \cdot (u(X) + u(Z)). \tag{14}$$

2.5 Symmetric Equilibrium

For equilibrium, producers choose inverse demand functions (9), (10) and maximize profits (12)). So First Order Conditions (FOC)

$$\frac{\partial \Pi}{\partial X} = 0, \qquad \frac{\partial \Pi}{\partial Z} = 0 \tag{15}$$

and Second Order Conditions (SOC)

$$\frac{\partial^2 \Pi}{\partial X^2} < 0, \qquad \frac{\partial^2 \Pi}{\partial X^2} \cdot \frac{\partial^2 \Pi}{\partial Z^2} - \left(\frac{\partial^2 \Pi}{\partial X \partial Z}\right)^2 > 0 \tag{16}$$

hold.

Like in the standard monopolistic competition framework, the firms enter into the market until their profit remains positive. Therefore, free entry implies the zero-profit condition

$$\Pi = 0. \tag{17}$$

By definition, the ***symmetric market equilibrium*** is a bundle

$$\left(X^*, Z^*, \Lambda^*, p\left(X, \Lambda^*\right), p\left(Z, \Lambda^*\right), N^*\right)$$

satisfying the following:

- the rational consumption conditions (9) and (10);
- the rational production conditions (15)–(16);
- the free entry condition (17);
- the labor balance condition (13).

Thus, the bundle

$$\left(X^*, Z^*, \Lambda^*\right)$$

satisfies the following equations:

$$\frac{\partial \Pi}{\partial X} = 0,$$

$$\frac{\partial \Pi}{\partial Z} = 0,$$

$$\Pi = 0.$$

3 The Case of Linear Costs

Let

$$V(Q) = c \cdot Q + f,$$

i.e., production costs are linear. Then

$$\frac{\partial \Pi}{\partial X} = L \cdot \left(\frac{R'(X)}{\Lambda} - c \right),$$

$$\frac{\partial \Pi}{\partial Z} = L \cdot \left(\frac{R'(Z)}{\Lambda} - c \cdot \tau \right),$$

Since (3) holds, then due to *FOC*

$$R'(X) > 0, \qquad R'(Z) > 0$$

and *FOC* is

$$\frac{R'(X)}{\Lambda} = c, \qquad \frac{R'(Z)}{\Lambda} = c \cdot \tau.$$

Further,

$$\frac{\partial^2 \Pi}{\partial X^2} = \frac{L}{\Lambda} \cdot R''(X),$$

$$\frac{\partial^2 \Pi}{\partial Z^2} = \frac{L}{\Lambda} \cdot R''(Z),$$

$$\frac{\partial^2 \Pi}{\partial X \partial Z} = 0.$$

Since (3) holds, then *SOC* is

$$R''(X) < 0, \qquad R''(Z) < 0.$$

3.1 The Local Comparative Statics w.r.t. τ

The local comparative statics with respect to transport costs τ is defined by the equations

$$\frac{d}{d\tau} \left(\frac{\partial \Pi}{\partial X} \right) = 0, \tag{18}$$

$$\frac{d}{d\tau} \left(\frac{\partial \Pi}{\partial Z} \right) = 0, \tag{19}$$

$$\frac{d\Pi}{d\tau} = 0, \tag{20}$$

i.e., due to *FOC*,

$$\frac{\partial^2 \Pi}{\partial X^2} \cdot \frac{dX}{d\tau} + \frac{\partial^2 \Pi}{\partial X \partial \Lambda} \cdot \frac{d\Lambda}{d\tau} = 0,$$

$$\frac{\partial^2 \Pi}{\partial Z^2} \cdot \frac{dZ}{d\tau} + \frac{\partial^2 \Pi}{\partial Z \partial \Lambda} \cdot \frac{d\Lambda}{d\tau} = -\frac{\partial^2 \Pi}{\partial Z \partial \tau},$$

$$\frac{\partial \Pi}{\partial \Lambda} \cdot \frac{d\Lambda}{d\tau} = -\frac{\partial \Pi}{\partial \tau},$$

i.e.,

$$\frac{L}{\Lambda} \cdot R''(X) \cdot \frac{dX}{d\tau} - \frac{L}{\Lambda^2} \cdot R'(X) \cdot \frac{d\Lambda}{d\tau} = 0,$$

$$\frac{L}{\Lambda} \cdot R''(Z) \cdot \frac{dZ}{d\tau} - \frac{L}{\Lambda^2} \cdot R'(Z) \cdot \frac{d\Lambda}{d\tau} = L \cdot c,$$

$$-\frac{L}{\Lambda^2} \cdot (R(X) + R(Z)) \cdot \frac{d\Lambda}{d\tau} = L \cdot c \cdot Z.$$

Due to Free Entry Condition, the last equation can be rewritten as

$$-(c \cdot Q + f) \cdot \frac{1}{\Lambda} \cdot \frac{d\Lambda}{d\tau} = L \cdot c \cdot Z.$$

Thus, the solution of (18), (19) and (20) is

$$\frac{dX}{d\tau} = \frac{1}{\tau} \cdot \frac{R'(X)}{R''(X)} \cdot E_\Lambda > 0,$$

$$\frac{dZ}{d\tau} = \frac{1}{\tau} \cdot \frac{R'(Z)}{R''(Z)} \cdot (E_\Lambda + 1),$$

$$E_\Lambda = -\frac{L \cdot c \cdot \tau \cdot Z}{c \cdot Q + f} < 0,$$

where

$$E_\Lambda = \frac{\tau}{\Lambda} \cdot \frac{d\Lambda}{d\tau}$$

is the elasticity of Λ with respect to transport costs τ.

Moreover, since

$$Q = L \cdot (X + \tau \cdot Z),$$

$$N = \frac{L}{c \cdot Q + f},$$

one has

$$\frac{dQ}{d\tau} = L \cdot \left(\frac{dX}{d\tau} + \tau \cdot \frac{dZ}{d\tau} + Z \right),$$

$$\frac{dN}{d\tau} = -L \cdot \frac{c \cdot \dfrac{dQ}{d\tau}}{(c \cdot Q + f)^2}.$$

Further, social welfare (14) is

$$W = L \cdot N \cdot (u(X) + u(Z)).$$

Therefore,

$$\frac{dW}{d\tau} = L \cdot \left(\frac{dN}{d\tau} \cdot (u(X) + u(Z)) + N \cdot \left(u'(X) \cdot \frac{dX}{d\tau} + u'(Z) \cdot \frac{dZ}{d\tau}\right)\right),$$

i.e.,

$$\frac{dW}{d\tau} \cdot \frac{1}{L^2}$$

$$= \frac{-c \cdot \frac{dQ}{d\tau} \cdot (u(X) + u(Z)) + (c \cdot Q + f) \cdot \left(u'(X) \cdot \frac{dX}{d\tau} + u'(Z) \cdot \frac{dZ}{d\tau}\right)}{(c \cdot Q + f)^2}.$$

Further, by direct calculations it can be shown that

$$\frac{dW}{d\tau} = 0$$

$$\Longleftrightarrow \left(-(u(X) + u(Z)) + u'(Z) \cdot \frac{R'(X)}{R'(Z)}\right) \cdot \frac{1}{R''(Z)}$$

$$= \left(\frac{u'(X) \cdot R'(X)}{R''(X)} + \frac{u'(Z) \cdot R'(Z)}{R''(Z)}\right) \cdot \frac{\frac{R(X)}{R'(X)} + \frac{R(Z)}{R'(Z)}}{R(X) + R(Z)} \cdot \frac{R'(X)}{R'(Z)} \cdot Z.$$

3.2 The Case $u(\xi) = 1 - e^{-\xi}$

Let

$$u(\xi) = 1 - e^{-\xi}.$$

Then

$$u'(\xi) = e^{-\xi},$$

$$R(\xi) = u'(\xi)\,\xi = e^{-\xi}\xi,$$

$$R'(\xi) = e^{-\xi}(1 - \xi),$$

$$R''(\xi) = -e^{-\xi}(2 - \xi).$$

Thus

$$\frac{dW}{d\tau} = 0$$

$$\Longleftrightarrow \left(-\left(1 - e^{-X} + 1 - e^{-Z}\right) + e^{-Z} \cdot \frac{1}{\tau}\right) \cdot \frac{1}{e^{-Z}(Z-2)}$$

$$= \left(\frac{1}{\tau} \cdot \frac{1}{X-2} + \frac{1}{Z-2}\right) \cdot \frac{e^{-X}X + e^{-Z}Z \cdot \frac{1}{\tau}}{e^{-X}X + e^{-Z}Z} \cdot Z.$$

Let us note that due to FOC

$$\frac{1}{\tau} = \frac{e^{-X}(1-X)}{e^{-Z}(1-Z)}.$$

Therefore,

$$\frac{dW}{d\tau} = 0 \Longleftrightarrow A_1(X,Z) + A_2(X,Z) = F \cdot B(X,Z),$$

where

$$A_1(X,Z) = (\Gamma(X,Z) - 1)\left(\frac{X}{2-Z} - \frac{Z}{2-X}\right),$$

$$A_2 = HK - J,$$

$$B(X,Z) = \frac{1-(1-Z)\Gamma(X,Z)}{2-Z},$$

$$F = \frac{f}{cL},$$

$$\Gamma(X,Z) = \frac{2e^{X+Z} - \left(e^X + e^Z\right)}{Xe^Z + Ze^X},$$

$$H(X,Z) = \Gamma(X,Z)XZ,$$

$$K(X,Z) = \frac{X-Z}{(2-X)(2-Z)},$$

$$J(X,Z) = \frac{Z}{1-X}\left(2e^X - 1 - e^{X-Z}\right).$$

Lemma 1. *Functions $A_1(X,Z)$ and $A_2(X,Z)$ increase with respect to τ, while function $B(X,Z)$ decreases with respect to τ.*

Proof. The proof of the Lemma is rather technical, let us omit it.

Corollary 1. *The equation*

$$\frac{dW}{d\tau} = 0$$

has the unique solution.

Corollary 2. *In equilibrium, Social Welfare W is unimodal with respect to τ.*

4 Conclusion

The paper investigates in the structure of monopolistic competition a homogeneous model of international trade of the type Dixit-Stiglitz-Krugman with additively separable utility functions. The only production factor is the labor.

Main attention is paid to the concept of market equilibrium, i.e., optimization of producer behavior:

- firms maximize profits by using inverse demand functions (provided to them by representative consumers),
- free entry conditions (firms enter the market until their profits are positive),
- labor balances (in each country, the total costs are equal to total labor).

Moreover, the results can help to generalize the results of [14] on the behavior of social welfare with respect to transport costs near "free trade" and near "autarky" not only to the case of market equilibrium, but also to the case of social optimality [15–18]. Let us recall that Social optimality means the optimization of the behavior of the state in the interests of consumers: the state (more precisely, all countries together) maximizes the social welfare function under the only constraint – the labor balances in each country.

In addition, it is of interest to see how the results obtained can be transferred to the case of models with retail, see [19,20], etc.

Finally, it is interesting to investigate models in which utility is not additively separable from the same perspective, see [21], etc.

Acknowledgements. The author is grateful to many colleagues for useful discussions and valuable comments. Especially I am indebted to Evgeny Zhelobodko (1973–2013) and Sergey Kokovin, who sparked my interest in monopolistic competition models. Moreover, many thanks to my student Ruslan Shikhov (Novosibirsk State University, Russia) who took part in some calculations.

References

1. Chamberlin, E.H.: The Theory of Monopolistic Competition: A Re-orientation of the Theory of Value, 1st edn. Harvard University Press, Cambridge (1933)
2. Robinson, J.: The Economics of Imperfect Competition, 1st edn. Macmillan, London (1933)
3. Dixit, A.K., Stiglitz, J.E.: Monopolistic competition and optimum product diversity. Am. Econ. Rev. **67**(3), 297–308 (1977)
4. Krugman, P.: Increasing returns, monopolistic competition, and international trade. J. Int. Econ. **9**(4), 469–479 (1979)
5. Krugman, P.: Scale economies, product differentiation, and the pattern of trade. Am. Econ. Rev. **70**(5), 950–959 (1980)
6. Brander, J., Krugman, P.: A reciprocal dumping model of international trade. J. Int. Econ. **15**(3–4), 313–321 (1983)
7. Melitz, M.J.: The impact of trade on intra-industry reallocations and aggregate industry productivity. Econometrica **71**(6), 1695–1725 (2003)

8. Melitz, M.J., Ottaviano, G.I.P.: Market size, trade, and productivity. Rev. Econ. Stud. **75**(1), 295–316 (2008)
9. Behrens, K., Murata, Y.: General equilibrium models of monopolistic competition: a new approach. J. Econ. Theory **136**(1), 776–787 (2007)
10. Zhelobodko, E., Kokovin, S., Parenti, M., Thisse, J.-F.: Monopolistic competition: beyond the constant elasticity of substitution. Econometrica **80**(6), 2765–2784 (2012)
11. Arkolakis, C., Costinot, A., Rodríguez-Clare, A.: New trade models, same old gains? Am. Econ. Rev. **102**(1), 94–130 (2012)
12. Arkolakis, C., Costinot, A., Donaldson, D., Rodríguez-Clare, A.: The elusive pro-competitive effects of trade. Rev. Econ. Stud. **86**(1), 46–80 (2019)
13. Bykadorov, I., Gorn, A., Kokovin, S., Zhelobodko, E.: Why are losses from trade unlikely? Econ. Lett. **129**, 35–38 (2015)
14. Kokovin, S., Molchanov, P., Bykadorov, I.: Increasing returns, monopolistic competition, and international trade: revisiting gains from trade. J. Int. Econ. **137**, 103595 (2022)
15. Dhingra, S., Morrow, J.: Monopolistic competition and optimum product diversity under firm heterogeneity. J. Polit. Econ. **127**(1), 196–232 (2019)
16. Dhingra, S.: Trading away wide brands for cheap brands. Am. Econ. Rev. **103**(6), 2554–2584 (2013)
17. Corchón, L.C.: Monopolistic competition: equilibrium and optimality. Int. J. Ind. Organ. **9**(3), 441–452 (1991)
18. Bykadorov, I.: Social optimality in international trade under monopolistic competition. Commun. Comput. Inf. Sci. **1090**, 163–177 (2019)
19. Perry, M.K., Groff, R.H.: Resale price maintenance and forward integration into a monopolistically competitive industry. Q. J. Econ. **100**(4), 1293–1311 (1985)
20. Choi, S.C.: Price competition in a channel structure with a common retailer. Mark. Sci. **10**(4), 271–296 (1991)
21. Ottaviano, G.I.P., Tabuchi, T., Thisse, J.-F.: Agglomeration and trade revised. Int. Econ. Rev. **43**(2), 409–436 (2002)

Mathematical Modeling in Forecasting the Development of the Construction Industry in the Russian Federation

Victor Orlov[1](✉), Tatiana Ivanova[2], Natalia Kazakova[3], Valentina Pavlova[4], and Iraida Ivanitskaya[5]

[1] Moscow State University of Civil Engineering, Yaroslavskoye Shosse, 26, Moscow 129337, Russia
orlovvn@mgsu.ru

[2] Institute of Economics of the Russian Academy of Sciences, Nakhimovsky Avenue, 32, Moscow 117218, Russia

[3] Cheboksary Institute (branch) of Moscow Polytechnic University, K.Marx Street, 54, Cheboksary 428000, Russia

[4] Russian University of Economics named after G.V. Plekhanov, Stremyanny Lane, 36, Moscow 117997, Russia
Pavlova.VV@rea.ru

[5] Chuvash State University named I.N. Ulyanov, Moskovsky Prospect, 15, Cheboksary 428015, Russia

Abstract. In the paper, the authors present a method for assessing the pace of development of the construction industry and their forecasting in the Russian Federation, which allows considering factors with different units of measurement. For this purpose, an index analysis of nine indicators was carried out (divided into three blocks: the volume of work and the number of employees; fixed assets and investments in fixed capital; commissioning of buildings and public utilities), an integral indicator was calculated that reflects their changes, and the values of the rate development of the construction industry in the Russian Federation. Then, on the basis of mathematical modeling, a multi-factor mathematical model was developed and verified in real time, with the help of which it is possible to carry out a qualitative prediction of the indicator under study. The analysis of forecasts, at the next stage, makes it possible to substantiate the importance of adjusting certain indicators that have the maximum impact on the process under study. Thus, we obtain a mathematical justification for making a managerial decision in the development of strategies and programs in order to develop the construction industry in the Russian Federation.

Keywords: Forecasting · Construction Industry · Assessment Methodology · Integral Indicator · Mathematical Modeling

N. Olenev et al. (Eds.): OPTIMA 2023, CCIS 1913, pp. 119–130, 2024.
https://doi.org/10.1007/978-3-031-48751-4_9

1 Introduction

The construction industry is a capital-forming industry, playing an important role in the socio-economic development of any country and is able to improve the infrastructure of the regions, raising the material and cultural level of their inhabitants. For the successful functioning and development of the construction industry in the Russian Federation, it is necessary to increase the effectiveness of interaction between all stakeholders, including through the use of mathematical justification in predicting emerging trends. The aim of the work is to develop a method for assessing the development of the construction industry in the Russian Federation using a mathematical model to obtain a high-quality predictable result that provides a positive trend in the development of the industry based on statistical information with a heterogeneous metric. To achieve the goal of the study, the following methods were used: index and integral calculation of indicators, correlation and regression analysis, the theory of multivariate regression and, based on analysis of variance, forecasting technology. Analysis of foreign works in recent years [7,9,12–14,19,25,27,29,30], indicates that they use research technology based on the analysis of the original information without the possibility of forecasting. In the works of [4–6,8,15–18,24,26,28], mathematical models of the processes under study are presented, but there is no technology for obtaining a forecast and the approach of accounting for indicators with different units of measurement is not taken into account. Based on the developed assessment technology: the impact of socio-economic factors on the reproduction of human resources in agriculture [10,22,23], innovative development of the economy [10], as well as forecasting reproductive processes in agriculture, in general [21], and the pace of agricultural production, in particular [20], the author's methodology for assessing the pace of development of the construction industry and its forecasting in the Russian Federation is recommended for use.

2 Materials and Methods

At the initial stage, information was prepared on the most significant indicators for assessing the pace of development of the construction industry, divided into three blocks: the amount of work and the number of employees; fixed assets and investments in fixed assets; commissioning of buildings and public utilities (Table 1).

Based on the analysis of the data in Table 1, it can be noted that the volume of work performed by the type of activity "Construction" increased from 2010 to 2019 by 4677.9 billion rubles, or 2.05 times; the average monthly nominal accrued wages of employees increased by 2.01 times, investments in fixed assets – by 1.91; fixed assets – by 2.06, the total area of buildings put into operation – by 1.60 times.

An analysis of the commissioning of production capacities through construction indicates their slight increase in conventional units from 9630.6 in 2010 to 10805.8 in 2019 (Rosstat, 2020) [3]. In total, 83 parameters were taken into

Table 1. The main indicators characterizing the pace of development of the construction industry of the Russian Federation in 2010–2019.

Indicators	Years									
	2010	2011	2012	2013	2014	2015	2016	2017	2018	2019
VWE – indicators of the volume of work and the number of employees										
Amount of work performed by type of activity “Construction”, billion rubles	4454.2	5140.3	5714.1	6019.5	6125.2	7010.4	7213.5	7579.8	8470.6	9132.1
Profitability of sold goods, products (works, services), %	4.5	4.3	5.0	8.3	3.4	3.8	4.0	3.8	3.9	4.5
Average annual number of employees, thousand people	6153.0	5473.6	5641.9	5711.9	5664.1	6383.5	6204.8	6318.9	6390.8	6416.3
Average monthly nominal accrued wages of employees, rubles	21172	23682	25951	27701	29354	29960	32332	33678	38518	42630
FAI – indicators of fixed assets and investments in fixed assets										
Investments in fixed assets aimed at development construction, billion rubles	342.1	337.0	348.6	438.1	469.3	401.2	443.7	511.5	638.4	653.7
Availability of fixed assets, billion rubles	1499.9	1490.2	1581.9	1676.9	1774.7	2049.4	2084.4	2191.1	2718.5	3094.1
Depreciation of fixed assets (at the end of the year), %	48.3	49.0	49.0	50.0	51.2	50.4	48.4	48.4	46.1	48.2
BPU – indicators of commissioning of buildings and public utilities										
The total area of buildings put into operation – total, mln sq. m, including:	91.5	98.9	110.1	117.8	138.6	139.4	135.8	137.3	132.7	146.7
– residential purpose	70.3	77.2	82.0	87.1	104.4	106.2	103.4	104.6	101.8	111.7
– non-residential purpose	21.2	21.6	28.1	30.7	34.2	33.2	32.4	32.7	30.9	35.0
Commissioning of utility facilities, km, including: – water networks	2234.4	2590.1	2340.1	2618.4	2401.2	2697.0	1675.2	1397.0	1499.0	1083.8
– gas networks	14016.7	14158.7	15905.1	12091.6	12020.3	9680.1	8724.0	7501.3	8043.4	7094.4
– sewer networks	496.3	447.7	428.8	566.9	481.7	471.8	445.8	437.8	363.5	335.3
– heating networks	204.3	439.6	227.5	247.2	222.3	106.1	170.1	213.8	173.5	182.6

Source – (Rosstat) [3]

account, of which: 8 – mining and processing capacities, 26 – production capacities, 44 – built, 1 – electrified, 4 – work carried out. The degree of depreciation of fixed assets decreased over the analyzed period by 0.1% points, while the profitability of sold goods, products (works, services) remained unchanged – 4.5%. Therefore, the fact that the average annual number of people employed in the construction industry has changed slightly (6153.0 thousand people in 2010 and 6416.3 thousand people in 2019) can only be explained by a decrease in the commissioning of public utilities in Russian Federation. Thus, the commissioning of water supply networks decreased by 2.06 times, gas - by 1.98, sewer – by 1.5, heating networks – by 1.1 times, respectively.

Considering the above, we will exclude from the list of indicators those that have an insignificant impact on the pace of development of the construction industry, since they have remained almost unchanged for 10 years. These are: the average annual number of employees, the degree of depreciation of fixed assets, the profitability of goods sold, products, as well as the area of residential and non-residential buildings put into operation (we will use their total area in calculations).

The selected factors are brought to the index value (in % of the previous year) in order to be able to take them into account when calculating the integral indicator (Table 2). The index analysis method allows aggregating a fairly wide range of quantitative indicators for assessing the pace of development of the construction industry, which have different units of measurement and cannot be compared with each other without standardization of values. Based on Table 1, Table 2 is formed, reflecting the index values of indicators characterizing the pace of development of the construction industry of the Russian Federation in 2010–2019.

Taking into account the information in Table 2, according to formula 1 (author's development), an integral indicator is calculated that characterizes the amount of work and the number of people employed in the construction of the Russian Federation in 2010–2019, (II_{VWE}), in %:

$$II_{VWE}=\sqrt[2]{I_{WC}*I_{AMW}}, \quad (1)$$

where I_{WC} – index of changes in the volume of work performed by the type of activity "Construction", %;
where I_{AMW} – index of change in average monthly nominal accrued wages, %.

Further, based on the information in Table 2, formula 2 is developed that calculates an integral indicator that characterizes fixed assets and investments in fixed capital of the construction industry of the Russian Federation in 2010–2019,(II_{FAI}), in %:

$$II_{FAI}= \sqrt[2]{I_{IFA}*I_{FA}}, \quad (2)$$

where I_{IFA} – index of change in investments in fixed assets, %;
I_{FA} – index of change in the availability of fixed assets, %.

Also, based on the data in Table 2, an integral indicator is calculated that characterizes the commissioning of buildings and public utilities in 2010–2019, according to formula 3 (author's development), (II_{BPU}), in %:

Table 2. Dynamics of changes in indicators characterizing the pace of development of the construction industry of the Russian Federation in 2010–2019, in% of the previous year.

Indicators	Years										Medium pace
	2010	2011	2012	2013	2014	2015	2016	2017	2018	2019	
II_{VWE} – index indicators of the volume of work and the number of employees											
Index of changes in the volume of work performed by the type of activity “Construction”	91.9	115.4	111.2	106.0	101.8	114.4	102.9	105.1	111.7	107.8	106.9
Index of change in average monthly nominal accrued wages	92.8	111.9	109.6	103.0	106.0	102.1	107.9	104.2	114.4	110.7	106.8
II_{FAI} – index indicators of fixed assets and investments in fixed assets											
Index of change in investment in fixed assets	93.4	98.5	103.4	101.9	107.1	85.5	110.6	115.3	124.8	102.4	107.4
Index of changes in the availability of fixed assets	97.1	99.4	106.2	99.7	105.8	115.5	101.7	105.1	124.1	113.8	107.8
II_{BPU} – index indicators of the commissioning of buildings and public utilities											
Index of change in the total area of buildings put into operation	93.3	108.1	111.3	96.9	117.7	100.6	97.4	101.1	96.6	110.6	104.6
Index of change in the commissioning of public utilities, including: – water supply networks	92.6	115.9	90.3	111.9	91.7	112.3	62.1	83.4	107.3	72.3	94.0
– gas networks	99.5	101.0	112.3	76.0	99.4	80.5	90.1	86.0	107.2	88.2	94.0
– sewer networks	104.4	90.2	95.8	132.2	85.0	97.9	94.5	98.2	83.0	92.2	97.4
– heating networks	63.3	215.2	51.8	108.7	89.9	47.7	160.3	125.7	81.2	105.2	104.9

Source – Compiled by the authors.

$$II_{BPU} = \sqrt[5]{I_{BPIO} * I_{WSN} * I_{CGN} * I_{CSN} * I_{CHN}}, \tag{3}$$

where I_{BPIO} – index of change in the total area of buildings put into operation, %;

I_{WSN} – index of change in the commissioning of water supply networks, %.

I_{CGN} – index of change in the commissioning of gas networks, %.

I_{CSN} – index of change in the commissioning of sewer networks, %.

I_{CHN} – index of change in the commissioning of heat networks, %.

Analysis of II_{VWE} (Table 3) indicates that for 2010–2019. The average value of the indices of changes in the volume of work performed by the type of activity "Construction" and the nominal accrued wages was almost identical - 106.9 and 106.8%, that is, both had the same effect on the integral indicator. II_{FAI}, shows that for the analyzed period, the average value of the indices of changes in investments in fixed assets and the availability of fixed assets was also almost the same – 107.4 and 107.9%, respectively. At the same time, the second indicator had a more significant impact on the integral indicator. Analyzing II_{BPU} it can be seen that the maximum average value of the indices is characteristic of the commissioning of heat networks – 105.0% and the total area of buildings put into operation – 104.5%; the other three are in decline.

Further, the values of formulas 1–3 are substituted into formula 4 to calculate the pace of development of the construction industry of the Russian Federation ($P_{D.C.I.}$), in %:

$$P_{D.C.I.} = \frac{II_{VWE} + II_{FAI} + II_{BPU}}{3}, \tag{4}$$

where II_{VWE} – an integral indicator characterizing the volume of work and the number of people employed in construction, %;
II_{FAI} – integral indicator characterizing fixed assets and investments in fixed assets of the construction industry, %;
II_{BPU} – integral indicator characterizing the commissioning of buildings and public utilities, %.

Values of the indicator of the pace of development of the construction industry of the Russian Federation ($P_{D.C.I.}$), are entered in Table 3.

Table 3. The pace of development of the construction industry of the Russian Federation in 2010–2019, in %.

Indicators	Years										Medium pace
	2010	2011	2012	2013	2014	2015	2016	2017	2018	2019	
II_{VWE}	93.6	113.6	110.4	106	103.8	108.1	105.4	104.6	113.1	109.2	106.8
II_{FAI}	100.5	98.9	104.8	115.4	106.5	99.4	106.1	110.1	124.4	108.0	107.4
II_{BPU}	89.3	121.6	89.4	103.4	96.1	84.3	96.3	97.8	94.4	92.7	96.5
$P_{D.C.I.}$	94.5	111.4	101.5	108.3	102.1	97.3	102.6	104.2	110.6	103.3	103.6

Source – Compiled by the authors.

The data in Table 3 indicate that the maximum impact on the pace of development of the construction industry in the Russian Federation in 2010–2019 was provided by integral indicators characterizing fixed assets and investments in fixed capital, as well as the commissioning of buildings and public utilities. Accordingly, the first indicator had the maximum impact (with an average rate of 107.4%), the second - the minimum (96.5%). This fact will be additionally substantiated below when constructing the correlation matrix of factors. The average rate of development of the construction industry in the Russian Federation for the analyzed period was 103.6%.

3 Results

At the next stage, in order to further refine the problem under study, it is proposed to use the mathematical modeling tools successfully implemented in the works [11, 20–22]. Consider a variant of the three-factor model:

$$Y = a_0 + a_1 x_1 + a_2 x_2 + a_3 x_3, \tag{5}$$

where $x_1 = II_{VWE}; x_2 = II_{FAI}; x_3 = II_{BPU}$ (Table 3) and analyze the correlation matrix (Table 4).

We calculate the correlation matrix (Table 4) according to the initial information in Table 3.

Table 4. Correlation matrix of factors presented in Table 3.

	Column 1	Column 2	Column 3	Column 4
Column 1	1			
Column 2	−0.316822	1		
Column 3	0.01026245	−0.8061103	1	
Column 4	−0.7888729	0.3673122	−0.2473968	1

In Table 4: column 1 corresponds to the designation Y, column 2 –x_1, column 3 – x_2, column 4 – x_3.

Analysis of the correlation matrix (Table 4) indicates the correlation of the first and second factors. Meaning of Interfactorial Partial Correlation $R_{x_1 x_2}$ in absolute value exceeds the value of 0.7 (column 2, Table 4). We carry out the adjustment (Table 3) by factors and get Table 5.

Table 5. Adjusted rate of development of the construction industry in the Russian Federation in 2010–2019, in%.

Indicators	Years										Medium pace
	2010	2011	2012	2013	2014	2015	2016	2017	2018	2019	
II_{FAI}	100.5	98.9	104.8	115.4	106.5	99.4	106.1	110.1	124.4	108.0	107.4
II_{BPU}	89.3	121.6	89.4	103.4	96.1	84.3	96.3	97.8	94.4	92.7	96.5
$P_{D.C.I.}$	94.5	111.4	101.5	108.3	102.1	97.3	102.6	104.2	110.6	103.3	103.6

Adjusted notation of values (Table 5): $Y - P_{D.C.I.}; x_1 - II_{FAI}; x_2 - II_{BPU}$.

Reanalysis of the correlation matrix get in Table 6.

For the data in Table 5, we build a multifactorial model of the form:

$$Y = a_0 + a_1 x_1 + a_2 x_2, \tag{6}$$

where $a_0 = 24.382128; a_1 = 0.48907; a_2 = 0.280009$. The presented multivariate regression equation has a multiple determination coefficient $R^2 = 0.86432$. The

Table 6. Correlation matrix of factors presented in Table 5.

	Column 1	Column 2	Column 3
Column 1	1		
Column 2	0.73123498	1	
Column 3	0.90031505	0.36145431	1

Designations in Table 6: column 1 corresponds to the designation Y, column 2 – x_1, column 3 – x_2. The values of Table 6 indicate the absence of correlation between the factors in Table 5.

significance of the regression equation is confirmed by the Fisher test at the level of significance $\alpha = 0.05$ and $F_{CRIT}(2;7;0.05) = 4.74$; $F_{OBSERV} = 22.2959$. A competing hypothesis is accepted $H1 : R^2 \neq 0$, asserting the significance of the constructed multifactorial model (6). Figure 1 shows the geometric interpretation of the model.

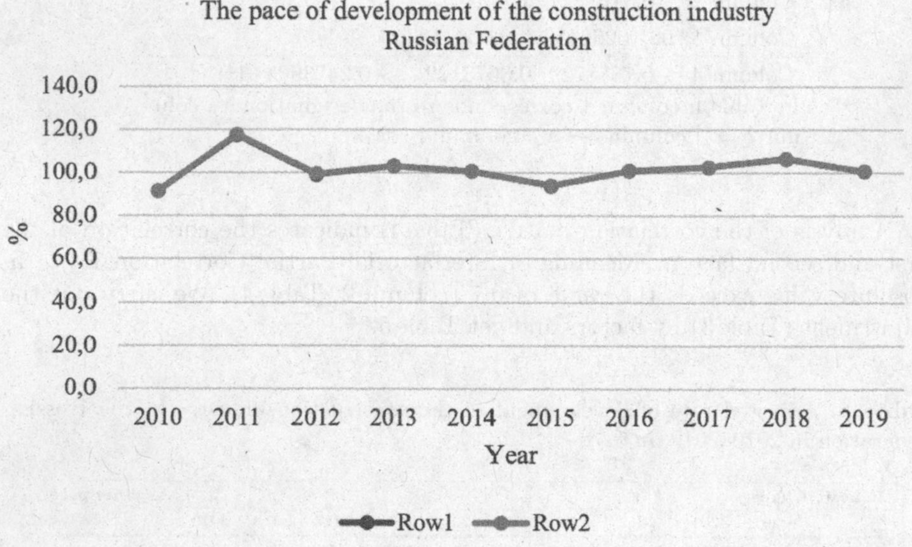

Fig. 1. Geometric interpretation of the multifactorial model of the rate of development of the construction industry in the Russian Federation, where row 1 is the initial information, row 2 is the multifactorial model.

This model has been tested for data from 2010 to 2018. The variant of the tested model has the form

$$Y = 24.382128 + 0.489070x_1 + 0.280009x_2. \tag{7}$$

We get the coefficient of multiple determination $R^2 = 0.864351$. Similarly, based on Fisher's statistical test, with a significance level $\alpha = 0.05$, as well as the values: $F_{CRIT} = 5.14$ and $F_{OBSERV} = 19.1159$ we obtain confirmation of the significance of the multiple coefficient of determination based on the acceptance of a competing hypothesis $H1 : R^2 \neq 0$. The geometric interpretation of the model is shown in Fig. 2.

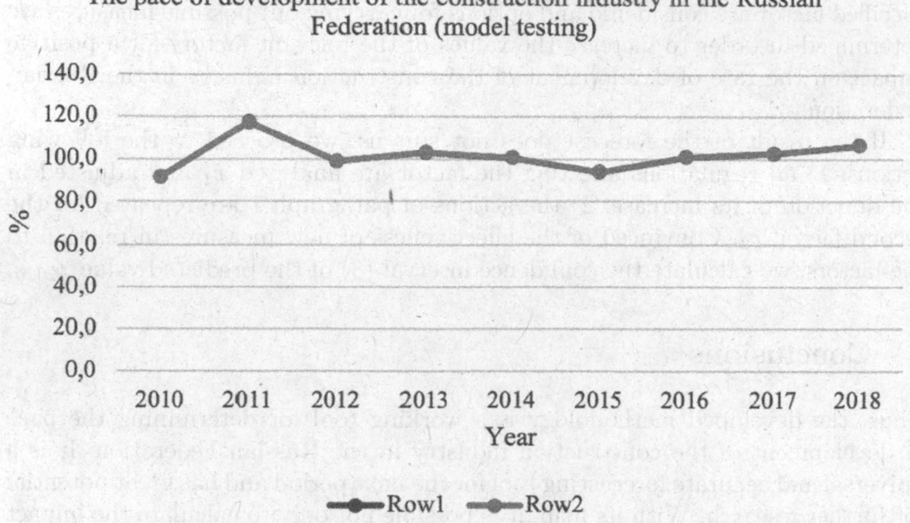

Fig. 2. Geometric interpretation of the multifactorial model of the rate of development of the construction industry in the Russian Federation, where row 1 is the initial information, row 2 is the multifactorial model.

Based on predicted factor values: $x_1 = 108.6$ and $x_2 = 92.7$ we obtain from the expression of the mathematical model (7) the confidence interval of the predicted value Y_{2019}: $89.4188 \leq Y_{2019} \leq 114.4465$ according to the formula:

$$Y_{pointforecast} R^2 \leq Y_{forecast} \leq Y_{pointforecast}(2 - R^2), \quad (8)$$

which covers the real value of the forecast $Y_{2019} = 103.3$. Wherein $Y_{pointforecast}$ calculated according to the model (7) for the values of the factors for 2019, Table 5.

To calculate the predicted value of the rate of development of the construction industry in the Russian Federation Y_{2020} according to model (6), it is necessary to know the factors x_1 and x_2 for 2020. One of the options for obtaining these factor values is based on the construction of a mathematical model for each factor, based on the theory of time series. For example, based on the dynamics of the development of the factors of the model (6), consider the values of the factors: $x_1 = 103.1; x_2 = 93.5$. From expression (8) and model (6) we obtain a confidence interval for the predicted value of the indicator of the rate of development of the construction industry of the Russian Federation $Y_{2020} P_{D.C.I.}$ for 2020: $87.2750 \leq P_{D.C.I.2020} \leq 114.6757$.

An analysis of the multifactorial model (6) shows that the rate of development of the construction industry in the Russian Federation Y_{2020} significant impact is exerted by the second integral indicator - II_{FAI}, characterizing fixed assets and investments in fixed capital of the construction industry. This is reflected in the values of the coefficients of partial correlation, the values of the coefficients of the structure of the multifactorial model (6). At the next stage, all indicators of the specified factor are considered and options for carrying out possible measures are determined in order to increase the values of the relevant factors for a positive impact on the rate of development of the construction industry in the Russian Federation.

If the result of the forecast does not suit us, we proceed to the following actions: 1) all regulations affecting the factor are analyzed x_1 and adjusted in the direction of its increase; 2) the actions of paragraph 1 are repeated for the second factor x_2. Convinced of the effectiveness of new measures in relation to the factors, we calculate the confidence interval (8) of the predicted value Y_{2020}.

4 Conclusions

Thus, the developed methodology is a working tool for determining the pace of development of the construction industry in the Russian Federation. It is a universal and accurate forecasting tool for the next period and has great potential for further research. With its help, it is possible not only to calculate the impact of certain indicators on the development of the construction industry of the state, but also to substantiate the most effective options for the implementation of measures of state targeted programs when assessing their regulatory impact.

So, analyzing the pace of development of the construction industry in the Russian Federation, based on Table 3, we can see a decrease in this indicator in 2010 (94.5%) and in 2015 (97.3%). This is connected, in our opinion, with the fact that in 2010 the implementation of the federal target program "Housing for 2002–2010" was completed. In 2011 there is a sharp jump (111.4%), since on December 17, 2010, a new program "Housing for 2011–2015" was approved by the Government of the Russian Federation No. 1050. By 2015 it was completing its implementation, which caused a reduction in the pace of development of the construction industry, but the program "Housing for 2015–2020" was adopted. Decree of the Government of the Russian Federation of August 25, 2015 No. 889 (FCP, 2022) [2]. On January 1, 2018, it was terminated ahead of schedule, and the implementation of its activities began to be carried out as part of the state program "Providing affordable and comfortable housing and utilities for citizens of the Russian Federation" Decree of the Government of the Russian Federation of December 30, 2017 No. 1710 (EFLNTD, 2022) [1]. This circumstance caused an additional jump in the pace of development of the construction industry in 2018 to 110.6%. Therefore, the proposed methodology really allows mathematically to prove the relationship between the adoption of state programs and the pace of development of the construction industry in the country.

References

1. Electronic fund of legal and normative-technical documents. https://docs.cntd.ru/document/556184998?marker=7DM0K9
2. Federal target programs of russia. https://fcp.economy.gov.ru/cgi-bin/cis/fcp.cgi/Fcp/ViewFcp/View/2007/382
3. Materials of the federal state statistics service. https://rosstat.gov.ru
4. Boyadjiev, C.: Industrial processes stability modeling. Math. Model. **3**(2), 40–45 (2019). https://stumejournals.com/journals/mm/2019/2/40
5. Cha, H., Lee, D.: Determining the risk value for evaluating building renovation projects using probability-based fuzzy set theory. J. Asian Arch. Struct. Eng. **17**(1), 63–70 (2018). https://doi.org/10.3130/jaabe.17.63
6. Chen, T.Y.: Interval-valued pythagorean fuzzy trade-off approach with correlation-based proximity indices for multicriteria decision analysis of bridge construction methods. Complexity, 6463039 (2018). https://doi.org/10.1155/2018/6463039
7. Díaz-Cuevas, P., Camarillo-Naranjo, J.M., Pérez-Alcántara, J.P.: Relational spatial database and multi-criteria decision methods for choosing the optimal location of photovoltaic power plants in the province of Seville (Southern Spain). Clean Technol. Environ. Policy **20**, 1889–1902 (2018). https://doi.org/10.1007/s10098-018-1587-2
8. Gungor, A., Akyuz, A.O., Sirin, C., Tuncer, A.D., Zaman, M., Gungor, C.: Importance of mathematical modeling in innovation. Math. Model. **3**(1), 32–34 (2019). https://stumejournals.com/journals/mm/2019/1/32
9. Hilty, L., Aebischer, B.: ICT Innovations for Sustainability. Springer, Heidelberg (2015). https://doi.org/10.1007/978-3-319-09228-7
10. Ivanova, T., Ivanova, A.: Formation of an effective system for managing the reproduction of personnel potential in agriculture in the region. In: IOP Conference Series: Earth and Environmental Science, vol. 613, p. 012049. IOP Publishing (2020). https://doi.org/10.1088/1755-1315/613/1/012049
11. Ivanova, T., Kadyshev, E., Ladykova, T., Brenchagova, S., Nemtsev, V., Ivanova, A.: Forecasting agricultural production as a tool for effective industry management (on the example of the chuvash republic). Smart Innov. Syst. Technol. **247**, 393–402 (2022). https://doi.org/10.1007/978-981-16-3844-2
12. Kahraman, C., Otay, İ (eds.): Fuzzy Multi-criteria Decision-Making Using Neutrosophic Sets. SFSC, vol. 369. Springer, Cham (2019). https://doi.org/10.1007/978-3-030-00045-5
13. Lee, R., Lee, J.H., Garrett, T.C.: Synergy effects of innovation on firm performance. J. Bus. Res. **99**, 507–515 (2019)
14. Lee, S.J., Lee, E.H., Oh, D.S.: Establishing the innovation platform for the sustainable regional development: Tech-valley project in Sejong city, Korea. World Technopolis Rev. **6**(1), 75–86 (2017)
15. Lennert, J.: Complex spatial modelling possibilities of the socio-economic changes of hungary - potential approaches and methods. Math. Model. **2**(4), 160–162 (2018). https://stumejournals.com/journals/mm/2018/4/160
16. Liang, H., Zhang, S., Su, Y.: Evaluation of the effectiveness of the industrialization process in prefabricated residential buildings using the method of making fuzzy multi-criteria decisions. Math. Prob. Eng. 6078490 (2017). https://doi.org/10.1155/2017/6078490
17. Mavi, R.K., Standing, C.: Critical success factors for sustainable project management in construction: a fuzzy dematel-anp approach. J. Clean. Prod. **194**, 751–765 (2018). https://doi.org/10.1016/j.jclepro.2018.05.120

18. Mirahadi, F., Zayed, T.: Construction performance prediction based on simulation using neural network fuzzy reasoning (2016). https://doi.org/10.1016/j.autcon.2015.12.021
19. Njos, R., Jakobsen, S.E.: Cluster policy and regional development: scale, scope and renewal. Reg. Stud. Reg. Sci. **3**(1), 146–169 (2016)
20. Orlov, V., Ivanova, T., Kadyshev, E., Chernyshova, T., Prokopiev, A., Ivanova, A.: Mathematical modeling in forecasting reproduction processes in agriculture. In: Beskopylny, A., Shamtsyan, M. (eds.) XIV International Scientific Conference "INTERAGROMASH 2021". LNNS, vol. 246, pp. 330–338. Springer, Cham (2022). https://doi.org/10.1007/978-3-030-81619-3_37
21. Orlov, V., Ivanova, T., Sokolova, G., Arkhipova, V., Ivanova, A.: Assessment of innovative development of the russian economy and mathematical modelling. In: VIII International Scientific and Practical Conference "Current Problems of Social and Labor Relations (ISPC-CPSLR 2020)": Advances in Social Science, Education and Humanities Research, vol. 527, pp. 501–507 (2021). https://doi.org/10.2991/assehr.k.210322.166
22. Orlov, V.N., Ivanova, T.V., Arkhipova, V.A., Ivanitskaya, I.P.: Assessment of the influence of social factors on reproduction of personnel potential in agriculture of Russia, pbwosce-2018: business technologies for sustainable urban development. In: E3S Web of Conference, vol. 110 (2019). https://doi.org/10.1051/e3sconf/201911002143
23. Orlov, V.N., Ivanova, T.V., Brechagova, S.A., Rumbayeva, N.N.: Mathematical modeling of economic factors impact: reproduction of personnel potential in agriculture sector of Russia. In: IOP Conference Series: Earth and Environmental Science, vol. 433, p.012012. IOP Publishing (2020). https://doi.org/10.1088/1755-1315/433/1/012012
24. Pan, Y., Zhang, L.: The role of artificial intelligence in design and construction management: a critical review and future trends. Build. Autom. **122**, 103517 (2021). https://doi.org/10.1016/j.autcon.2020.103517
25. Patanacul, P., Kwak, Y.H., Zwickael, O., Liu, M.: What influences the effectiveness of large-scale government projects? Int. J. Project Manag. **34**(3), 452–466 (2016). https://doi.org/10.1016/j.ijproman.2015.12.001
26. Rahimi, Y., Tavakkoli-Moghaddam, R., Iranmanesh, S.H., Vaez-Alaei, M.: A hybrid approach to risk management of construction projects while using FMEA/ISO 31000/evolutionary algorithms: an empirical optimization study. J. Struct. Eng. Manag. **144**(6), 4018043 (2018). https://doi.org/10.1061/(ASCE)CO.1943-7862.0001486
27. Rezai, J.: Best-worst multicriteria decision-making method. Omega **53**, 49–57 (2015). https://doi.org/10.1016/j.omega.2014.11.009
28. Samantra, C., Datta, S., Mahapatra, S.S.: A fuzzy-based risk assessment module for a capital construction project: an empirical study. Artif. Intell. Eng. Appl. **65**, 449–464 (2017). https://doi.org/10.1016/j.engappai.2017.04.019
29. Yoon, D.: The regional-innovation cluster policy for R&D efficiency and the creative economy: with focus on daedeok innopolis. J. Sci. Technol. Policy Manag. **8**(2), 206–226 (2017)
30. Zavadskas, E.K., Vilutiene, T., Turskis, Z., Saparauskas, J.: Multi-criteria analysis of the effectiveness of projects in construction. Arch. Civil Mech. Eng. **14**(1), 114–121 (2014). https://doi.org/10.1016/j.acme.2013.07.006

Position Auctions for Sponsored Search in Marketplaces

Vladimir Yankovskiy[1,2] and Yuriy Dorn[3,4,5](✉)

[1] High School of Economics, Doctoral School of Mathematics, Budapest, Hungary
vladimir.vank0vskiy@yandex.com
[2] Yandex, Moscow, Russia
[3] MSU Institute for Artificial Intelligence, Lomonosov Moscow State University, Moscow, Russia
[4] Moscow Institute of Physics and Technology, Dolgoprudny, Russia
[5] Institute for Information Transmission Problems of the Russian Academy of Sciences, Moscow, Russia
dornyv@yandex.ru

Abstract. In recent years, online shopping has become increasingly concentrated in marketplaces, which now have a similar number of daily users as searching platforms and social networks. As a result, marketplaces have started adopting monetization mechanics from search platforms, with sponsored search being the main example. However, compared to searching platforms, where "search" is the main source of income and the only decision point (to show sponsored results or not), marketplaces have more complex mechanics. Each seller has at least two decisions that affect marketplace income: what price to set for their goods and how much to spend on search boost. This complexity makes the development of position auctions for sponsored searches more challenging. Currently, most marketplaces still use auctions that were designed for searching platforms, like VCG and GSP auctions. In this work, we consider the properties of VCG and GSP auctions for a simplified marketplace model. In the marketplace setting, those auctions are well studied, but all previous works focused on auction fairness, search costs, and so on, which are usually important in the late stages, i.e., when companies already use auctions and want to make them work better. On the other hand, there is a lack of research on possible upper and lower bounds on the revenue of auctions for marketplaces when information is limited. The question "how much revenue can it generate?" is critical in the early stages of auction implementation. In this work, we try to answer this question and get upper and lower bounds on revenue generated by VCG and GSP auctions for marketplaces.

Keywords: Mechanism design · Sponsored search · Position auctions

1 Introduction

In recent years, online marketplaces have become increasingly popular as a platform for buying and selling goods. With the rise of marketplaces, there has

N. Olenev et al. (Eds.): OPTIMA 2023, CCIS 1913, pp. 131–144, 2024.
https://doi.org/10.1007/978-3-031-48751-4_10

been a growing interest in sponsored search and position auctions as a means of generating revenue. Sponsored search is a form of advertising where advertisers bid for the right to have their ads displayed alongside search results. Position auctions, on the other hand, determine the order in which ads are displayed in search results.

In his seminal paper [21] Hal Varian discusses the use of position auctions in sponsored search and their effectiveness in generating revenue. However, while position auctions have been extensively studied in the context of search engines ([1–3,5,6,10,13,16,18–20,22]), their application to marketplaces is less well understood. This is due to the unique challenges that marketplaces face, such as the need to balance the interests of multiple sellers and the complexity of pricing decisions.

In recent years, the use of position auctions in online marketplaces has garnered significant attention. As a result, there has been a large number of articles examining the use of position auctions in various contexts.

However, most of these articles have focused on two key aspects: fairness and searching costs for customers ([9,11,15,17,24]). Fairness is a critical concern in marketplaces, as sellers compete for ads space and customers seek the most relevant results. Searching costs, on the other hand, can be a significant barrier for customers, as they must navigate through a large number of ads to find what they are looking for. There was also works, that research sellers incentives and assortment selection ([14]). But all of this usually involve specific seller cost model (which is usually unknown and not stochastic) and unrealistic assumptions regarding customer choice. Contrary to this in practice even huge marketplaces usually prefer simple, robust and easy to implement auctions. Another important remark is that the main focus of each marketplace is still to generate revenue.

In this paper, we primarily focus on position auctions and sponsored searches for marketplaces, with revenue optimization as the main objective. Similar to the sponsored search problem for search engines ([19]), this problem can be formulated as an optimization problem ([4]). While there have been some works proposing new models and algorithms for auction design specifically for marketplaces (e.g., [7] or [8] and their references), in our experience, most smaller marketplaces are more interested in adopting well-known mechanisms such as VCG (Vickrey-Clarke-Groves, [23]) and GSP (Generalized Second Price, [12]), which are widely recognized. Equilibrium solutions for GSP and VCG auctions for sponsored search are well studied ([19]), and equilibrium solutions for position auctions in marketplaces for simple models follow directly.

On the other hand, after a marketplace implements a particular auction, switching to the new auction becomes a very hard task. Thus, it is desirable to analyze possible revenue generated by auction before actual implementation, i.e., for business analysis purposes, it is good to have upper and lower bounds on possible revenue that can be estimated beforehand. The main result of this work is such bounds for VCG and GSP auctions. To the best of our knowledge, this is the first result for VCG and GSP auctions on online marketplaces.

2 General Setting of Position Auctions with Preference

Suppose we have N participants and N slots. To any slot corresponds a number α_j which determines the conversion rate of that slot. Let's consider $\alpha_1 \geq \alpha_2 \geq \alpha_3 \geq ... \geq \alpha_N \geq 0$.

To any participant corresponds a pair (v_i, c_i) ($v_i \geq 0$, $c_i \geq 0$), where v_i is the conversion value of that participant or themselves and c_i is the conversion value of that participant for the organizer.

The participants are then distributed by slots according to some mechanism (two such mechanisms will be discussed in this article). Let's define $\sigma \in S_N$ as such an ordering, that

Then $\sigma(i)$-th participant will gain $v_{\sigma(i)}\alpha_i$ and the organizer will gain $\sum_{i=1}^{N} c_{\sigma(i)}\alpha_i$

Within this setting we will consider two mechanisms and the games that correspond to them. First, let's recall the definition of a mechanism.

Definition 1 (*[19]*). *A mechanism for n players is given by*

- *players type spaces* $T_1, ..., T_N$
- *players action spaces* $X_1, ..., X_N$
- *alternative set* A
- *players valuation functions* $V_i : T_i \times A \to \mathbb{R}$
- *outcome function* $a : X_1 \times ... \times X_N \to A$
- *players payment functions* $p_i : X_1, ..., X_N\mathbb{R}$

The game induced by the mechanism is given by using T_i *as type spaces,* X_i *as action spaces and utilities*

$$u_i(t_i, x_1, ..., x_N) = V_i(t_i, a(x_1, ..., x_N)) - p_i(x_1, ..., x_N)$$

In our case the types are pairs (v_i, c_i) set of alternatives is the distribution of slots between participants and actions are bids.

3 VCG Position Auctions with Preference

First, consider a somewhat "canonical" class of mechanisms:

Definition 2 (Vickrey-Clarke-Growes mechanism, *[19]*, **p.218).** *A mechanism is called a VCG mechanism if*

- $X_i = V_i(t_i, A)$
- *If* $x_i = V_i(t_i, a(x_1, ..., x_N))$ *for all* i *then* $a(x_1, ..., x_N) = argmax_a \sum_{i=1}^{N} V_i(t_i, a)$
- *There exists a set of functions* $h_1, ..., h_N$, $h_i : X_1 \times ... \times X_{i-1} \times X_{i+1} \times ... \times X_N \to \mathbb{R}$ *such that* $p_i(x_1, ..., x_N) = h_i(x_1, ..., x_{i-1}, x_{i+1}, ..., x_N) - \sum_{i \neq j} V_i(t_i, a(x_1, ..., x_N))$

Now let's determine how Vickrey-Clarke-Growes mechanism looks for this type of auction. Note that values of participants are uniquely determined by v_i, as c_i and α_i are common knowledge. First of all, if i-th participant is assigned to i-th slot for all i-s, then the revenue of i-th participant is $v_i\alpha_i$ and the revenue of organizer (without payments from participants) is $\sum_{i=1}^{N} c_i\alpha_i$. Therefore, total welfare is equal to $\sum_{i=1}^{N}(v_i + c_i)\alpha_i$. It is maximised if the participants are sorted according to $v_i + c_i$.

Also, if we remove j-th participant all participants behind them move one place up and the new total welfare will be

$$\sum_{i=1}^{j-1}(v_i + c_i)\alpha_i + \sum_{i=j}^{N}(v_i + c_i)\alpha_{i+1}$$

Therefore VCG mechanism in this setting will take the following form:

Definition 3 (VCG positional auction with preference). *The participants place bids b_j and then are given places $\sigma^{-1}(j)$ so, that $b_{\sigma(i)} + c_{\sigma(i)}$ is non-increasing. Then j-th participant pays*

$$\left(\sum_{i=1}^{j-1}(b_{\sigma(i)} + c_{\sigma(i)})\alpha_i + \sum_{i=j+1}^{N}(b_{\sigma(i)} + c_{\sigma(i)})\alpha_{i-1}\right) - \left(\sum_{i=1}^{N}(b_{\sigma(i)} + c_{\sigma(i)})\alpha_i\right) + b_{\sigma(j)}\alpha_j$$

$$= \sum_{i=j}^{N-1}(b_{\sigma(i+1)} + c_{\sigma(i+1)})(\alpha_i - \alpha_{i+1}) - c_{\sigma(j)}\alpha_j$$

The profit of the participant on j-th slot is

$$(v_{\sigma(j)} + c_{\sigma(j)})\alpha_j - \sum_{i=j+1}^{N}(b_{\sigma(i)} + c_{\sigma(i)})(\alpha_{i-1} - \alpha_i)$$

and the profit of the organizer is

$$\sum_{j=1}^{N}\sum_{i=j+1}^{N}(b_{\sigma(i)} + c_{\sigma(i)})(\alpha_{i-1} - \alpha_i) = \sum_{i=1}^{N}(b_{\sigma(i+1)} + c_{\sigma(i+1)})\alpha_i$$

Lemma 1. *The combination of bids forms a Nash Equilibrium for VCG auction in and only if*

$$b_{\sigma(k+1)} + c_{\sigma(k+1)} \leq v_{\sigma(k)} + c_{\sigma(k)} \leq b_{\sigma(k-1)} + c_{\sigma(k-1)}$$

Proof. One can notice, that if the participant on k-th slot changes their bid to move to the j-th slot, then their profit will change to $(v_{\sigma(k)} + c_{\sigma(k)})\alpha_j - \sum_{i=j}^{k-1}(b_{\sigma(i)} + c_{\sigma(i)})(\alpha_i - \alpha_{i+1}) - \sum_{i=k+1}^{N}(b_{\sigma(i)} + c_{\sigma(i)})(\alpha_{i-1} - \alpha_i)$ in case $j < k$ or to $(v_k + c_k)\alpha_j - \sum_{i=j+1}^{N}(b_{\sigma(i)} + c_{\sigma(i)})(\alpha_{i-1} - \alpha_i)$ otherwise.

Therefore, a sequence of bids $b_1, \ldots, b_N$ forms a Nash equilibrium if and only if

$$(v_{\sigma(k)} + c_{\sigma(k)})(\alpha_j - \alpha_k) \leq \sum_{i=j}^{k-1}(b_{\sigma(i)} + c_{\sigma(i)})(\alpha_i - \alpha_{i+1}) \leq$$

$$\leq \frac{\sum_{i=j}^{k-1}(b_{\sigma(k-1)} + c_{\sigma(k-1)})(\alpha_i - \alpha_{i+1})}{\alpha_j - \alpha_k} = b_{\sigma(k-1)} + c_{\sigma(k-1)}$$

for all $j < k$ and

$$(v_{\sigma(k)} + c_{\sigma(k)})(\alpha_k - \alpha_j) \geq \sum_{i=k+1}^{j}(b_{\sigma(i)} + c_{\sigma(i)})(\alpha_{i-1} - \alpha_i) \geq$$

$$\geq \sum_{i=k+1}^{j}(b_{\sigma(k+1)} + c_{\sigma(k+1)})(\alpha_{i-1} - \alpha_i) = b_{\sigma(k+1)} + c_{\sigma(k+1)}$$

for $k < j$

Lemma 2. *In any Nash equilibrium for VCG* $(v_{\sigma(i)} + c_{\sigma(i)}) \geq (v_{\sigma(j)} + c_{\sigma(j)})$ *for all* $j > i + 1$

Proof.

$$v_{\sigma(i)} + c_{\sigma(i)} \geq b_{\sigma(i+1)} + c_{\sigma(i+1)} \geq b_{\sigma(j-1)} + c_{\sigma(j-1)} \geq v_{\sigma(j)} + c_{\sigma(j)}$$

Theorem 1. *The set of possible organizer revenues in a Nash equilibrium in VCG is a segment*

$$[A_1; \sum_{i=1}^{N}(v_{\sigma(i)} + c_{\sigma(i)})\alpha_i]$$

where A_i *is defined by recurrence*

$$A_{N+1} = 0$$

$$A_N = (v_{\sigma(N)} + c_{\sigma(N)})\alpha_N$$

$$A_i = \min((v_{\sigma(i+2)} + c_{\sigma(i+2)})\alpha_i + A_{i+1}, (v_{\sigma(i+2)} + c_{\sigma(i+2)})\alpha_{i+1} + (v_{\sigma(i+3)} + c_{\sigma(i+3)})\alpha_i + A_{i+2})$$

and where σ *sorts* $(v_{\sigma(i)} + c_{\sigma(i)})$ *in non-increasing order.*

Proof. Let's construct the maximum possible combination of bids that forms a Nash equilibrium.

By Lemma 1 for a fixed σ it is achieved exactly when

$$b_{\sigma(k)} = v_{\sigma(k+1)} + c_{\sigma(k+1)} - c_{\sigma(k)}$$

Because the organizer's revenue monotonously depends on bids, their revenue with this set of bids will be maximal. And it will be equal to

$$\sum_{i=1}^{N}(b_{\sigma(i+1)} + c_{\sigma(i+1)})\alpha_i = \sum_{i=1}^{N}(v_{\sigma(i)} + c_{\sigma(i)})\alpha_i$$

Now, let's construct a minimum possible combination of bids that forms an envy-free equilibrium.

By Lemma 1 under fixed σ it is achieved exactly when

$$b_{\sigma(k)} = v_{\sigma(k-1)} + c_{\sigma(k-1)} - c_{\sigma(k)}$$

Because the organizer's revenue monotonously depends on bids, their revenue with this set of bids will be minimal. And it will be equal to

$$\sum_{i=1}^{N}(b_{\sigma(i+1)} + c_{\sigma(i+1)})\alpha_i = \sum_{i=1}^{N}(v_{\sigma(i+2)} + c_{\sigma(i+2)})\alpha_i$$

The recurrent relation demonstrates the minimum of this function with respect to Lemma 2.

Finally let's show that this is a segment. As the revenue is a continuous function of bids, it is sufficient to prove that the set of Nash equilibria is connected.

First note, that for any σ the set of corresponding equilibria is a convex polytope. Now let's consider $s_i : S_N \to S_N$ to be a function on orderings that swaps $\sigma(i)$-th and $\sigma(i+1)$-th elements, while leaving others intact. Now, suppose σ is a permutation, that is compliant with Lemma 2, suppose exists i such that $s_i(\sigma)$ is also compliant and suppose $b_{\sigma(j)}$ is a Nash equilibrium for σ. Then define $b'_{\sigma(j)} = b_{\sigma(j)}$ if $j \notin \{i, i+1\}$ and $b'_{\sigma(j)} = \frac{\max(b_{\sigma(i+2)}, v_{\sigma(i+2)}) + \min(b_{\sigma(i-1)}, v_{\sigma(i-1)})}{2}$ if $j \in \{i, i+1\}$ (in case of edge conditions consider $b_{N+1} = v_{N+1} = 0$, $b_0 = v_0 = b_1 + v_1$)

One can see, that $b'_{\sigma(j)}$ are a Nash equilibrium both in case when the final order is σ and in case when it is $s_i(\sigma)$. Therefore, the equilibrium polytopes for σ and $s_i(\sigma)$ share a point.

However one can notice that any ordering compliant with Lemma 2 can be achieved from the ordering for which $(v_{\sigma(i)} + c_{\sigma(i)})$ are non decreasing by a combination of s_i for various i-s such that all intermediate results are also compliant with Lemma 2. Therefore, the total set of Nash equilibria is indeed connected.

4 Generalized Second Price Auctions with Preference

Consider yet another type of rules:

Definition 4 (GSP positional auction with preference). *The participants place bids b_j and then are given places $\sigma^{-1}(j)$ so, that $b_{\sigma(i)} + c_{\sigma(i)}$ is non-increasing. Then i-th participant pays $(b_{\sigma(i+1)} + c_{\sigma(i+1)} - c_{\sigma(i)})\alpha_i$.*

If all c_i are 0 then this auction turns into a classical GSP position auction.

The profit of j-th participant will be $(v_{\sigma(j)} - b_{\sigma(j+1)} - c_{\sigma(j+1)} + c_{\sigma(j)})$ and the profit of the organizer will be $\sum_{i=1}^{N}(b_{\sigma(i+1)} + c_{\sigma(i+1)})\alpha_i$ - the same as in VCG.

Lemma 3. *The combination of bids forms a Nash Equilibrium for GSP auction in and only if for all i holds:*

$$\max_{i<j;\alpha_j\neq\alpha_i} \frac{(b_{\sigma(i+1)} + c_{\sigma(i+1)})\alpha_i - (b_{\sigma(j+1)} + c_{\sigma(j+1)})\alpha_j}{\alpha_i - \alpha_j} \leq v_{\sigma(i)} + c_{\sigma(i)} \leq$$

$$\leq \min_{j<i;\alpha_j\neq\alpha_i} \frac{(b_{\sigma(j)} + c_{\sigma(j)})\alpha_j - (b_{\sigma(i+1)} + c_{\sigma(i+1)})\alpha_i}{\alpha_j - \alpha_i}$$

Proof. One can also notice, that if the participant on i-th slot changes their bid to move to the j-th slot, then their profit will change to $(v_{\sigma(i+1)} - b_{\sigma(j)} - c_{\sigma(j)} + c_{\sigma(i)})\alpha_j$ in case $i > j$ or to $(v_{\sigma(i)} - b_{\sigma(j+1)} - c_{\sigma(j+1)} + c_{\sigma(i)})\alpha_j$ otherwise.

Therefore, a sequence of bids $b_1, \ldots, b_N$ (with participants sorted by $b_{\sigma(i)} + c_{\sigma(i)}$) forms a Nash equilibrium if and only if

$$(v_{\sigma(i)} - b_{\sigma(j)} - c_{\sigma(j)} + c_{\sigma(i)})\alpha_j \leq (v_{\sigma(i)} - b_{\sigma(i+1)} - c_{\sigma(i+1)} + c_{\sigma(i)})\alpha_i$$

for all $i > j$ and

$$(v_{\sigma(i)} - b_{\sigma(j+1)} - c_{\sigma(j+1)} + c_{\sigma(i)})\alpha_j \leq (v_{\sigma(i)} - b_{\sigma(i+1)} - c_{\sigma(i+1)} + c_{\sigma(i)})\alpha_i$$

for $j > i$

These conditions are equivalent to

$$(v_{\sigma(i)} + c_{\sigma(i)})(\alpha_j - \alpha_i) \leq (b_{\sigma(j)} + c_{\sigma(j)})\alpha_j - (b_{\sigma(i+1)} - c_{\sigma(i+1)})\alpha_i \quad .$$

for all $i > j$ and

$$(v_{\sigma(i)} + c_{\sigma(i)})(\alpha_i - \alpha_j) \geq (b_{\sigma(i+1)} + c_{\sigma(i+1)})\alpha_i - (b_{\sigma(j+1)} - c_{\sigma(j+1)})\alpha_j$$

for $j > i$

In case of GSP it seems to be quite hard to describe the actual set of possible revenues. However, this problem seems to be more tangible for more refined types of equilibria.

5 Envy-Free Equilibria

Definition 5. *Let's call a sequence of bids $b_1, ..., b_N$ in the positional auction with preference setting an envy-free equilibrium if no participant can benefit from exchanging both slot and payment with another participant.*

Lemma 4. *The combination of bids forms an envy-free equilibrium for VCG auction in and only if*

$$b_{\sigma(k+1)} + c_{\sigma(k+1)} \leq v_{\sigma(k)} + c_{\sigma(k)} \leq b_{\sigma(k)} + c_{\sigma(k)}$$

Proof. Follows from the formulas of participant revenues in each position. The proof is basically the same as in Lemma 1.

Lemma 5. *The combination of bids forms an envy-free equilibrium for GSP auction in and only if*

$$\max_{j>k;\alpha_j \neq \alpha_k} \frac{(b_{\sigma(k+1)} + c_{\sigma(k+1)})\alpha_i - (b_{\sigma(j+1)} - c_{\sigma(j+1)})\alpha_j}{\alpha_j - \alpha_k} \leq v_{\sigma(k)} + c_{\sigma(k)} \leq$$

$$\leq \min_{j<k;\alpha_j \neq \alpha_k} \frac{(b_{\sigma(j+1)} + c_{\sigma(j+1)})\alpha_j - (b_{\sigma(k+1)} - c_{\sigma(k+1)})\alpha_k}{\alpha_j - \alpha_k}$$

Proof. Follows from the formulas of participant revenues in each position.

Lemma 6. *In any envy-free equilibrium for VCG auction the values $(v_{\sigma(i)} + c_{\sigma(i)})$ are monotonously non-increasing.*

Proof.

$$v_{\sigma(i)} + c_{\sigma(i)} \geq b_{\sigma(i+1)} + c_{\sigma(i+1)} \geq v_{\sigma(i+1)} + c_{\sigma(i+1)}$$

by Lemma 4.

Lemma 7. *In any envy-free equilibrium for GSP auction the values $(v_{\sigma(i)} + c_{\sigma(i)})$ are monotonously non-increasing.*

Proof.

$$(v_{\sigma(i)}+c_{\sigma(i)}) \geq \frac{(b_{\sigma(i+1)} + c_{\sigma(i+1)})\alpha_i - (b_{\sigma(i+2)} - c_{\sigma(i+2)})\alpha_{i+1}}{\alpha_i - \alpha_{i+1}} \geq (v_{\sigma(i+1)}+c_{\sigma(i+1)})$$

by Lemma 5.

Lemma 7 allows us to improve the necessary and sufficient conditions of envy-free equilibria for GSP.

Lemma 8. *The combination of bids forms an envy-free equilibrium for GSP auction in and only if*

$$\frac{(b_{\sigma(k+1)}+c_{\sigma(k+1)})\alpha_k-(b_{\sigma(k+2)}+c_{\sigma(k+2)})\alpha_{k+1}}{\alpha_k-\alpha_{k+1}}\le v_{\sigma(k)}+c_{\sigma(k)}\le$$

$$\le\frac{(b_{\sigma(k)}+c_{\sigma(k)})\alpha_{k-1}-(b_{\sigma(k+1)}+c_{\sigma(k+1)})\alpha_k}{\alpha_{k-1}-\alpha_k}$$

Proof. Suppose $j>k$ and the aforementioned conditions are true. Then

$$(v_{\sigma(k)}+c_{\sigma(k)})(\alpha_k-\alpha_j)=\sum_{i=k+1}^{j}(v_{\sigma(k)}+c_{\sigma(k)})(\alpha_{i-1}-\alpha_i)\ge$$

$$\ge\sum_{i=k+1}^{j}(v_{\sigma(i-1)}+c_{\sigma(i-1)})(\alpha_{i-1}-\alpha_i)\ge$$

$$\ge\sum_{i=k+1}^{j}((b_{\sigma(i)}+c_{\sigma(i)})\alpha_{i-1}-(b_{\sigma(i+1)}-c_{\sigma(i+1)})\alpha_i)=$$

$$=(b_{\sigma(k+1)}+c_{\sigma(k+1)})\alpha_i-(b_{\sigma(j+1)}-c_{\sigma(j+1)})\alpha_j$$

Suppose $j<k$ and the aforementioned conditions are true. Then

$$(v_{\sigma(k)}+c_{\sigma(k)})(\alpha_j-\alpha_k)=\sum_{i=j+1}^{k}(v_{\sigma(k)}+c_{\sigma(k)})(\alpha_{i-1}-\alpha_i)\le$$

$$\le\sum_{i=j+1}^{k}(v_{\sigma(i)}+c_{\sigma(i)})(\alpha_{i-1}-\alpha_i)\le\sum_{i=j+1}^{k}(b_{\sigma(i)}+c_{\sigma(i)})\alpha_{i-1}-(b_{\sigma(i+1)}+c_{\sigma(i+1)})\alpha_i=$$

$$=(b_{\sigma(j+1)}+c_{\sigma(j+1)})\alpha_j-(b_{\sigma(k+1)}-c_{\sigma(k+1)})\alpha_k$$

One can see, that any envy-free equilibrium for VCG is a Nash equilibrium for VCG. Using Lemma 7 one can prove the same about GSP.

Lemma 9. *Any envy-free equilibrium for GSP is a Nash equilibrium for GSP*

Proof. Suppose $b_1,...,b_N$ forms an envy-free equilibrium. Then

$$\max_{j>k;\alpha_j\neq\alpha_k}\frac{(b_{\sigma(k+1)}+c_{\sigma(k+1)})\alpha_i-(b_{\sigma(j+1)}-c_{\sigma(j+1)})\alpha_j}{\alpha_j-\alpha_k}\le v_{\sigma(k)}+c_{\sigma(k)}\le$$

$$\le\min_{j<k;\alpha_j\neq\alpha_k}\frac{(b_{\sigma(j+1)}+c_{\sigma(j+1)})\alpha_j-(b_{\sigma(k+1)}-c_{\sigma(k+1)})\alpha_k}{\alpha_j-\alpha_k}\le$$

$$\le\frac{(b_{\sigma(j)}+c_{\sigma(j)})\alpha_j-(b_{\sigma(k+1)}-c_{\sigma(k+1)})\alpha_k}{\alpha_j-\alpha_k}$$

Theorem 2. *The possible revenues attainable by organizer in envy-free equilibria in VCG form a segment*

$$\left[\sum_{i=1}^{N}(v_{i+1}+c_{i+1})\alpha_i; \sum_{i=1}^{N}(v_i+c_i)\alpha_i\right]$$

where σ sorts $(v_{\sigma(i)}+c_{\sigma(i)})$ in non-increasing order.

Proof. Let's construct the maximum possible combination of bids that forms an envy-free equilibrium.

By Lemma 4 it is achieved exactly when

$$b_{\sigma(k)} = v_{\sigma(k+1)} + c_{\sigma(k+1)} - c_{\sigma(k)}$$

Because the organizer's revenue monotonously depends on bids, their revenue with this set of bids will be maximal. And it will be equal to

$$\sum_{i=1}^{N}(b_{\sigma(i+1)}+c_{\sigma(i+1)})\alpha_i = \sum_{i=1}^{N}(v_{\sigma(i)}+c_{\sigma(i)})\alpha_i$$

Now, let's construct a minimum possible combination of bids that forms an envy-free equilibrium.

By Lemma 4 it is achieved exactly when

$$b_{\sigma(k)} = v_{\sigma(k)}$$

Because the organizer's revenue monotonously depends on bids, their revenue with this set of bids will be minimal. And it will be equal to

$$\sum_{i=1}^{N}(b_{\sigma(i+1)}+c_{\sigma(i+1)})\alpha_i = \sum_{i=1}^{N}(v_{\sigma(i+1)}+c_{\sigma(i+1)})\alpha_i$$

Finally, let's notice, that because the order σ depends only on v_i and c_i the set of all envy-free equilibria forms a convex polytope. Therefore the set of all possible revenues of organizer is a segment, as its bounded projection onto 1-d space.

Theorem 3. *The possible revenues attainable by organizer in envy-free equilibria in GSP form a segment*

$$\left[\sum_{j=1}^{N-1}(v_{\sigma(j+1)}+c_{\sigma(j+1)})(\alpha_j-\alpha_{j+1}); \sum_{j=1}^{N-1}(v_{\sigma(j)}+c_{\sigma(j)})(\alpha_j-\alpha_{j+1})\right]$$

where σ sorts $(v_{\sigma(i)}+c_{\sigma(i)})$ in non-increasing order.

Proof. Let's construct the maximum possible combination of bids that forms an envy-free equilibrium.

Consider $b_{\sigma(N+1)} = 0$ and $c_{\sigma(N+1)} = 0$. Then Lemma 8 yields recurrent formula for the maximum possible combination of bids:

$$b_{\sigma(k)} = (b_{\sigma(k+1)} + c_{\sigma(k+1)})\frac{\alpha_k}{\alpha_{k-1}} + (v_{\sigma(k-1)} + c_{\sigma(k-1)})(1 - \frac{\alpha_k}{\alpha_{k-1}}) - c_{\sigma(k)}$$

From this recurrence it follows, that

$$b_{\sigma(k)} = -c_{\sigma(k)} + \frac{1}{\alpha_{k-1}} \sum_{i=k-1}^{N-1} (v_{\sigma(j)} + c_{\sigma(j)})(\alpha_i - \alpha_{i+1})$$

Because the organizer's revenue monotonously depends on bids, their revenue with this set of bids will be maximal and equal to

$$\sum_{i=1}^{N}(b_{\sigma(i+1)} + c_{\sigma(i+1)})\alpha_i = \sum_{i=1}^{N}\sum_{j=i}^{N-1}(v_{\sigma(j)} + c_{\sigma(j)})(\alpha_j - \alpha_{j+1}) =$$

$$= \sum_{j=1}^{N-1}(v_{\sigma(j)} + c_{\sigma(j)})(\alpha_j - \alpha_{j+1})$$

Now let's construct the maximum possible combination of bids that forms an envy-free equilibrium.

Consider $b_{\sigma(N+1)} = 0$ and $c_{\sigma(N+1)} = 0$. Then Lemma 8 yields recurrent formula for the minimum possible combination of bids:

$$b_{\sigma(k)} = (b_{\sigma(k+1)} + c_{\sigma(k+1)})\frac{\alpha_k}{\alpha_{k-1}} + (v_{\sigma(k)} + c_{\sigma(k)})(1 - \frac{\alpha_k}{\alpha_{k-1}}) - c_{\sigma(k)}$$

From this recurrence it follows, that

$$b_{\sigma(k)} = -c_{\sigma(k)} + \frac{1}{\alpha_{k-1}} \sum_{i=k-1}^{N-1} (v_{\sigma(j+1)} + c_{\sigma(j+1)})(\alpha_i - \alpha_{i+1})$$

Because the organizer's revenue monotonously depends on bids, their revenue with this set of bids will be maximal. And it will be equal to

$$\sum_{i=1}^{N}(b_{\sigma(i+1)} + c_{\sigma(i+1)})\alpha_i = \sum_{i=1}^{N}\sum_{j=i}^{N-1}(v_{\sigma(j+1)} + c_{\sigma(j+1)})(\alpha_j - \alpha_{j+1}) =$$

$$= \sum_{j=1}^{N-1} j(v_{\sigma(j+1)} + c_{\sigma(j+1)})(\alpha_j - \alpha_{j+1})$$

Finally, let's notice, that because the order σ depends only on v_i and c_i the set of all envy-free equilibria forms a convex polytope. Therefore the set of all possible revenues of organizer is a segment, as its bounded projection onto 1-d space.

6 Truthfulness

Now let's consider a following property of auction mechanisms:

Definition 6. *An auction mechanism is called truthfull if and only if the strategy profile where every participant bids their true value is a Nash equilibrium. That strategy profile is also called truthfull.*

For our VCG truthfulness is a direct consequence of the theorem that any VCG auction is truthfiull. However, this fact can be yet strengthened in our case.

Proposition 1. *The truthful strategy profile is an envy-free equilibrium in VCG.*

Proof. Follows from Lemma 4, as

$$v_{\sigma(k+1)} + c_{\sigma(k+1)} \leq v_{\sigma(k)} + c_{\sigma(k)} \leq v_{\sigma(k)} + c_{\sigma(k)}$$

However, for GSP the things happen to be more diverse: it is not always truthful. Still, a following fact holds:

Theorem 4. *The following three conditions are equivalent:*

- *GSP positional auction with preference is truthful*
- *The truthful strategy profile is an envy-free equilibrium.*
- $(v_{\sigma(k)} + c_{\sigma(k)} - v_{\sigma(k+1)} - c_{\sigma(k+1)})\alpha_k \geq (v_{\sigma(k)} + c_{\sigma(k)} - v_{\sigma(k+2)} - c_{\sigma(k+2)})\alpha_{k+1}$ *for all* $k < N$

Proof. The first condition follows from the second by Lemma 9.

The second condition follows from the third by Lemma 8, as

$$\frac{(v_{\sigma(k+1)} + c_{\sigma(k+1)})\alpha_k - (v_{\sigma(k+2)} + c_{\sigma(k+2)})\alpha_{k+1}}{\alpha_k - \alpha_{k+1}} \leq v_{\sigma(k)} + c_{\sigma(k)}$$

is equivalent to

$$(v_{\sigma(k)} + c_{\sigma(k)} - v_{\sigma(k+1)} - c_{\sigma(k+1)})\alpha_k \geq (v_{\sigma(k)} + c_{\sigma(k)} - v_{\sigma(k+2)} - c_{\sigma(k+1)})\alpha_{k+1}$$

and

$$v_{\sigma(k)} + c_{\sigma(k)} \leq \frac{(v_{\sigma(k)} + c_{\sigma(k)})\alpha_{k-1} - (v_{\sigma(k+1)} + c_{\sigma(k+1)})\alpha_k}{\alpha_{k-1} - \alpha_k}$$

is always true due to $v_{\sigma(k+1)} + c_{\sigma(k+1)} \le v_{\sigma(k)} + c_{\sigma(k)}$.

The third condition follows from the first one by Lemma 2, as

$$(v_{\sigma(i)} + c_{\sigma(i)}) \ge \max_{i<j;\alpha_j\neq\alpha_i} \frac{(b_{\sigma(i+1)} + c_{\sigma(i+1)})\alpha_i - (b_{\sigma(j+1)} + c_{\sigma(j+1)})\alpha_j}{\alpha_i - \alpha_j} \ge$$

$$\ge \frac{(b_{\sigma(i+1)} + c_{\sigma(i+1)})\alpha_i - (b_{\sigma(i+2)} + c_{\sigma(i+2)})\alpha_{i+1}}{\alpha_i - \alpha_{i+1}}$$

Acknowledgements. The research part done by Yuriy Dorn is supported by the Ministry of Science and Higher Education of the Russian Federation (Goszadaniye) 075-00337-20-03, project No. 0714-2020-0005.

References

1. Aggarwal, G., Feldman, J., Muthukrishnan, S., Pál, M.: Sponsored search auctions with Markovian users. In: Papadimitriou, C., Zhang, S. (eds.) WINE 2008. LNCS, vol. 5385, pp. 621–628. Springer, Heidelberg (2008). https://doi.org/10.1007/978-3-540-92185-1_68
2. Ashlagi, I., Braverman, M., Hassidim, A., Lavi, R., Tennenholtz, M.: Position auctions with budgets: existence and uniqueness. BE J. Theor. Econ. **10**(1) (2010)
3. Athey, S., Ellison, G.: Position auctions with consumer search. Q. J. Econ. **126**(3), 1213–1270 (2011)
4. Bai, G., Xie, Z., Wang, L.: Practical constrained optimization of auction mechanisms in e-commerce sponsored search advertising. arXiv preprint arXiv:1807.11790 (2018)
5. Bayir, M.A., Xu, M., Zhu, Y., Shi, Y.: Genie: an open box counterfactual policy estimator for optimizing sponsored search marketplace. In: Proceedings of the Twelfth ACM International Conference on Web Search and Data Mining, pp. 465–473 (2019)
6. Börgers, T., Cox, I., Pesendorfer, M., Petricek, V.: Equilibrium bids in sponsored search auctions: theory and evidence. Am. Econ. J.: Microecon. **5**(4), 163–187 (2013)
7. Choi, H., Mela, C.F.: Monetizing online marketplaces. Mark. Sci. **38**(6), 948–972 (2019)
8. Chu, L.Y., Nazerzadeh, H., Zhang, H.: Position ranking and auctions for online marketplaces. Manage. Sci. **66**(8), 3617–3634 (2020)
9. Dash, A., Chakraborty, A., Ghosh, S., Mukherjee, A., Gummadi, K.P.: When the umpire is also a player: bias in private label product recommendations on e-commerce marketplaces. In: Proceedings of the 2021 ACM Conference on Fairness, Accountability, and Transparency, pp. 873–884 (2021)
10. Deng, Y., Mao, J., Mirrokni, V., Zuo, S.: Towards efficient auctions in an auto-bidding world. In: Proceedings of the Web Conference 2021, pp. 3965–3973 (2021)
11. Derakhshan, M., Golrezaei, N., Manshadi, V., Mirrokni, V.: Product ranking on online platforms. Manage. Sci. **68**(6), 4024–4041 (2022)
12. Edelman, B., Ostrovsky, M., Schwarz, M.: Internet advertising and the generalized second-price auction: selling billions of dollars worth of keywords. Am. Econ. Rev. **97**(1), 242–259 (2007)

13. Edelman, B., Schwarz, M.: Optimal auction design and equilibrium selection in sponsored search auctions. Am. Econ. Rev. **100**(2), 597–602 (2010)
14. Etro, F.: Product selection in online marketplaces. J. Econ. Manage. Strategy **30**(3), 614–637 (2021)
15. Ferreira, K.J., Parthasarathy, S., Sekar, S.: Learning to rank an assortment of products. Manage. Sci. **68**(3), 1828–1848 (2022)
16. Jansen, B.J., Mullen, T.: Sponsored search: an overview of the concept, history, and technology. Int. J. Electron. Bus. **6**(2), 114–131 (2008)
17. Li, W., Qi, Q., Wang, C., Yu, C.: Optimally integrating ad auction into e-commerce platforms. In: Frontiers of Algorithmic Wisdom: International Joint Conference, IJTCS-FAW 2022, Hong Kong, China, August 15–19, 2022, Revised Selected Papers. pp. 44–58. Springer (2023)
18. Milgrom, P.: Simplified mechanisms with applications to sponsored search and package bidding. Tech. rep., Working paper, Stanford University (2007)
19. Nisan, N., Roughgarden, T., Tardos, E., Vazirani, V.V.: Algorithmic game theory. Cambridge University Press (2007)
20. Roughgarden, T.: Twenty lectures on algorithmic game theory. Cambridge University Press (2016)
21. Varian, H.R.: Position auctions. Int. J. Indust. Org. **25**(6), 1163–1178 (2007)
22. Varian, H.R.: Online ad auctions. Am. Econ. Rev. **99**(2), 430–434 (2009)
23. Vickrey, W.: Counterspeculation, auctions, and competitive sealed tenders. J. Financ. **16**(1), 8–37 (1961)
24. Watts, A.: Fairness and efficiency in online advertising mechanisms. Games **12**(2), 36 (2021)

Optimization in Economics and Finance

Agent-Based Model of Cultural Landscape Evolution in Euclidean Space

Nikolay Belotelov[1(✉)] and Fedor Loginov[2]

[1] Federal Research Center "Computer Science and Control" of the Russian Academy of Science, Moscow, Russia
belotel@mail.ru

[2] Moscow Institute of Physics and Technology, Moscow, Russia
loginov@phystech.edu

Abstract. The cultural landscape determines the cultural agent's diversity. It is believed that the cultures space can be represented as a Hilbert space, in which certain sub-spaces - cultural cones -correspond to different cultures. This allows the agents state to be described by a vector in Hilbert space in the first part of the article and in Euclidean space in the second. The numbers of agents belonging to certain cultures are determined by demographic processes and the educational process, as well as the intensity of intercultural contacts. Interaction between agents occurs within clusters, into which the entire set of agents is divided according to certain criteria. During interaction between agents according to a certain algorithm, the length and angle characterizing the state of the agent changes. It is believed that in the process of intercultural interactions, agents leave their pure cultural cones and form intercultural clusters that do not completely belong to any culture. The mathematical formalization of this process is presented in the first part of the article. We suppose that in such clusters as we described the interaction of agents occurs. The second part of the work is devoted to describing the dynamics of this process. In which the Kolmogorov equations were obtained. These equations describe changes in cultural diversity within these clusters.

Keywords: Euclidean space · Agent-based model · Markov chain · Kolmogorov equations · Queueing theory

1 Introduction

One of the problems that modern civilization is facing is the rapidly increasing migration processes, which significantly change the "cultural landscapes" of the regions, giving rise in some cases to problems of intercultural interactions. Apparently, these are inevitable phenomena associated with the growth in the number and mobility of the population, as well as with the growth of the information connectivity of the planet. Researchers use such a new concept as "anthropological transition", which, of course, will lead to global cultural transformations. A

N. Olenev et al. (Eds.): OPTIMA 2023, CCIS 1913, pp. 147–157, 2024.
https://doi.org/10.1007/978-3-031-48751-4_11

large number of humanitarian studies are devoted to the problems of cultural change, the so-called "cultures of transition" [1,2]. Conditions associated with changing cultural landscapes (the emergence of cultural uncertainties) precede the emergence of cultures of transition. The analysis of factors influencing the intensity of occurrence of such uncertainties is of considerable interest.

The speed with which the development of crisis phenomena is taking place urgently requires the intensification of interdisciplinary research using the methods of mathematical and computer modeling. Mathematical formalization of the inevitable assimilation processes that are naturally generated by migration and lead to cultural uncertainty is an extremely complex but relevant task. The unambiguity of definitions necessary for an objective description is achieved by using mathematical symbols, precisely because in this way the references to the conscious subject that pervade everyday language are avoided." [3]. Currently, computer and mathematical approaches to the analysis of social dynamics, the analysis of dynamic changes in social networks, the influence of information impacts on the formation of social heterogeneity are being intensively developed [4–6]. This article is in line with these studies.

The article proposes a formal mathematical scheme of an agent-based simulation model of intercultural interactions, in which an attempt is made to obtain an assessment of the dynamics of change in the "cultural landscape", interaction between different cultures, the depth of cultural memory, as well as the intensity of interaction between subjects. Here as a cultural landscape we consider set of agents that belong to different cultures. There is empirical consensus in presence of universal characteristics – general for all agents and unique characteristics - individual for each small group of the agents.

Here we assume that universal and unique characteristics are formed space with infinite dimension. Unfortunately, modern computational representation of any object can be only finite. We can assume that we have, generally speaking, an infinite set of "symbols-objects" that encode culture. Further, realizing that each culture reflects the world around us that existed, exists and the world that will be, we are forced to place our symbols - objects in an infinite dimensional space. Describing it, any model should provide a mechanism to select finite dimensions in advance. It could be done, as an example, in Hilbert space with dimensions reordering rule and cutting predefined finite number of dimensions, where each dimension is encode some "symbol-object". At the same time, the space itself must be provided with some structure so that it is possible to formally describe the interaction of cultural subjects that are the "carriers" of these "symbols-objects". A subject located in a cultural space and having certain beliefs - assessments in the first approximation can be characterized by a vector formed by the basic assessments of culture, that is, length and angle. The length of the vector characterizes the loading of the agent with some symbols-objects, as well as the potential activity of the agent, or, one might say, the viability of the agent. While the angle allows you to take into account cultural proximity to other agents. The similarity or closeness of the angles is interpreted in the proposed mathematical construction by the cultural similarity of the corresponding agents. That is why the Hilbert space was chosen as the simplest mathematical structure that has the corresponding properties. This space has lengths and angles. Further reasoning is based on the following humanitarian constructions [7]. Culture as a set of symbols that reflect the external and internal world of the subject, initially arises in a certain natural environment. With the development of civilization, it becomes more complex and develops. Therefore, in the most simplified way, we can consider that this is an expanding sphere in some infinite-dimensional space, separated by solid angles that correspond to different cultures. In the proposed model, culture is represented as a cone characterized by basic vectors, which we call the base of culture. These cones come out from some point. It is believed that in the process of cross-cultural interactions, agents leave their pure cultural cones and form cross-cultural clusters that do not completely belong to any culture. The mathematical formalization of this process is presented in the first part of the article. We suppose that in such clusters as we described the interaction of agents occurs. The second part of the work is devoted to describing the dynamics of this process.

The task is close to the problem of opinion dynamic modeling. The analysis of opinion dynamic distribution changing is an important task. The problem has a significant interest from scientists that created a lot of opinion dynamic models such as [8–10]. But provided papers do not split the behaviour between opinion dynamic in online - taking into account behaviour of neighbour agent changes in time and offline - with determine agents behaviour in advance. In the papers [11,12] authors investigate the opinion dynamics as it.

One of the problems is how to represent "cultural" features as a mathematical object. One of possible solutions is to reuse words representation in Euclidean space from [13].

We suggest to consider culture as opinion extension. Our work is based on assumptions that were derived in [14,15] about a cultural states dynamic in Hilbert space problem statement and agent states dimensions sampling to be computationally modelled.

In this paper we introduce an algorithm that allows to dynamic of agents set cultural distribution modeling. There is a similar approach developed by M. DeGroot [16] where authors suggest to use trust matrix for agent states recomputing. Further in the article [17] the approach is supplied by Markov chain assumption usage it for opinion dynamic modeling. In our work we suggest to use continuous-time Markov chain assumption for probability matrix recalculation and use probabilities for stochastic dynamic of agent states. We assume that the cultural landscape transformation is significantly changed for a period that is less than characteristic time of generation change. It that assumption we suggest do not consider a demographic process when we analyse a agents dynamic is separate group of agents.

2 Cluster Type Vector Model of Assimilation Dynamics

We reuse assumptions and definitions from our previous paper [14]. Let‘s introduce topological space P as a set of all elements describes any phenomenas symbols-objects. Consider a subset of $p_i \subsetneq P$ and call the subset as an agent i. Introduce an complete infinite Hilbert space H with basis $\{\psi_d\} \subset H$. Suppose we have a bijective function $l(\cdot) : \tau_P \to$ H reflected each subset p_i to H. Note that the construction of such a function is a task of interdisciplinary analysis and, generally speaking, is a separate difficult task. But, if such a function is constructed, then each element of the space $x_i \in$ H will be called the cultural state of the i-th agent in the space H. Then call H as a space of all cultural states of agents or just cultural space. Let's denote a metric as a $\rho(\cdot, \cdot) : \mathrm{H} \times \mathrm{H} \to \mathrm{R}^+ \cup \{0\}$. Consider a set $B_r(a) = \{x \in \mathrm{H} | 0 < \rho(x, a) = r\} \subset$ H. The set $B_r(a)$ is a sphere with radius r and center a.

Let's call a base of a culture as a pair $(h_i, \phi_i) \in \mathrm{H} \times \mathrm{R}^+ \cup \{0\}$. Here h_i is a center of a culture i and ϕ_i is a critical angle (Term angle is introduced from property of scalar product). This pair is natively introduces a term cultural cone, where h_i is an axis and phi_i is an angle between generatrix and axis of a cone. Let's denote the cone as $Y_i = \{x \in \mathrm{H} | \arccos((h_i, x)) < \phi_i\}$.

Let's introduce a finite-dimensional space H_0, which is a subspace of H. We assume that $\mathrm{H}_0 = \mathrm{R}^\mathrm{N} \subset$ H. Then, according to the projection theorem [18], for any $x \in$ H there is a unique projection onto H_0. This fact can be interpreted through the presence in all agents of common (for example, physiological) symbol-objects expressed through a state subvector from H_0, such as the need for food, the need for sleep, the instinct of reproduction, etc. Then there is an orthogonal complement to H_0, let's denote it as $\mathrm{H}^+ : \mathrm{H}_0 \times \mathrm{H}^+ = \mathrm{H}$ and call it the space of all unique character-objects inherent in different cultures. Denote the operator of the projection of the space H onto H_0 as $\pi(\cdot) : \mathrm{H} \to \mathrm{H}_0$. Then for agent i the projection of the vector x_i onto H_0 is $\pi(x_i)$. Consider the orthogonal complement H^+ to the space H. Since H_0 is finite-dimensional, it follows that H^+ is infinite-dimensional. Let $x_i^+ \in \mathrm{H}^+$, $x_i^+ = \pi(x_i) = x_i$ be the state of an agent from the orthogonal complement to the space of common object symbols. To further construct a projection from an infinite-dimensional space to a finite-dimensional one, it is necessary to define an algorithm for clustering agents in the space H_0.

Lemma 1. *Orthogonal complement X^+ of $X = Y_i \cap \mathrm{H}_0$ is not empty in* H^+.

Proof. Let's fix index i. $X \subset \mathrm{R}^n$. Notify that $\forall x \in \mathrm{H}^+ \to x \times 0 \in$ H and $\forall x \in \mathrm{H}_0 \to x \times 0 \in$ H. Here $\times$ is a Cartesian product. Orthogonal complement of X is $X^+ = \{x \in \mathrm{H}^+ | \forall \hat{x} \in X : (x, \hat{x}) = 0\}$. Here we consider a scalar product in therms of H space. Due to H is a Hilbert space, H^+ has a scalar product $(\mathrm{H}^+ \subset \mathrm{H})$. $\forall x_k, x_l \in$ H we can write

$$(x_k, x_l) = \sum_{j=1}^{\infty} x_k^j \cdot x_l^j = \sum_{j=1}^{n} x_k^j \cdot x_l^j + \sum_{j=n+1}^{\infty} x_k^j \cdot x_l^j \tag{1}$$

Here each infinite sum is converged due to property of Hilbert space. Let's assume that $X^+ = \varnothing$. It means that the both terms of 1 is zero $\forall x_k, x_l$. It means $\forall x_k, x_l \in \mathrm{H} (x_k, x_l) = 0$. That could be true only if $\mathrm{H} = \{0\}$ that is untrue in our assumption that H is complete metric space.

Consider a finite set of cultures $\{Y_i\}_{i=1}^{K}$. Lets introduce a cluster $X(x,r) = \{x_j | \forall x_j \in \cup_{i=1}^{K} Y_i : \arccos(\pi(x_j), x) < r\}$, where $\arccos(\cdot,\cdot) = \hat{\rho}(\cdot,\cdot)$ is an angle metric. [19]. Here $x \in \mathrm{H}$ is a center of cluster and r is a critical angle of cluster forming. The clustering procedure is carried out according to the states of agents from the subspace of universal values H_0. It is assumed that the center of the cluster is some agent i that does not belong to any of the clusters. We will also assume that the critical angle of cluster formation is less than critical angle of the i-th agent of the k-th culture: $r < \phi_k \forall k$.

Let's introduce a closure procedure for a cluster and denote it as $CX(x,r)$. $CX(x,r) = \{x_j \in x_i \in X(x,r) | \hat{\rho}(\pi(x_j), \pi(x_i)) < r\}$. This means that in the process of closing into a cluster $X(x,r)$ agents are added that do not belong to any of the clusters. We introduce a clustering algorithm through the use of a closure operation over $X(x,r)$ according to the above rule until the result of the closure $C^{m+1}X(x,r)$ will not match with $C^m X(x,r)$. Obviously, this procedure converges in a finite number of steps. The resulting set will be denoted as $X_i = C^m X(x,r)$. Algorithm allows to split all agents into disjoit clusters that means $\forall X_i, X_j : X_i \cup X_j = \varnothing$.

Let us introduce the projection $\pi(\cdot) : \mathrm{H}^+ \to \mathrm{R}^M$ of the orthogonal complement of states for the cluster X_k. We choose some natural number $M \in \mathrm{N}, M < \infty$. This number corresponds to the number of selected dimensions from the space of unique values H^+. We introduce a discrete uniform distribution, where each dimension will correspond to the probability of its choice, equal to $\frac{1}{M}$. We choose M dimensions from the space of unique values H^+ and denote the vector from R^M for the i-th agent from $\hat{X}_k$ as $\hat{x}_i$. Lets denote the projected vector for agent i at time t as $x_i^t = \pi(x_i) + \hat{x}_i$.

Under the dynamics of intercultural interactions, we mean the change in the cultural state of each agent over time in subspaces $Y_{i\,i=1}^{\;K}$. Cultural state will change through turnaround procedures and stretching (compression) of the state of the agent within the clusters. Using the projected state of agent i at time t: x_i^t. Let us introduce the following rule for the interaction of agents i and j within clusters separately:

$$x_{j,i}^{t+1} = x_{j,i}^t + \cos(\phi_{k_{j,i}} + \hat{\rho}(x_{i,j}^t, x_{j,i}^t))(x_{i,j}^t - x_{j,i}^t) \tag{2}$$

Here $\hat{\rho}(x_i^t, x_j^t) = \arccos((x_i^t, x_j^t))$ - angle metric.

2.1 Demography

The agent birth process is modeled as follows. For each culture k, the fertility parameter ξ_k is introduced, which describes the rate of emergence of new agents. Inside one cluster X_i all agents are randomly divided into pairs (if there is no pair, the agent does not participate in the birth process). In just one iteration,

$$NW_k = \xi_k \cdot \alpha \cdot N_k \tag{3}$$

new agents appear, where k is the culture, N_k is the abundance of the k-th culture, ξ_k is the fecundity of the k-th culture, and α is the fertility parameter, a macrocharacteristic of the entire population. Of all couples within the same culture using a discrete uniform distribution, NW_k pairs are selected, in each of which a new agent is born with a cultural state equal to the average for the pair. Let x_i be the cultural state of the new agent emerging from agents with cultural states x_{i_1}, x_{i_2}: $xi = \frac{x_{i_1}+x_{i_2}}{2}$ belonging to the Y_k culture.

Let us proceed to the description of the process of mortality. At each time step of all agents, that did not appear on the current cycle, using a discrete uniform distribution the proportion of agents equal to β is chosen, which is the mortality parameter. The resulting number of ND agents is

$$ND = \beta \cdot N \tag{4}$$

where N is the total number of agents. Next, the selected agents are removed from the model, and the process of demography ends at this cycle. In this model implementation, we consider that mortality does not depend on culture.

2.2 Education

The educational process in the model is taken into account as follows. Since the educational process means the process of transferring values to offspring from parents, then in the formalism constructed above this means that in the process of education, the cultural state approaches a certain neighborhood in which a new agent. In the current implementation, the memory depth is one cycle. All agents that appeared on a given cycle, change their cultural states to cultural states that are different from initial by a fixed angle ζ_k, for the k-th culture. Then education changes the cultural state of the new agent xi by the angle ζ_k towards h_k - the axis of Y_k culture. That is $x_i : \hat{\rho}(h_k, x_i) \leq \zeta_k$. However, the direction of change. In this case, the direction of the change in the cultural state of the agent occurs towards the center of the cluster in which the agent appeared.

The model considers an agents dynamic with different clusters interaction. But agents dynamic is separate cluster is no less interesting. Further we suggest to consider model described agent dynamic in separate cluster without taking into account demographic process in assumption of that the cultural landscape transformation is significantly changed for a period that is less than characteristic time of generation change.

3 Euclidean Model

3.1 Dimensional Sampling

We consider an agent k as a vector $x_k : x_k \in \mathrm{E}^{\bar{i}_k}$ that was sampled from H using $\bar{\epsilon}$ and m. Let's call E^m as a subspace of universal agents values. This subspace is general for all agents. Other subspace $\mathrm{E}^{\bar{i}_k - m} = \mathrm{E}^{\bar{i}_k}/\mathrm{E}^m \neq \varnothing$ is called as a space of unique agents values.

3.2 Interaction Into Sub-Spaces

Consider a set of agents $\{x_i(t)\}_{i=1}^{N}$ where $N < \infty$ and $x_i(t)$ - is a state of i-th agent in time t. We consider $\pi(\cdot) : \mathrm{E}^{m+\sum_{i=1}^{N}(\bar{i}_i - m)} \to \mathrm{E}^m$ as an projection operator to E^m sub-space. Corresponding to projection operator theory there is only one projection operator to subspace [18] [19]. Let's denote $\pi(x_i(t))$ as $x_i^{all}(t)$ and then $x_i - \pi(x_i(t))$ as $x_i^{self}(t)$.

We introduce interaction for universe sub-space in time as an ensemble averaging. Consider $M(t)$ as an average over agents in time t.

$$M(t) = \frac{\sum_{i=1}^{N} \pi(x_i(t))}{N} \tag{5}$$

Where $\pi(x_i(t)) = x_i^{all}(t)$ and N - is a total number of agents. Then we introduce an universe values sub-space interaction rule as

$$x_i^{all}(t + \delta t) = x_i^{all}(t) + \left(M(t) - x_i^{all}(t)\right) \cdot \delta t \tag{6}$$

That could be written in continuous form:

$$\frac{dx_i^{all}(t)}{dt} = \left(M(t) - x_i^{all}(t)\right) \tag{7}$$

Let's introduce interaction procedure for unique values sub-space. Firstly, we denote a unique values sub-space as $\mathrm{E}^s = \mathrm{E}^{\sum_{i=1}^{N}(\bar{i}_i - m)} = \mathrm{E}^{m+\sum_i^N(\bar{i}_i - m)}/\mathrm{E}^m$ and $s = \sum_{i=1}^{N}(\bar{i}_i - m)$

We consider interaction procedure for unique values sub-space as an dynamic of "actualized" dimensions. Here "actualized" means dimensions with non zero value. For pair of agents x_i and x_j we can define the number of actualized dimensions D_a as $D_a = \sum_{l=1}^{s} 1(x_i^l \cdot x_j^l)$. Here $1(\cdot)$ is indicator function.

Introduce the rule of interaction procedure for unique values sub-space. And denote $P_{lf}^{ji}(h) = P\left(x_f^{self_i}(t+h) | x_l^{self_j}(t)\right)$ as a probability of affection from l-th dimension of j-th agent to f-th dimension of i-th agent at the moment t. Let's write a balance equation by total probability formula for i-th agent

$$\sum_{j}^{N} \sum_{l,f}^{s} P_{lf}^{ji}(t) = 1 \tag{8}$$

Denote $P_l^i(t)$ as an affection probability to l-th dimension of i agent, then we can write according to law of total probability

$$P_l^i(t + \delta t) = \sum_{j \neq i}^{N} P_l^j(t) \sum_{f}^{s} P_{fl}^{ji}(\delta t) + P_l^i(t) \left(1 - \sum_{f}^{s} P_{fl}^{ii}(\delta t)\right) \tag{9}$$

Where $P_l^j(t)$ is a probability of l-th dimension for j-th agent at the moment t, and $P_{lf}^{ji}(\delta t)$ is a probability of affection from l-th dimension of j-th agent to f-th dimension of i-th agent during the time δt.

Let's assume that the probability $P_{lf}^{ji}(t)$ has a following view according to continuous-time Markov chain assumption:

$$P_{lf}^{ji}(\tau) = \lambda_{lf}^{ij} \cdot \tau + o(\tau) \tag{10}$$

Then we can write:

$$P_l^i(t+h) = h \cdot \sum_{j \neq i}^{N} P_l^j(t) \sum_{f}^{s} \lambda_{fl}^{ji} + P_l^i(t) \left(1 - h \cdot \sum_{f}^{s} \lambda_{fl}^{ji}\right) \tag{11}$$

$$P_l^i(t+h) - P_l^i(t) = h \cdot \sum_{j \neq i}^{N} P_l^j(t) \sum_{f}^{s} \lambda_{fl}^{ji} - h \cdot P_l^i(t) \cdot \sum_{f}^{s} \lambda_{fl}^{ji} \tag{12}$$

$$\frac{P_l^i(t+h) - P_l^i(t)}{h} = \sum_{j \neq i}^{N} P_l^j(t) \sum_{f}^{s} \lambda_{fl}^{ji} - P_l^i(t) \sum_{f}^{s} \lambda_{fl}^{ji} \tag{13}$$

When we direct $h \to 0$ as a result we have:

$$\frac{dP_l^i(t)}{dt} = \sum_{j \neq i}^{N} P_l^j(t) \sum_{f}^{s} \lambda_{fl}^{ji} - P_l^i(t) \sum_{f}^{s} \lambda_{fl}^{ji} \tag{14}$$

In general parameters λ_{fl}^{ji} are depend on time and state of the agents. Let's consider the case when one agent depends on the next in some order: $\lambda_{fl}^{ji}(t) = \lambda \in [0,1]$ if $f = l+1, j = i+1$; $\lambda_{fl}^{ji}(t) = 1 - \lambda$ if $f = l, j = i$ and $\lambda_{fl}^{ji}(t) = 0$ otherwise. Then Eq. 14 will have the following form for:

$$\frac{dP_l^i(t)}{dt} = \lambda \sum_{j \neq i}^{N} P_l^j(t) + (1-\lambda) P_l^i(t) \tag{15}$$

Form a random vector $\xi_i(t)$ from the distribution built on $P_l^i(t)$ along k dimension: $\xi_i(t) \sim \{P_l^i(t)\}$. Then we suggest to consider a unique feature sub-vector dynamic:

$$\frac{dx_i^{self}(t)}{dt} = x_i^{self}(t) \odot \xi_i(t) \tag{16}$$

where $\odot$ - is the Hadamard product [20]. Hole process of $x_i^{self}(t)$ evolution is split into 2 parts: firstly we solve an Eq. 15. After that with formed vector $\xi_i(t)$ we solve an Eq. 16.

3.3 Education

The educational process in the model is taken into account as follows. Since the educational process means the process of transferring values to offspring from parents, then in the formalism constructed above this means that in the process of education, the cultural state approaches a certain neighborhood in which a new agent.

As described in [14] educational process is based on attraction of some agent cultural vector to some cultural state. Denote $F^i(t)$ as a educational state for i-th agent. Then probability vector of $P^i_F(t)$ is forming as uniform distribution over nonzero dimensions. As an example for educational vector $F^i(t) = (0, 1, 0, 0.5, 0.01)$ probability vector will be $P^i_F(t) = \left(0, \frac{1}{3}, 0, \frac{1}{3}, \frac{1}{3}\right)$. Then we can update the system of equation on probabilities Eq. 15:

$$\frac{dP^i_l(t)}{dt} = w\left(\lambda \sum_{j \neq i}^{N} P^j_l(t) + (1-\lambda) P^i_l(t)\right) + (1-w)\, P^i_F(t) \tag{17}$$

Here $w \in [0, 1]$ is a weight factor to control educational process intensity. Let's notice that if $w << 1$ then only educational part of the equation will affect on unique values dynamic.

Finally, for i-th agent state $x_i(t)$ dynamic that could be split into two parts: $x_i^{self}(t)$ and $x_i^{all}(t)$ we have a following system:

$$\begin{cases} \frac{dx_i^{self}(t)}{dt} = x_i^{self}(t) \odot \xi_i(t) \\ \frac{dx_i^{all}(t)}{dt} = \left(M(t) - x_i^{all}(t)\right) \end{cases} \tag{18}$$

4 Conclusion

The problem of mathematical formalization of cross-cultural interactions has great practical interest. Proposed model is based on [14,21]. And allows to analyse and estimate a dynamic of assimilation processed that appears in the evolution process of cultural landscapes. In the current work we eliminate inaccuracies such as cultural cone and orthogonal complement non-emptiness and cross-cultural clusters appearing description. Additionally we introduce new reinterpretation of education affection to agents dynamic. In the previous paper we considered only 3-dimensional space and didn't observe a behaviour of agents from highly-dimensional spaces. Also we apply relaxation time estimation for education parameter from low to high changing that leads to decreasing the number of cross-cultural clusters - cultural uncertainties elimination.

This approach is based on the assumption and metrizability of an infinite-dimensional cultural space. That allows us to introduce lengths and angles between agent states. The interaction between agents is built on the principles of localization of the cultural state in the cultural space and the calculation of the distance between two vectors through angles. To describe intercultural interactions, an algorithm was proposed based on the formation of clusters. These clusters allow us to consider general cultural trends within the cultural space. And also to take into account the cultural characteristics of each agent.

We consider education and the depth of cultural memory to be the main factor in increasing cultural uncertainty. And the main result of the article is a description of these factors.

In proposed algorithm we consider an finite set of agents in Euclidean space that divided to universal and unique characteristics. Where in the universal characteristics sub-space there is an ensemble averaging. In the unique characteristics sub-space Markov chain over finite dimensions is introduced. As the result we derive Kolmogorov equations system and consider the simple example of the connection graph for the Markov chain.

Also our assumption that the cultural landscape transformation is significantly changed for a period that is less than characteristic time of generation change could be wrong. So in that case we should consider not only interaction process with dynamic distribution changing, but also we need to modify the system Eq. 17 with demographic process including and make it solvable - find a border conditions.

References

1. Geertz, C.: The interpretation of cultures. Library of Congress (1973)
2. Van Gennep, A.: The Rites of Passage. Psychology Press (2004)
3. Bohr, N.: Atomic Physics and Human Knowledge. Dover Publications (2010)
4. Gubanov, D., Petrov, I.: Multidimensional model of opinion polarization in social networks. In: Twelfth International Conference Management of large-scale system development (2019)
5. Kozitsin, I.V., et al.: Modeling political preferences of Russian users exemplified by the social network Vkontakte. Math. Models Comput. Simul. **12**, 185–194 (2020)
6. Mikhailov, A.P., Petrov, A.P., Proncheva, O.G.: A model of information warfare in a society with a piecewise constant function of the destabilizing impact. Math. Models Comput. Simul. **11**, 190–197 (2019)
7. Lotman, Yu.M.: Semiosphere (in Russian). Art-SPb (2000)
8. Flache, A., et al.: Models of social influence: towards the next frontiers. J. Artif. Societies, Social Simul. **20**(4) (2017)
9. Mäs, M.: Challenges to simulation validation in the social sciences. a critical rationalist perspective. In: Beisbart, C., Saam, N.J. (eds.) Computer Simulation Validation. SFMA, pp. 857–879. Springer, Cham (2019). https://doi.org/10.1007/978-3-319-70766-2_35
10. Peralta, A.F., Kertesz, J., Iniguez, G.: Opinion dynamics in social networks: from models to data (2022)
11. Kozitsin, I.V.: Opinion dynamics of online social network users: a microlevel analysis. J. Artif. Soc. Soc. Simul. **47**(1), 1–41 (2021)
12. Kozitsin, I.V.: A general framework to link theory and empirics in opinion formation models. Sci. Reports. **12**(1) (2022)
13. Mikolov, T., Sutskever, I., Chen, K., Corrado, S. G., Dean, J.: Distributed representations of words and phrases and their compositionality. In: Proceedings of Workshop at ICLR. arXiv:1310.4546 (2013)

14. Belotelov, N.V., Loginov, F.V.: The agent model of intercultural interactions: the emergence of cultural uncertainties. Comput. Res. Model. **14**(5), 1143–1162 (2022)
15. Belotelov, N.V., Loginov, F.V.: Cross-cultural interactions model. AIP Conf. Proc. **2425**(1), 420030 (2022). https://doi.org/10.1063/5.0081727
16. DeGroot, M.: Reaching a consensus. J. Am. Stat. Assoc. **69**(345), 118–121 (1974). https://doi.org/10.2307/2285509
17. Jackson, M.O.: Social and Economic Networks. Princeton University Press (2008)
18. Gilbarg, D., Trudinger, N.S.: Elliptic partial differential equations of second order. Nauka, Moscow. (in Russian), pp. 464–465 (1989)
19. Burago, D., Burago, Y., Ivanov, S.: A Course in metric geometry. Rhode Island: American Mathematical Society Providence, Example 1.2.25. (2001)
20. Chandler, D.: The norm of the Schur product operation. Numer. Math. **4**, 343–344 (1962)
21. Belotelov, N.V., Loginov, F.V.: Agent-based approach of cross-cultural interactions in hilbert space. Proc. Future Technol. Conf. **2021**(1), 252–263 (2022)

Development of an Agent-Based Optimization Model for the Human Capital Market

I. G. Kamenev(✉)

FRC CSC of RAS, Moscow, Russia
igekam@gmail.com

Abstract. The article considers a multi-agent optimization model of human capital markets: education, healthcare, recreation, labor. The agent's problem of a Firm and a Household are studied considering the State funding. It is shown that the joint equilibrium of human capital markets depends on agents' predictions of each other's actions. Proposed micro-economic basis and design of the general equilibrium model with the sector of investment in human capital.

Keywords: Human capital · Education · Healthcare · Labor market · Social policy · Agency task · Optimal choice

1 Introduction

This article considers the problem of interaction between workers (c), employers (b) and the state (g) in the human capital markets (H) (education, healthcare, recreation). Traditionally, the main focuses of research in this area are the problem of property rights specification, labor mobility of workers, and the role of human capital in the firms' competition [1].

Human capital as a factor of production in the last 30 years has been actively considered as a conceptual replacement for the classical interpretation of labor as a purely demographic quantitative indicator. The number of workers employed in the economy together with the number of the unemployed of working age allowed a fairly accurate simulation of mass industrial production, in which the majority of workers performed simple standardized operations.

However, in the modern information economy, the qualitative characteristics of a person (health, qualifications, mobility, etc.) also have a significant impact on labor productivity both at the firm level and at the level of the national economy. Human capital as a concept implies, foremost, that these qualitative characteristics are not random, so it is incorrect to simply average them out: on the contrary, investments in human capital increase labor productivity.

The research was supported by RSCF grant No. 22-21-00746 "Models, methods and software to support the modeling of socio-economic processes with the possibility of forecasting and scenario calculations".

N. Olenev et al. (Eds.): OPTIMA 2023, CCIS 1913, pp. 158–169, 2024.
https://doi.org/10.1007/978-3-031-48751-4_12

Human capital is understood as a stock (unobservable, studied by external signs) inherent in a human being that makes his work more productive than the work of other people. The generalised mathematical formulation is as follows: if one (identical, marginal) workplace would be filled by one of two workers with different human capital, then they would show a different increase in the company's product, and the difference will be proportional to the difference in the human capital of workers:

$$TP(K, L, H), MPH_1/MPH_2 = H_1/H_2 \quad (1)$$

K is capital,
L is labor (quantity of workers),
H is human capital,
TP is total product,
MP is marginal product.

This concept can be mathematically formalized in different ways, but reflects the following ideas: human capital is inherent in a particular individual (as his asset, phase variable); may be accumulated; have positive return on investment; manifests itself in any conditions (for any capital-labor ratio, firm size, etc., the return on human capital stays positive); cannot completely replace other factors of production.

At the macroeconomic level, human capital is usually seen as an additional factor that explains economic growth and inequality [2]. In the Inada Conditions [3] models, sustainable growth is possible only if there are two growing factors of production (otherwise, diminishing returns to scale on a factor of production cause at a certain level investment in it not being able to compensate amortization). Considering declining population and labor resources in developed countries, economic growth in them cannot be explained by investment in material capital alone. Accordingly, different models of economic growth with human capital (see, for example, [4–6]) offer different mathematical formalization of growth based on investments in material and human capital. Alternative approaches consider information and innovation, social capital, etc. as the second growth factor [7]. (in general, all these approaches are close and represent one or another way of technological progress endogenous modeling).

However, such models leave without consideration the very market mechanism of human capital markets. If human capital is an asset in which economic agents invest, then there are important problems of property rights specification, economic rationality of investment, and the return on investment. Thus, the mechanism of the human capital market (markets) is an important link between the micro-economic description of human capital for the firm level and the macroeconomic problem of searching for sustainable growth sources.

Thus, the purpose of this publication is to develop the architecture of a mathematical model of interaction between workers, firms and the state in human capital markets, which would meet key requirements: it would describe the decisions of all agents regarding investments in human capital and their influence on each other, and also allow study of various models of agents' expectations regarding the actions of other agents.

2 Theoretical Approaches to Human Capital Markets

The human capital markets that shape the human being and ensure its productivity include education, health care and recreation. The education market increases human productivity (the ability to produce more product per unit of time). The health care market increases the number of effective hours at the disposal of a person (if a person is unhealthy, then part of the available time is spent on severe illnesses or part of the time is permanently lost due to disability). The recreation market provides labor amortization, worker's readiness to work productively (motivational component and psychological health). (Note that in macroeconomic modelling it is standard to count amortisation not as a diminishment of the asset but as a costs that are spent on a asset's replenishment). Of these three markets, the education market remains the most studied, with a number of classic works in human capital theory devoted to it [8,9]. It should be noted that the approach proposed in these works reflects the economic realities of the second half of the 20-th century and in many aspects is outdated, but this issue is not the main subject of this study. Most importantly, in these classic works, the education market is seen as part of the micro-basis for the parenting household problem. As a result, they aren't describing the displacement effect, arising from the fact that investment in this asset might be done by three different agents (types of agents) that receive a return on human capital.

This classical approach to economics agent modeling assumes that each agent acts in his own interests, maximizing his own net benefit (considering information asymmetry and other rationality constraints). In particular, for the Firm, this means maximizing profits or (in more complex formulation) the discounted dividend flows. However, this formulation contradicts the practice of the modern labor market, in which firms in many countries (though not everywhere) finance expensive social programs. There are various explanations for this phenomena in terms of the contract between the worker and the employer. In general, they come down to the hypothesis that social programs are part of the wages in the form of services preferred by workers, purchased at wholesale prices. This interpretation, however, leads to a systematic underestimation of the benefits that Firm itself receive from this social programs: through increased workers' productivity.

The role of the health care market in the human capital formation is much less studied. From a macroeconomic point of view, it works the same way: services are purchased that create investments in the asset (human capital), and the accumulation of this asset increases the overall productivity of the economy. At the micro-economic level, the issue is much more complicated: it seems that the lost health capital is very difficult to restore, and it is more fair for a particular person to describe it exclusively as the (in)full amortization. For a society as a whole, this can be mitigated by the fact that different people are at different stages of the life cycle, but unexpected effects might be found in the information asymmetry (people might not get information about their health deterioration until it is too late to invest in). The problems of population aging and the difference between life expectancy and healthy life expectancy, the impact of the pension system on the economic activity of the population at an older age deserve a separate discussion as well.

Finally, the recreation market is the least studied, and the question of delimiting the recreation of human capital and ordinary consumption remains open.

In general, the current state of research on human capital markets can be described as follows. These markets correspond to three sectors of the economy that are clearly observed in macroeconomic statistics (not taking into account issues with the recreation market). In these markets, investments are made in the asset (human capital), which should cover its amortization (including the loss of part of human capital due to aging, professional deformation, obsolescence of professional education, etc.). These industries consume the social product, and, accordingly, to describe them in general equilibrium models, at least a three-sector model (with a consumer product, investment in material capital and investment in human capital) is required. The supply in these industries can be considered highly competitive (for most market economies), but on the demand side there is a serious deviation from the perfectly competitive state: the question of whether large firms have market power (the ability to dictate prices) remains debatable, and the state in most countries acts as an active market player, so everyone is forced to adapt to social policies.

3 Agents Basis Mathematical Formalization

3.1 Model of Return on Human Capital Investment

The return on human capital arises from the growth of worker productivity. Returns to scale are assumed to be negative (marginal product is decreasing):

$$R_t = Pq(H_t), \frac{dq(H_t)}{dH_t} > 0, \frac{d^2q(H_t)}{dH_t^2} < 0, H_t > 0 \tag{2}$$

The resulting return on human capital is divided between the employer (additional profit, h_b), the employee (bonus salary, h_c), and the government (increased tax revenue from the firm and employee, h_g). The proportion is determined by tax legislation and equilibrium in the labor market. For the sake of simplicity (while not constructing a special model of the labor market), we will assume that the proportion is constant:

$$R_t = h_c R_t + h_b R_t + h_g R_t \tag{3}$$

Investments in human capital can be made by the household-worker, the firm-employer and the state. At the same time, human capital becomes obsolete, wears out, etc., which is manifested in its amortization. The corresponding investment values (h_c, h_b, h_g) are the agents' controls.

$$\frac{dq(H_t)}{dH_t} = \frac{Ic_t + Ib_t + Ig_t}{Pht} \tag{4}$$

3.2 Optimization Tasks

A household can invest in its human capital in the long run based on the difference between the expected human capital and the desired human capital in terms of the consumer's task:

$$U = U(C_t) \rightarrow \max \tag{5}$$

Its budget constraint includes (in its simplest setting without financial savings) labor income w_0 and rents on human capital, which are distributed between consumption and investment in human capital:

$$w_0 + h_c R_t = Ic_t + C_t \tag{6}$$

A firm, likewise, may invest in the human capital of its employees if it further increases its dividend. In the simplest setting, the firm invests part of the income received as a return on human capital (to build a complete model of the firm, a demographic development model L_t and a capital market model K_t are needed, the introduction of which also requires the construction of a complete household problem with savings).

$$D = D(d_t) \rightarrow \max, d_t = h_b R_t - Ib_t \tag{7}$$

Let us remind that the discounted dividend flow maximization task is a dynamic analogue of the static profit maximization task. In a static setting (standard, without considering the agent-principal problem), the firm acts in the interests of the owner, maximizing profits, which are paid to the owner in dividends. In a dynamic setting, the owner makes inter-temporal choice: whether to extract all the profits from the company through dividends, or to leave part of the profits in the company (investing it in material and human capital) hoping this will increase future earnings and dividends.

Similarly, a government can reinvest part of the returns it receives from human capital into its markets:

$$Ib_t \leq h_g R_t \tag{8}$$

3.3 Mutual Expectations of Players in the Human Capital Markets

The interaction of agents of all three types in the human capital market can be considered from the standpoint of game theory. Assuming a plurality of firms and households, neither ones nor the others are capable of coordinated strategic behavior. Thus they only react to the actions of the opposite side and the state. Accordingly, households consider Ib_t and Ig_t as exogenous constants of Ib and Ig. Firms, respectively, consider Ic and Ig as constants. In relation to the state, however, the task of the leading player can be set (predicting the behavior of other players and turning the game problem into a optimization one). Controls of households, firms and the state (the amount of investment in human capital) can be presented in various settings, for example, as a percentage of the benefits received:

$$Ic = x_c h_c R_t, Ib = x_b h_b R_t, Ig = x_g h_g R_t \tag{9}$$

The study of the problem posed shows that the logic of the optimal choice of agents (firms and households) includes the construction of their response functions. The main features of the model:

- since human capital amortization is linear, then (given the value of other factors of production) there is a threshold level of human capital, above which the return on human capital is less than the cost of its depreciation, even if all agents direct the entire return on human capital to investments in human capital;
- firms and households make additional investments in human capital if the increase in income in the future, taking into account inter-temporal preferences, compensating its decrease in the present;
- if the national market structure of human capital markets assumes the absence of investment by other agents (for example, investments are made only by parental households, or only by firms), then the model degrades into a classical problem of inter-temporal choice;
- if the national model allows co-financing of human capital, then a displacement effect occurs: the presence of investments from other agents reduces the marginal return on the agent's own investments, reducing the attractiveness of financing human capital.

Presumably, these features are preserved when the model is expanded with the addition of a capital market and bringing all prices into real form. At the same time, however, the joint equilibrium of the market for goods (including consumer goods, human and material capital) becomes a rather difficult task from the point of view of the formation of equilibrium prices. The most promising direction of development seems to be the division of human capital into the markets of education, healthcare and recreation, because traditionally they have a different institutional structure of the transaction even within the same country. In the future, this will make it possible to explain the imbalances within human capital that are characteristic of the economic growth of individual countries, and to develop new approaches to the conditions for balanced growth.

4 Statement of the General Equilibrium Model Taking Into Account the Micro-basis of Human Capital Markets

To study approaches to a general equilibrium model with human capital markets design, it is necessary to build the appropriate agency macro-models and formulate prerequisites about the nature of interaction between agents (see Fig. 1).

Such a macroeconomic model will include a dynamic optimization problem for a firm and a household. Note that taxes collected by the government are not included in this scheme because complete model of the state budget requires modelling beyond the considered segment of the general equilibrium. For the same reason, the optimization task of the state is not formulated.

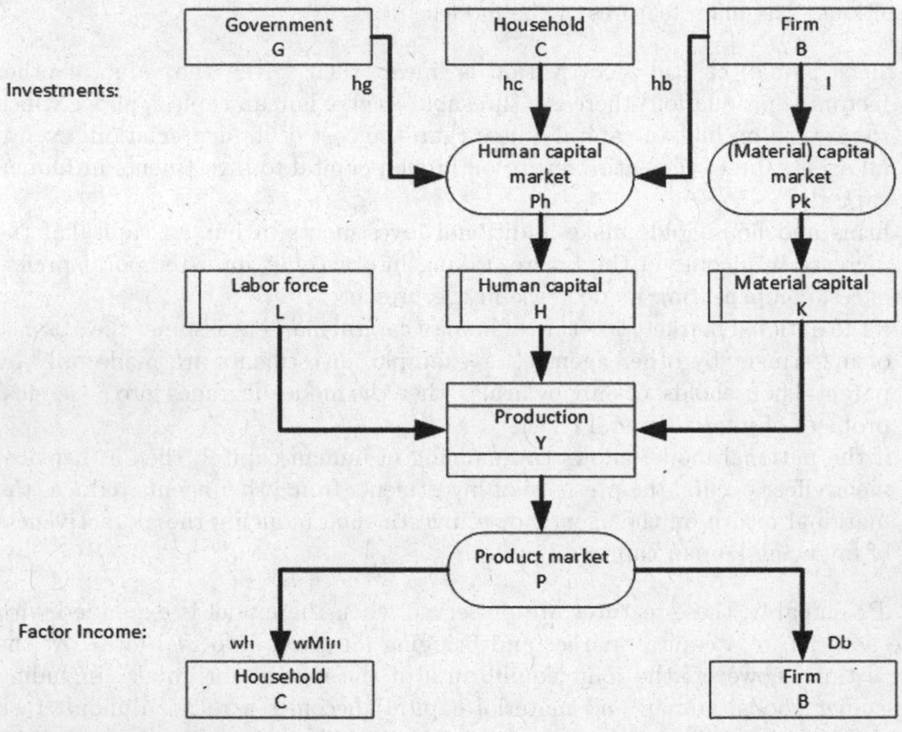

Fig. 1. General scheme of interaction between the household, the firm and the state in a human capital market in general equilibrium model

4.1 Household Task

Household functional:

$$Uc = \int_{t_0}^{T} (Cc(t)^{0.5} e^{-zt}) dt \to \max : \tag{10}$$

Uc: consumer functional
t_0: current time
$Cc(t)$: consumption (control)
T: planning horizon

t: time

z: discount factor.

The household maximizes utility, its objective function reflects diminishing returns to scale and the household's inter-temporal preferences (prioritizing the present over the future). This task is solved under a budget constraint, which includes labor and non-labor income, consumption expenditures and investments in human capital.

Conditions:

budget balance:

$$Cc(t) = wMin_e(t) + Db_e(t) + wh_e(t)H(t) - hc(t) \tag{11}$$

$hc(t) >= 0$: household investment in human capital (control)

Amh: human capital amortization factor

$H(t) >= 0$: human capital

Household assets (transfer dynamic differential equations):

human capital transfer equation:

$$\frac{dH(t)}{dt} = ((hc(t) + hb_e(t) + hg_e(t))/Ph_e(t)) - AmhH(t) \tag{12}$$

Household informational exogenous variables which it has to predict:

$Ph_e(t)$: the price of human capital formation

$wMin_e(t)$: wage

$wh_e(t)$: human capital rent

$Db_e(t)$: dividends (autonomous income)

$hb_e(t)$: forecast of firm investment in human capital

$hg_e(t)$: forecast of state investment in human capital.

4.2 Firm Task

Firm functional:

$$Ub = \int_{t_0}^{T} (Db(t)^{0.5}(\exp -zt))dt \rightarrow \max \tag{13}$$

Ub: firm owner's functional

$Db(t)$: dividends (control, redundant, because is redundant determined by two other controls).

The firm maximizes the discounted dividend flow (including diminishing returns to scale and inter-temporal preferences). The functional configuration is similar to that of the household, because it reflects the interests of the people who own companies in the national economy.

This problem is solved taking into account the company's budget constraint, which includes the necessity to produce goods for sale (production function), revenue obtaining and spending it on factors of production payments, and distribution of the remaining profit.

Conditions:
production function

$$Y(t) = L(K(t)^a)(H(t)^{1-a}) \tag{14}$$

budget distribution balance

$$Rr(t) = Db(t) + hb(t) + I(t) \tag{15}$$

budget balance

$$Rr(t) = P_e(t)Y(t) - wMin_e(t)L - wh_e(t)H(t) \tag{16}$$

$Rr(t)$: revenue flow
$hb(t)$: firm investment in human capital (control)
Amk: material capital amortization factor
$Pr(t)$: profit
$I(t)$: firm investment in material capital (control)
$K(t)$: material capital
$Y(t)$: output (total product)
L: labour force (constant, full employment)
a: scale factor
Firm assets (transfer dynamic differential equations):
human capital transfer equation

$$\frac{dH(t)}{dt} = (hc_e(t) + hb(t) + hg_e(t))/Ph_e(t) - AmhH(t) \tag{17}$$

material capital transfer equation:

$$\frac{dK(t)}{dt} = ((I(t)/Pk_e(t)) - AmkK(t)) \tag{18}$$

It should be noted that for the firm, both material and human capital are assets financed by it, despite the fact that the human capital does not belong to the firm. Although there are no transfers of workers between firms at this level of aggregation, there is still the possibility that the household will decide to work less. Despite this, the firm may be interested in financing human capital if its volume is insufficient.

Firm informational exogenous variables which it has to predict:
$P_e(t)$: production price
$Ph_e(t)$: the price of human capital formation
$Pk_e(t)$: the price of material capital formation
$wMin_e(t)$: labour price
$wh_e(t)$: human capital rent
$hc_e(t)$: forecast of household investment in human capital
$hg_e(t)$: forecast of state investment in human capital.

As for the state, its influence can be described at a later stage using taxes spent on social programs (thus investing in human capital). At this stage, however, it is sufficient to consider public investment in human capital as an exogenous constant. We also omit the conditions for the equilibrium of goods market and pricing mechanisms (this next step in the consideration of the model is beyond a limited scope of paper).

4.3 Expectations Model

A more fundamental question is whether agents are aware of the future dynamics of information variables, including (most importantly) other agents' investment in human capital. Traditional general equilibrium models assume that agents are able to use the model in which they exist to predict the future. If we assume that there are many firms and households, and the state policy is constant, then agents have information about the information variables vectors.

However, there are also general equilibrium models in which agents are considered naive: making predictions of information variables not on the basis of the model, but on the basis of their previous dynamics or (in the simplest case) their observed value at a given moment. It is difficult to study such models by analytical methods, because divergence between the predicted values of information variables and the actual ones will force the agents to recalculate their decisions every moment of time. The natural way out is to discretize the model and turn it into an imitation one.

The principal source of naivete in the model is the price of human capital formation. This variable can only be felt at the moment, while the construction of a series of observations and even more so model calculations for it are not available to economic agents. This gives us reason to believe that it should be introduced into the agents' calculations as a constant: Ph. In turn, an error in the calculation of one of the prices through the conditions of market equilibrium inevitably means the agent's naivety with respect to all other informational variables. The question of the significance of these distortions and their significance from the point of view of a balanced growth trajectory can be investigated by simulation methods.

Finally, the most complex expectation model type that can be used to describe expectation is a forecast model. Agents can consider it being random and determine the distribution proceeding from the history. However, this is a far more complicated type of models (turning it into a stochastic one). Stochastic general equilibrium models are a currently actively developed, but their study is associated with additional difficulties both at the construction stage, optimality conditions expression, and their study by quantitative methods. Accordingly, the introduction of stochastic expectations is possible only after a model with adaptive expectations has been explored.

5 Optimality Conditions

Using the open system of agent-based modeling "Ecomod" in the Pyton language [1], we studied the optimality conditions for household and the firm in this model. For example, in the version with full information:

Firm optimality conditions:

$$-\frac{0.5\lambda_0 e^{t(-z)}}{\left(LH_B^{1-a}(t)K_B^a(t)P_e(t) - LwMin_e(t) - H_B(t)wh_e(t) - I_B(t) - hb_B(t)\right)^{0.5}} = \\ = \frac{\alpha_0(t)}{Ph_e(t)} \tag{19}$$

$$-\frac{0.5\lambda_0 e^{t(-z)}}{\left(LH_B^{1-a}(t)K_B^a(t)P_e(t) - LwMin_e(t) - H_B(t)wh_e(t) - I_B(t) - hb_B(t)\right)^{0.5}} = \\ = \frac{\alpha_1(t)}{Pk_e(t)} \tag{20}$$

Household optimality conditions:

$$-\frac{0.5\lambda_0 e^{t(-z)}}{\left(Db_e(t) + H_C(t)wh_e(t) - hc_C(t) + wMin_e(t)\right)^{0.5}} = \\ = \frac{\alpha_0(t)}{Ph_e(t)} \tag{21}$$

It can be seen that the firm's optimality conditions correspond to the classical choice between two factors of production in proportion to their prices (golden rule $MPx/Px = MPy/Py$). In turn, the task of the household is a classic intertemporal choice, similar to that which arises in models with savings (the fundamental difference, however, is the presence of external factors in the dynamics of human capital).

6 Conclusions and Perspectives

We considered the general requirements for building a model of human capital markets in which households, firms and the state invest in the development of human productivity. It is shown that such a model can be built in the standard form of a general equilibrium model, which allows one to analyze its properties analytically (under complete information) or to discretize and study the market response to external shocks.

The proposed model architecture makes it possible to correlate observed investments in human capital (which can be estimated using industry-wide macroeconomic statistics) with the choice of households, firms, and with social policy. It allows us to consider different models of mutual expectations of agents, which, after further research, should help determine which of the possible models of expectations allow us to more accurately reproduce actual interaction in these industries.

Potentially, this type of model is suitable for explaining endogenous economic growth in developed countries, and, unlike existing growth models with human capital, investment in it is also explained endogenously by the choice of agents. However, in order for the model to be comparable with macroeconomic statistics, identifiable and calibrated for a particular country, it is necessary to close it in the financial market and form a condition for the equilibrium of the goods market.

References

1. Crook, T.R., et al.: Does human capital matter? A meta-analysis of the relationship between human capital and firm performance. J. Appl. Psychol. **3**(96), 443–456 (2011)
2. Galor, O., Zeira, J.: Income distribution and macroeconomics. Rev. Econ. Stud. **1**(60), 35–52 (1993)
3. Uzawa, H.: On a two-sector model of economic growth. Rev. Econ. Stud. **30**(2), 10–118 (1963)
4. Rebelo, S.: Long run policy analysis and long run growth. J. Polit. Econ. **99**(3), 500–521 (1991)
5. Romer, P.M.: Increasing returns and long-run growth. J. Polit. Econ. **94**(5), 1002–1037 (1986)
6. Lucas, R.E.: On the mechanics of economic development. J. Monet. Econ. **1**(22), 3–42 (1988)
7. Romer, P.M.: Endogenous technological change. J. Polit. Econ. **5**(98), 71–102 (1990)
8. Becker, G.: Human Capital: Theoretical and Empirical Analysis, with Special Reference to Education, N.Y. (1964)
9. Mincer, J.: Investment in human capital and personal income distribution. J. Polit. Econ. **66**(4), 281–302 (1958)
10. Schultz, T.: Investment in Human Capital: The Role of Education and of Research, N.Y. (1971)
11. Iusup-Akhunov, B.B., Zhukova, A.A., Kamenev, I.G.: Support system for modeling socio-economic processes based on an open source platform [Sistema podderzhki modelirovaniya social'no-ekonomicheskih processov na osnove platformy s otkrytym iskhodnym kodom]. Proc. Moscow Inst. Phys. Technol. [Trudy MFTI] **4**(56), 69–83 (2022)

Applications

Exact Algorithm for Generating H-Cores in Simplified Lattice-Based Protein Model

Andrei Ignatov(✉)

HSE University, Moscow, Russia
aignatov@hse.ru

Abstract. Modeling protein folding, which is the process by which a protein obtains its spacial shape, still remains a challenging problem. Protein geometry might be simplified by using the coarse-grained models. The highest level of simplification is achieved in HP-models where only polarity of amino acid residues is considered, and the unified monomers are located in nodes of some discrete lattice. One possible way of predicting the spacial structure of a protein in this model implies creating a maximally dense hydrophobic core (H-core), and fitting a protein into it afterwards. The paper proposes setups of Linear Programming (LP) problems for constructing both maximally dense H-cores and H-cores with the predefined number of contacts. Setups are developed for two lattices – Pseudo-triangular and Face-Centered Cubic. Results of the conducted experiments show that the proposed methodology is efficient enough to be utilized in the protein structure prediction process.

Keywords: Protein folding · HP-Models · H-Core · Discrete Optimization · Linear Programming

1 Introduction

Protein folding is the physical process by which a protein obtains its spacial shape. Modeling this process still remains a challenging problem. Numerous researchers worldwide have been making advances to solving this problem for the last 60 years. However, the key aspects of this process are still not studied fully.

A protein is a chain of amino acid residues. Generally, there exist only 20 types of amino acids that can be present in a protein. These amino acids have the same *backbone* – an amino group, a carboxylic group, and a carbon which is named C_α for clarity (see Fig. 1). C_α atom has a *side-chain* bound to it. Side-chains are different for every type of amino acid.

When amino acids get bound to each other with their backbones, a strong peptide bond is created. OH group from the carboxylic group of one amino acid and H atom from the amino group of another amino acid get removed and constitute into water. The remaining structures of amino acids are called amino acid **residues**.

Generally, the problem of predicting the spacial structure of a protein is an extremely challenging task. This is a caused by a huge number of degrees of

N. Olenev et al. (Eds.): OPTIMA 2023, CCIS 1913, pp. 173–187, 2024.
https://doi.org/10.1007/978-3-031-48751-4_13

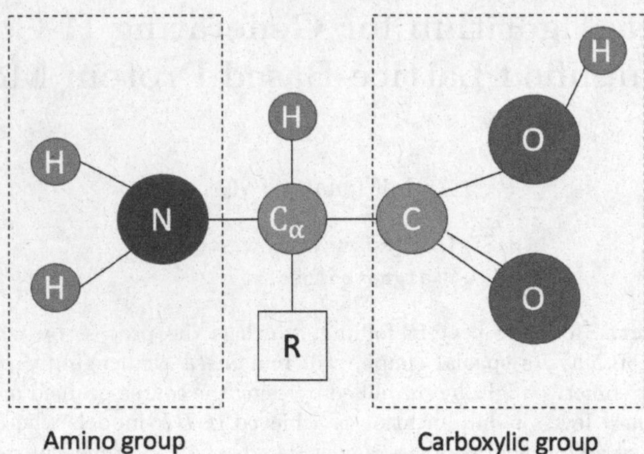

Fig. 1. The general structure of an α-amino acid. N, C_α, and C are the main backbone atoms, R represents the side chain, H are the hydrogen atoms

freedom, which results in an enormous number of variables. As it has been proposed [1], the native (folded) structure of a protein corresponds to some minima (possibly not global) of its potential energy. Unfortunately, minimization of all-atom potentials (e.g. OPLS [2]) faces a tremendous computational complexity. For that reason, their minimization can be managed only with local optimization methods [3].

In order to reduce the number of variables, a number of coarse-grained models [4–7] have been proposed. These models combine several atoms and/or their groups in one unified atom, which results in a significant simplification of the protein geometry. Usually, these models are accompanied by their own potentials, minimization of which is supposed to produce a structure similar to the native one.

In the paper, we focus on one of the most significant mechanisms of protein folding – hydropathy of side-chains. Some side-chains (*polar*) are able to contact the solvent around the protein well, while *hydrophobic* side-chains tend to 'hide' inside the protein to minimize their ASA (accessible surface area). This key idea became a foundation for the coarse-grained HP-model [7]. In the HP-model, every residue is marked with one of two types – H (hydrophobic) or P (polar). The model suggests a strong simplification where the whole residue is replaced with only one unified atom (monomer). These unified atoms can be located only in nodes of some 2D- or 3D-lattice. The model is built on the principle of maximizing the number of contacts between hydrophobic monomers, where two monomers are in contact if they are located in neighboring nodes of the lattice. Such optimization results in a dense hydrophobic core surrounded by polar monomers. This result reflects the real situation as most proteins have an explicitly observed hydrophobic core [8].

Strong simplification of the protein geometry proposed in the HP-model lets convert the complex problem of protein folding modeling into a discrete problem. Solving this problem implies finding a mapping of residues to the nodes of the lattice. Several approaches to discovering this mapping have been suggested. The standard approach [9] suggests building a maximally dense H-core and fitting the protein into it afterwards. If fitting a protein into all maximally dense H-cores fails, the number of contacts must be reduced, and the generating of H-cores is repeated.

In the paper, we focus on the problem of generating H-cores (both maximally dense and with the predefined number of contacts). Previously, H-core generation was performed with Branch and Bound techniques [9], genetic and random-walk algorithms [10], memetic algorithms [11,12], and constraint satisfaction techniques [13]. Some of these approaches are able to construct the maximally dense H-cores, but not the ones with the predefined number of contacts. Unfortunately, fitting the target protein into the maximally dense H-core does not succeed for many real protein chains. Additionally, utilizing these approaches may lead to exhaustive search, especially, if using Branch and Bound and genetic algorithms.

To resolve these issues, we suggest using Linear Programming (LP) techniques for H-core generation. The main contributions of the paper are setups of Linear Programming problems for generating H-cores. The setups are proposed both for generating the maximally dense H-cores and the ones with the predefined number of contacts. The problem statements have been developed for the 2D Pseudo-triangular [14] and the 3D Face-Centered Cubic (FCC) [16] lattices. Solving these problems with the simplex method results into exact solutions with the global maximum in the number of contacts [15]. According to the results of the conducted experiments, the developed algorithms are fast enough to be utilized in protein structure prediction process.

The rest of the paper is organized as follows. Section 2 contains the background of the considered problem. In Sects. 3 and 4, the proposed exact algorithm is described in detail. Section 5 contains the results of the algorithm testing and their analysis. Finally, Sect. 6 summarizes and concludes the research.

2 Problem Statement

In this section, basic concepts of the problem are discussed, and the general formal statement of the H-core generation problem is given.

In the HP-model, the amino acid sequence is replaced with an HP-sequence where the i'th letter is H if the i'th residue of the protein is hydrophobic, and P otherwise. In the current study, only hydrophobic monomers (H-monomers) are considered. Let N be the number of H-monomers in the protein chain, n_S be the length of the protein chain, S_i be the i'th monomer.

Monomers in the HP-model can only be located in the nodes of some 2D or 3D **lattice**. Formally, a **Lattice** L is a countable set of points such that $u+v \in L$ for any $u, v \in L$. Then, a minimal set of vectors N_L can be derived for a lattice.

Linear combination of these vectors with non-negative integral coefficients would represent any point from L:

$$L = \{u \in \mathbb{R}^{dim} | u = \sum_{v \in N_L} c_v \cdot v, c_v \in \mathbb{Z}^+\},$$

where dim is the space dimension (2 for planar lattices, and 3 for spacial). N_L sets of vectors for the 2D Pseudo-triangular, the 2D Square, and the 3D Face-centered Cubic (FCC) lattices are presented in Table 1.

Table 1. N_L sets for 2D Pseudo-triangular, 2D Square, and 3D FCC lattices

Lattice	N_L vectors
Pseudo-triangular	(1, 0), (−1, 0), (0, 1), (0, −1), (−1, 1), (1, −1)
Square	(1, 0), (−1, 0), (0, 1), (0, −1)
Face-Centered Cubic	(0, 1, 1), (0, −1, 1), (0, 1, −1), (0, −1, −1) (1, 0, 1), (−1, 0, 1), (1, 0, −1), (−1, 0, −1) (1, 1, 0), (−1, 1, 0), (1, −1, 0), (−1, −1, 0)

When constructing the maximally dense H-core, the aim is to maximize the number of $H - H$ contacts. Formally, two monomers i and j at coordinates M_i and M_j, respectively, $M_i \neq M_j$, are in contact if

$$M_i - M_j \in N_L.$$

Let us denote the set of indices of H-monomers in the HP-sequence S as h_S, i.e. $h_S = i \, for \, i = 1, 2, ..., n_S \, if \, S_i =$ 'H'. Let $M = (M_1, M_2, ..., M_N)$ be the set of lattice points selected to be present in the target H-core. The selected points must be unique, i.e. for any $i, j = 1, 2, ..., N, i \neq j \; M_i \neq M_j$.

Then, the goal function for the problem of generating the maximally dense H-core can be defined as follows:

$$E_S(M) = \sum_{1 \leq i < j \leq N} \Delta(M_i, M_j), \text{ where}$$

$$\Delta(M_i, M_j) = \begin{cases} 1, \text{if } M_i - M_j \in N_L, \\ 0, \text{otherwise.} \end{cases}$$

Having the goal function, we can formulate the problem of creating the maximally dense H-core as follows:

$$\begin{cases} E_S(M) \longrightarrow \max, \\ M_i \neq M_j, i, j = 1, ..., N, \; i \neq j. \end{cases} \tag{1}$$

In some cases, all maximally dense H-cores may turn out to be inappropriate for the target HP-sequence S, i.e. the target HP-sequence cannot be fit into the generated H-core, so the number of contacts in the newly generated H-cores must be reduced. Inappropriateness of an H-core may be revealed during fitting the HP-sequence into these cores, or at the step of HP-sequence analysis [17].

Let us denote the upper limit for the number of contacts in the target H-core as K_{lim}. Then, another constraint must be added to the Problem (1), which results in another problem statement:

$$\begin{cases} E_S(M) \longrightarrow \max, \\ E_S(M) \leq K_{lim} \\ M_i \neq M_j, i, j = 1, ..., N, \ i \neq j. \end{cases} \tag{2}$$

3 Linear Programming-Based Algorithm: 2D Case

In the paper, applying Linear Programming (LP) techniques to Problems (1) and (2) is proposed.

The main idea of the proposed approach is to split the lattice nodes into layers, and calculate the number of points in each layer with LP techniques. After the numbers of points in each layer are discovered, the shifts of layers that do not change the number of contacts can be derived. The latter step produces several H-cores of the same flavor, i.e. H-cores with the same number of nodes and contacts.

First, let the Pseudo-triangular lattice be considered. Let all the nodes of the target H-core be distributed among horizontal layers. Then, for each layer, one should consider the internal contacts and contacts with the previous layer (if the layer is not the first).

Generally, there may be up to N layers in the H-core with N nodes. Let x_i be the number of nodes in the i'th layer, $i = 1, 2, ..., N$. Let the layers be built from up to down, i.e. the y-coordinate of all nodes in the i'th layer is equal to $(N - i)$. Then, let us denote the number of contacts between the i'th and the $(i-1)$'th layers as w_i, $i = 2, 3, ..., N$.

3.1 Inter-layer Contacts

The number of inter-layer contacts w_i is maximized when the overlay of layers is maximal, i.e. the number of nodes (x, i) for which a node $(j, i-1)$ exists should be maximized, $j = 1, ..., x_i$. This effect can be seen for H-cores at $\Delta = 1, 2$ in Fig. 2.

Let m and k be the nonzero numbers of nodes in the current and the previous layer, respectively. The number of inter-layer contacts depends mostly on the number of nodes in the overlay $ov(x_{i-1}, x_i, \delta)$, where $\delta = (-m+1), (-m+2), ..., -1, 0, 1, ..., k-1$ is the horizontal shift of the i'th layer.

$$ov(k, m, x_0) = \begin{cases} \min(k, x_0 + m) - \max(0, x_0), x_0 = (-m+1), (-m+2), ..., (k-1) \\ 0, \text{otherwise.} \end{cases}$$

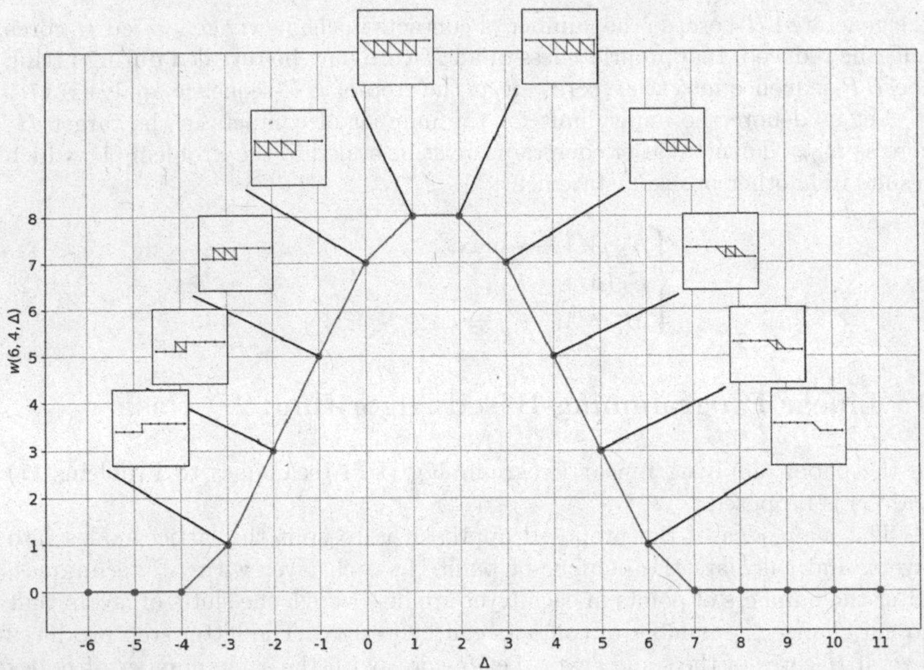

Fig. 2. Numbers of contacts between layers with $k = 6$ and $m = 4$ nodes depending on the shift Δ of the second layer from the first. The respective cores in the Pseudo-triangular lattice are presented.

Generally, every node in the overlay gives two contacts. However, this might be wrong in boundary cases, as seen in Fig. 2. If summarized, the number of contacts $w(k, m, x_0)$ between two layers can be expressed as

$$w(k, m, x_0) = 2 \cdot ov(k, m, x_0) - \begin{cases} 1, x_0 \leq 0 \\ 0, \text{otherwise} \end{cases} + \begin{cases} 1, k - m < x_0 \leq k \\ 0, \text{otherwise} \end{cases}, \tag{3}$$

where k and m are the numbers of nodes in the previous and the current layers, respectively, and x_0 is the horizontal shift of the current layer from the previous, $x_0 = (-m + 1), (-m + 2), ..., k$.

During H-core generation, the number of contacts between layers must be maximized. From Formula 3, the following evaluation of the maximal number of contacts between layers with k and m nodes can be derived:

$$w_{max}(k, m) = 2 \cdot \min(k, m) - \begin{cases} 1, k = m \\ 0, \text{otherwise} \end{cases} \tag{4}$$

In the proposed LP problem statement, maximization of the total number of inter-layer contacts is represented as follows. Let us introduce the variables:

- $v_i = 0, 1, ..., i = 2, 3, ..., N$. v_i is is intended to represent the minimum of x_i and x_{i-1}.
- $p_i \in \{0, 1\}, i = 2, 3, ..., N$. p_i must be 1 if $x_i = x_{i-1}$, and 0 otherwise.
- $t_i \in \{0, 1\}, i = 2, 3, ..., N$. According to Formula 4, 1 contact must be subtracted from the sum if $x_i = x_{i-1}$. But this is not valid if $x_i = x_{i-1} = 0$, which might be possible if the predefined number of layers is equal to N. Then, this case must be managed explicitly, and t_i is expected to be 0 if $x_i = x_{i-1} = 0$, and 1 otherwise.
- $w_i = 0, 1, ..., i = 2, 3, ..., N$. w_i is the number of contacts between the i'th and the $(i-1)$'th layers.

For the described variables, the following constraints are introduced. For every $i = 2, 3, ..., N$:

- $v_i \leq x_i$;
  $v_i \leq x_{i-1}$.
  When v_i is maximized, these constraints make it to be the maximum of x_i and x_{i-1}. Then, v_i is the maximal possible overlay of x_i and x_{i-1} nodes.
- $x_i + x_{i-1} - 2v_i \geq p_i$.
  If $x_i == x_{i-1} = 0$, the left part of this inequality is zero, and is greater than zero otherwise. Then, maximization of p_i means that it would be equal to 0 if $x_i == x_{i-1} = 0$, and to 1 otherwise.
- $2(x_i + x_{i-1}) - 2v_i \leq M \cdot t_i$, where $M >> 1$.
  Minimization of t_i along with this constraint ensures that t_i is 0 if $x_i = x_{i-1} = 0$, and 1 otherwise.
- $w_i = 2v_i + p_i - t_i$.
  This constraint makes w_i be the maximal number of contacts between layers i and $(i-1)$, as discussed above.

Introducing these constraints lets define the maximal number of contacts between layers i and $(i-1)$. Sum of $w_i, i = 2, 3, ..., N$ variables is the total number of inter-layer contacts K_{inter} in the generated H-core.

3.2 Intra-layer Contacts

The number of contacts in a layer with k nodes is $(k-1)$ if $k \geq 1$, and 0 otherwise. Summation over all layers gives the following formula for the total number of intra-layer contacts K_{intra}:

$$K_{intra} = \sum_{i=1}^{N} \max(x_i - 1, 0)$$

The formula above can be reduced to:

$$K_{intra} = N - n_{non-empty} \tag{5}$$

Let us introduce the binary variables $z_i, i = 1, 2, ..., N$ in the LP problem formulation. Then, for every $i = 1, 2, ..., N$ the following constraint can be formulated:

$$M \cdot z_i \geq x_i,$$

where $M >> 1$. When z_i is minimized, it would be equal to 1 if $x_i \geq 1$, and 0 otherwise. Then, the sum of all $z_i, i = 1, 2, ..., N$ variables gives the total number of non-empty layers.

The target function of the LP problem is the total number of contacts (both intra- and inter-layer):

$$K = \sum_{i=2}^{N} w_i + N - \sum_{i=1}^{N} z_i \tag{6}$$

Maximization of the variable K leads to the optimal solution for Problem (1). For solving Problem (2), an additional constraint on the number of contacts must be added to the LP formulation:

$$K \leq K_{lim}$$

3.3 Generation of *H*-Core Mutations

Once the LP problem formulation is ready, a Linear Programming solver is run. The solution contains the optimal values $x_i^*, i = 1, 2, ..., N$ and $w_i^*, i = 2, 3, ..., N$.

In the next step, possible offsets of layers are derived from x_i^* and w_i^* values. For every $i = 2, 3, ..., N$, all Δ values are selected such that $w(x_{i-1}, x_i, \Delta) = w_i^*$. In result, the following list of shift options R is derived:

$$R = [\{\Delta | w(x_{i-1}, x_i, \Delta) = w_i^*\}, i = 2, 3, ..., N] \tag{7}$$

During creation of R, all empty layers are ignored.

All combinations of shift values from R are derived. In result, a list of *mutations* is generated, i.e. a list of combinations of shifts of layers that lead to the same number of contacts in the produced H-cores.

For every combination R^c, the node coordinates are generated. Nodes of the first non-empty layer i_0 have coordinates $(x, 0), x = 1, 2, ..., x_{i_0}$. Nodes the i'th layer, $i = i_0 + 1, i_0 + 2, ..., n_{non-empty}$ have coordinates $(d_i + x, i_0 - i), x = 1, 2, ..., x_i$, where $d_i = \sum_{j=1}^{i-i_0} R_j^c$. Figure 3 presents the maximally dense H-cores for $N = 10$.

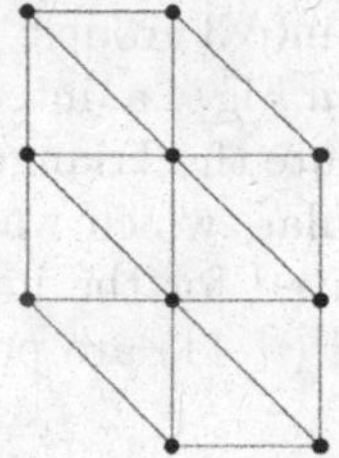

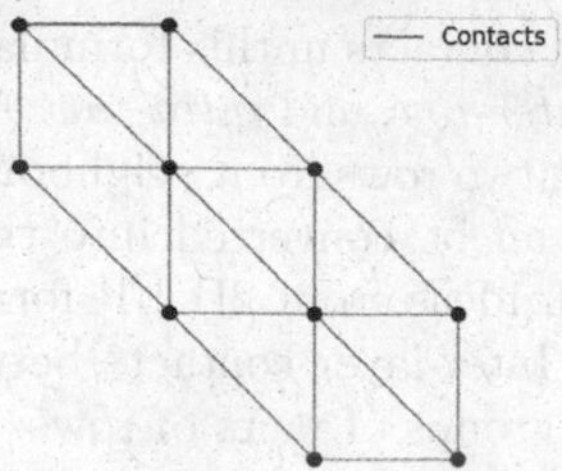

Fig. 3. Maximally dense H-cores in the Pseudo-triangular lattice generated for $N = 10$ with the exact LP-based algorithm

4 Linear Programming-Based Algorithm: 3D Case

In the spacial case (i.e., for the 3D FCC lattice), the LP problem formulation becomes significantly more complex. First, let the coordinate system for a spacial H-core be defined as presented in Fig. 4. Then, let a *layer* be a set of nodes with the same z coordinate value, and a *row* be a set of nodes with the same coordinate values along y and z axes.

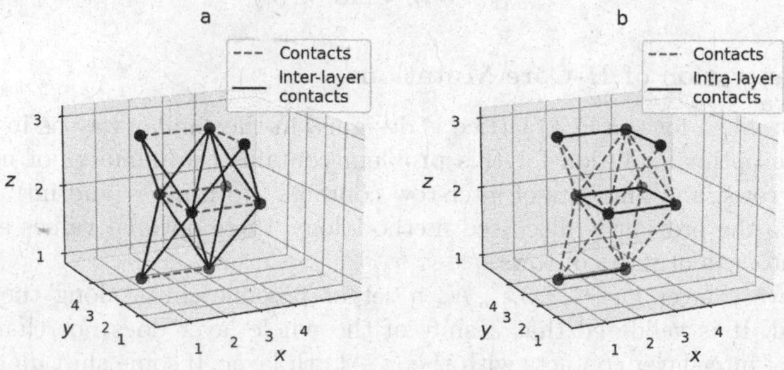

Fig. 4. Division of contacts in the modified FCC lattice into (a) inter-layer and (b) intra-layer contacts

Let all nodes of the generated H-core be distributed among N layers, where every layer has N rows. Let the H-core be built by layers from down to up. Then, in the LP formulation, let $x_{i,j}, i = 1, 2, ..., N,\ j = 1, 2, ..., N$ be the number of nodes in the row with coordinates (i, j), i.e. the row in the i'th layer and its j'th row.

In Fig. 4, it can be noticed the nodes in every next layer are horizontally shifted along the x and y axes as compared to the previous layer. This horizontal shift of rows in neighboring layers implies that the first non-empty row of a layer may have either even or odd index along the y axis. Let $j_i^0 = i \mod 2$ be the flag implying the parity of the index of the first non-empty row in the i'th layer.

Generally, contacts in this formulation can be split into 3 groups: *inter-layer*, *intra-layer inter-row*, and *intra-row*. As it can be seen in Fig. 4.a, inter-layer contacts between two rows from neighboring layers constitute the Triangular lattice. This lattice can be converted into the Pseudo-triangular, which was discussed in Sect. 3. All ideas from 2D LP formulation are derived for the inter-layer in the 3D case. Inter-layer contacts between layers i and $(i-1)$ are possible only between two groups of pairs of rows:

- (i, j) and $(i-1, j-1)$, $j = (2+j_i^0), (4+j_i^0), ..., (N-j_i^0)$;
- (i, j) and $(i-1, j+1)$, $j = j_i^0, (2+j_i^0), ..., (N-2+j_i^0)$.

Similarly, intra-row contacts follow the same logic as in the 2D case. The total number of intra-row contacts is equal to $N - n_{non-empty}$, where $n_{non-empty}$ is the number of non-empty rows in all layers.

All nodes inside one layer constitute the Square lattice, as seen in Fig. 4.b. In the i'th layer, inter-row contacts are found only between rows with indices (i, j) and $(i, j+2)$, $j = (2-j_i^0), (4-j_i^0), ..., (N-2-j_i^0)$. Here, the number of contacts between rows (i, j) and $(i, j+2)$ is maximal when the overlay of their nodes is maximized. Then, in the Square lattice, the maximal number of contacts $w_{Sq}(k, m, x_0)$ between two rows can be expressed as

$$w_{max}(k, m) = \min(k, m) \tag{8}$$

4.1 Generation of *H*-Core Mutations

An LP problem for the FCC lattice is designed in the similar way as in the 2D case. The optimal solution for this problem contains the numbers of nodes in different rows, and numbers of inter-row contacts (both inter- and intra-layer). Similar to the previously discussed methodology, these optimal values are converted into sets of shifts of rows.

For every layer $i, i = 2, 3, ..., N$, a set of possible shifts along the y axis is defined. It is validated that a shift of the whole layer does not change the number of inter-layer contacts with the $(i-1)$'th layer. If some shift dy changes the number of inter-layer contacts, it is ignored. This step of mutation generation is skipped for the first layer that is considered to be fixed.

Then, for every appropriate dy value, horizontal shifts of rows are determined. The first non-empty layer is considered to be fixed, and possible offsets are defined for all other rows with greater indices.

Finally, a set of mutations is derived in the same way as in the 2D case (see Sect. 3.3). For every mutation, an H-core is generated. In order to conform the FCC lattice requirements, all nodes of the resulting H-core are rotated in O_{xy} projection by $-\frac{\pi}{4}$. The z coordinate of all nodes is not mutated.

5 Results and Discussion

The developed H-core generation algorithms were implemented in the Python 3 programming language. All experiments were executed on a MacBook Pro 2020 machine that has a 2 GHz Intel Core i5 processor and 16 Gb of memory. For solving the Linear Programming problems, the PuLP Python module [20] was utilized.

5.1 Time Evaluations

The proposed exact algorithm for the Pseudo-triangular lattice faces the exponential rise in the elapsed time at large values of N, as shown in Fig. 5. This might be explained by setting a too high number of layers in the initial LP problem formulation. Generally, maximally dense H-cores present relatively small numbers of non-empty layers, but N layers are considered for N nodes by default. This results in a higher number of variables and in more extensive computation process.

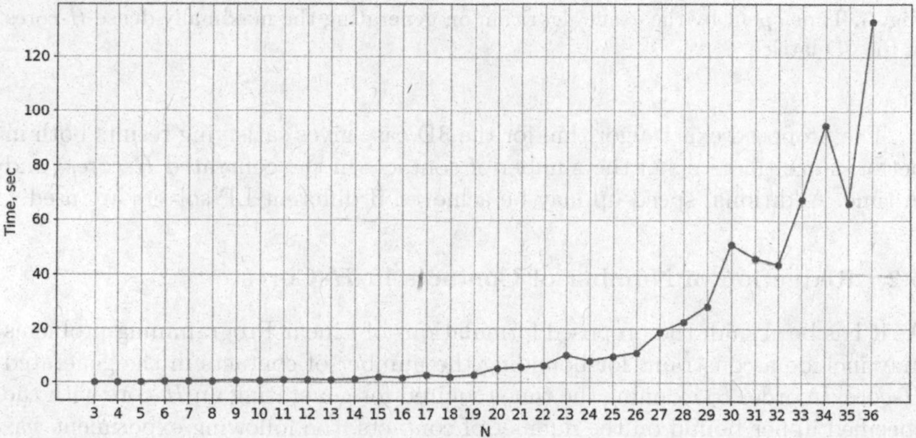

Fig. 5. Time spent by the exact algorithm on generating the maximally dense H-cores in the 2D lattice

The situation is different in the 3D case where we can suppose that all nodes of the maximally dense H-core are located within some 3D box. In Sect. 4, it was proposed to create N vertical layers in the LP formulation, but such runs lead to unreasonable running time values even at relatively small values of N. Thus, it was proposed to limit the number of vertical layers in the maximally dense H-core for N nodes with the value $\lceil \sqrt[3]{N} \rceil + 1$. Usage of such limit leads to a significant speed-up of the LP problem solving process. Time evaluations for these runs are shown in Fig. 6. Such time values are considered satisfying, which makes this algorithm applicable to the real world practice.

It can be concluded that the developed exact algorithms for generating the maximally dense H-cores are efficient enough. Performance of the algorithm for the 2D case could be improved if the number of non-empty layers is efficiently evaluated. This is supposed to be done in the future work.

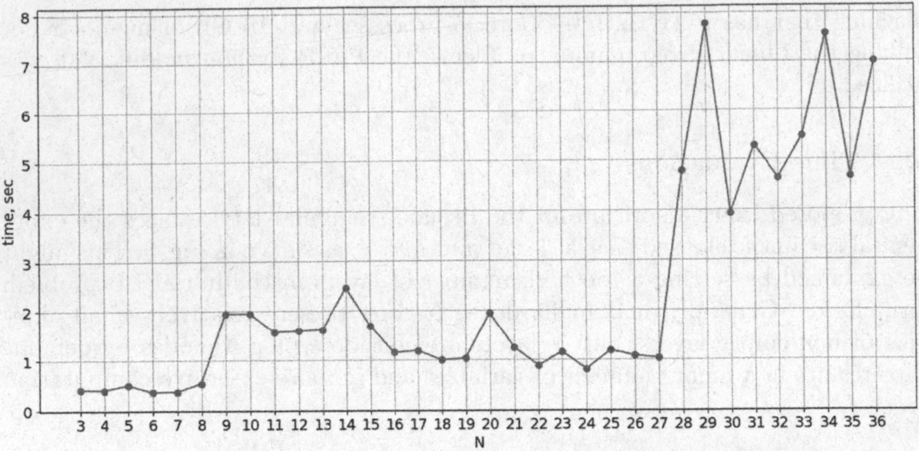

Fig. 6. Time spent by the exact algorithm on generating the maximally dense H-cores in the 3D lattice

The proposed exact algorithm for the 3D case gives satisfying results both in performance (in terms of the number of contacts in the generated H-cores) and in time. Additional speed-up may be achieved if different LP solvers are used.

5.2 Reduction of Number of Contacts in H-Cores

As it has been said, the proposed formulations of Linear Programming problems may include a constraint for bounding the number of contacts in the generated H-core. In order to examine the time required for generating an H-core with the specified upper bound on the number of contacts, the following experiment was conducted. N was set to 33, and the maximally dense H-core was constructed with the exact algorithm for the Pseudo-triangular lattice. Then, the discovered number of contacts in the maximally dense H-core K_{max} was fixed. After that, Problem (2) was solved for the same N and the predefined maximal numbers of contacts $K_{lim} = (K_{max} - 1), (K_{max} - 2), ..., (N - 1)$. For every run, the elapsed time and the number of layers in the generated H-cores were recorded.

Evaluations of the elapsed time are presented in Fig. 7. In can be seen that considering the upper bound on the number of contacts, i.e. solving Problem (2) instead of Problem (1), leads to reasonably less elapsed time. This was also proved true for $K_{lim} = K_{max}$: in this case, solving Problem (2) takes only 17 s instead of 42 s when solving Problem (1).

A similar experiment was conducted for the 3D case with the reduced number of vertical layers, as discussed in Sect. 5.1. Figure 8 presents time evaluations for solving Problem (2) for $N = 33$ and $K_{lim} = (K_{max} - 1), (K_{max} - 2), ..., (N - 1)$. First, it can be noticed that this plot contains several outlying values (at $K_{lim} = 93, 108, 112, 114, 117$). Except for them, the plot demonstrates the decreasing trend when K_{lim} decreases.

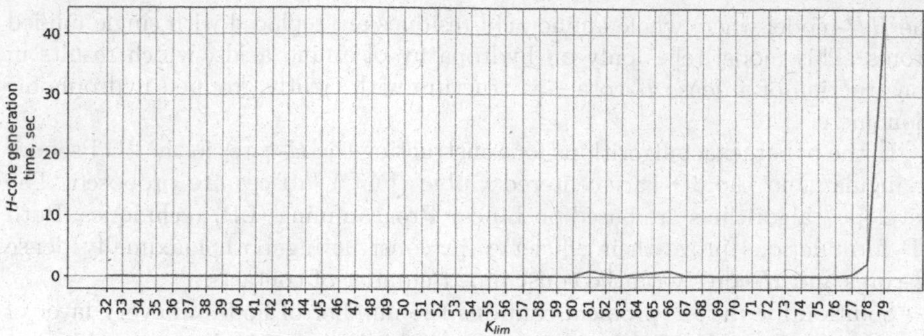

Fig. 7. Time required for solving Problem (2) for $N = 33$ and different values of K_{lim} in the 2D case

It can be concluded that the developed exact algorithms present the satisfying results when solving Problem (2). Another remarkable result is that a resulting configuration can be found reasonably faster if the upper bound of the number of contacts is known in advance. Thus, advanced analytical techniques for filtering H-cores by their basic features are worth developing.

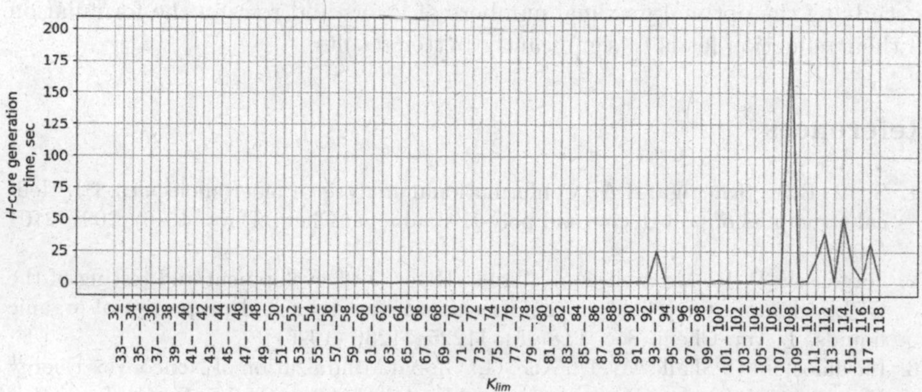

Fig. 8. Time required for solving Problem (2) for $N = 33$ and different values of K_{lim} in the 3D case

6 Conclusion

Modeling the spacial structure of protein molecules is a challenging problem that includes a huge number of degrees of freedom. Coarse-grained models propose an efficient way of decreasing the protein geometry, which results in a significant simplification of the overall problem. The strongest simplification is achieved in the HP-model where whole amino acid residues are replaced with single unified atoms. This model relies only on hydropathy of amino acids, which results in construction of a dense H-core – a structure with tightly grouped hydrophobic monomers.

In the paper, exact algorithms for constructing the H-cores in the 2D Pseudo-triangular and the 3D Face-centered Cubic (FCC) lattices are proposed. The developed algorithms are based on Linear Programming (LP) techniques. Both LP formulations prove their efficiency in constructing both maximally dense H-cores and H-cores with the constrained number of contacts.

Solution for the LP problem contains the number of nodes in every layer of the H-core. Possible offsets of layers are derived from this data, and different combinations of them are generated. These sets of offsets called *mutations* result in several H-cores with the same number of contacts. Thus, the algorithm manages to generate several analogous H-cores instead of only one. This might be useful for further steps of protein structure prediction where the protein sequence must be fit into the generated H-cores.

If the maximal possible number of contacts is predefined, H-core generation is shown to be performed faster. Thus, advanced rules of analytical filtering of H-cores are worth developing. If the maximal possible number of contacts can be derived for the target protein, this would provide a significant boost in time to the HP-model application. Another direction of possible algorithm improvement is studying the optimal maximal numbers of layers and rows in the formulation as their reduction leads to significantly better results.

References

1. Scott, R.A., Scheraga, H.A.: Conformational analysis of macromolecules. III. Helical structures of polyglycine and poly-L-alanine. J. Chem. Phys. **45**(6), 2091–2101 (1966)
2. Jorgensen, W.L., Maxwell, D.S., Tirado-Rives, J.: Development and testing of the OPLS all-atom force field on conformational energetics and properties of organic liquids. J. Am. Chem. Soc. **118**(45), 11225–11236 (1996)
3. Pardalos, P.M., Shalloway, D., Xue, G.: Global Minimization of Nonconvex Energy Functions: Molecular Conformation and Protein Folding: Molecular Conformation and Protein Folding: DIMACS Workshop, vol. 23, 20–21 March 1995 (1996)
4. Levitt, M.: A simplified representation of protein conformations for rapid simulation of protein folding. J. Mol. Biol. **104**(1), 59–107 (1976)
5. Koliński, A.: Protein modeling and structure prediction with a reduced representation. Acta Biochimica Polonica **51** (2004)
6. Kmiecik, S., Gront, D., Kolinski, M., Wieteska, L., Dawid, A.E., Kolinski, A.: Coarse-grained protein models and their applications. Chem. Rev. **116**(14), 7898–7936 (2016)
7. Lau, K.F., Dill, K.A.: A lattice statistical mechanics model of the conformational and sequence spaces of proteins. Macromolecules **22**(10), 3986–3997 (1989)
8. Munson, M., et al.: What makes a protein a protein? Hydrophobic core designs that specify stability and structural properties. Protein Sci. **5**(8), 1584–1593 (1996)

9. Mann, M., Backofen, R.: Exact methods for lattice protein models. Bio-Algorithms Med.-Syst. **10**(4), 213–225 (2014)
10. Rashid, M.A., Hoque, M.T., Newton, M.A.H., Pham, D.N., Sattar, A.: A new genetic algorithm for simplified protein structure prediction. In: Thielscher, M., Zhang, D. (eds.) AI 2012. LNCS (LNAI), vol. 7691, pp. 107–119. Springer, Heidelberg (2012). https://doi.org/10.1007/978-3-642-35101-3_10
11. Nazmul, R., Chetty, M.: A knowledge-based initial population generation in memetic algorithm for protein structure prediction. In: Lee, M., Hirose, A., Hou, Z.-G., Kil, R.M. (eds.) ICONIP 2013. LNCS, vol. 8227, pp. 546–553. Springer, Heidelberg (2013). https://doi.org/10.1007/978-3-642-42042-9_68
12. Nazmul, R., Chetty, M., Chowdhury, A.R.: An improved memetic approach for protein structure prediction incorporating maximal hydrophobic core estimation concept. Knowl.-Based Syst. **219**, 104395 (2021)
13. Will, S.: Constraint-based hydrophobic core construction for protein structure prediction in the face-centered-cubic lattice. In: Biocomputing 2002, pp. 661–672 (2001)
14. Irbäck, A., Sandelin, E.: Local interactions and protein folding: a model study on the square and triangular lattices. J. Chem. Phys. **108**(5), 2245–2250 (1998)
15. Dantzig, G.B.: Origins of the simplex method. In: A History of Scientific Computing, pp. 141–151 (1990)
16. Wilson, M.S., Landau, D.P.: Thermodynamics of hydrophobic-polar model proteins on the face-centered cubic lattice. Phys. Rev. E **104**(2), 025303 (2021)
17. Ignatov, A., Posypkin, M.: Techniques for speeding up H-core protein fitting. In: Olenev, N.N., Evtushenko, Y.G., Jaćimović, M., Khachay, M., Malkova, V. (eds.) OPTIMA 2021. LNCS, vol. 13078, pp. 336–350. Springer, Cham (2021). https://doi.org/10.1007/978-3-030-91059-4_24
18. Böckenhauer, H.-J., Dayem Ullah, A.Z.M., Kapsokalivas, L., Steinhöfel, K.: A local move set for protein folding in triangular lattice models. In: Crandall, K.A., Lagergren, J. (eds.) WABI 2008. LNCS, vol. 5251, pp. 369–381. Springer, Heidelberg (2008). https://doi.org/10.1007/978-3-540-87361-7_31
19. Wu, L.C., Kim, P.S.: A specific hydrophobic core in the α-lactalbumin molten globule. J. Mol. Biol. **280**(1), 175–182 (1998)
20. PuLP: PyPI. (n.d.). https://pypi.org/project/PuLP/. Accessed 5 Aug 2022

Graph Density and Uncertainty of Graphical Model Selection Algorithms

Valeriy Kalyagin(✉) and Ilya Kostylev

Laboratory of Algorithms and Technologies for Network Analysis, HSE University, Nizhny Novgorod, Russia
vkalyagin@hse.ru, idkostylev@edu.hse.ru
https://nnov.hse.ru/en/latna/

Abstract. Graphical models became a popular tool in machine learning and data analysis. Graphical Model Selection Problem is a problem to recover a specific graph structure (graphical model) from a set of data. In many cases the data are given by a sample of observations of some multivariate distribution. In this setting any reconstruction algorithm can be evaluated by uncertainty of identification of the hidden graphical model by observations. In the present paper we study uncertainty of identification of so-called concentration graph which represents a dependence structure for the components of multidimensional random vector. We introduce and discuss different measures of uncertainty appropriate for the concentration graph identification problem and compare on this basis different identification algorithms, including optimization (graphical lasso) algorithm and a family of known multiple hypotheses testing algorithms. Novelty of our approach is in the study of dependence of uncertainty on the graph density. Some new and interesting phenomena are observed and discussed.

Keywords: Graphical model · Concentration graph · Uncertainty · Optimal identification algorithm · Graph density

1 Introduction

In general, graphical model selection problem is to recover a hidden graph structure (graphical model) from a set of data. Graphical model of a particular interest is the probabilistic graphical model (Markov field, Bayesian network), where the hidden graph structure is a *concentration graph* which represents a conditional dependence structure of the components of a multidimensional distribution (components of a random vector). Set of data for graphical model selection problem in this case is given by a sample of observations of this multivariate distribution. Probabilistic graphical model has a variety of applications in different fields, and especially in bio-informatics and bio-genomics (see for example [1,2]).

Chapter 1 was prepared within the framework of the Basic Research Program at the National Research University Higher School of Economics (HSE University), results of the Chaps. 3–7 are obtained with a support from RSF grant 22-11-00073.

N. Olenev et al. (Eds.): OPTIMA 2023, CCIS 1913, pp. 188–201, 2024.
https://doi.org/10.1007/978-3-031-48751-4_14

Increasing number of publications is devoted to this problem. Different algorithms to recover (identify) the concentration graph by observations were proposed and investigated. Recent reviews can be found in [3,4]. Important class of identification algorithms is related with multiple hypothesis testing [5]. Another approach is connected with graphical LASSO type algorithms, based on maximum likelihood optimization with a regularization term (see original paper [6] and recent advances in [7]).

Important question for an identification algorithm is how close are concentration graphs identified by the algorithm from observations to a true concentration graph of the model (hidden graph structure). This difference measures a quality of the algorithm. Existing study of the quality of concentration graph identification algorithms are focused on the case where the associated concentration graph is sparse. Sparsity is observed in some applications but is not only the case of interest for the concentration graph identification. Therefore it is important to understand how the quality of identification algorithms is related with the concentration graph density. This issue in our knowledge is not well investigated in the literature. The main contribution of the present study is a detailed investigation of this question.

Quality of algorithms for the concentration graph identification can be measured by some statistical measure of error. From the Machine Learning point of view, graphical model selection problem can be considered as a binary classification problem. Therefore, to measure uncertainty of identification algorithms one can use measures of error from the binary classification. In our study to measure uncertainty of graphical model selection algorithms we use the following measures of error from binary classification: TPR (True Positive Rate), FOR (False Omission Rate), FDR (False Discovery Rate), TNR (True Negative Rate), and generalized characteristics such as BA (Balanced Accuracy), F1 score, and MCC (Matthews Correlation Coefficient).

In the present paper we compare uncertainty of popular multiple hypotheses testing algorithms and graphical lasso (optimization) algorithm using different these measures of quality. Our main focus in this comparison is to study the quality (uncertainty) of algorithms as a function of the concentration graph density. First, we suggest to consider concentration graph selection problem in a more general setting: identification of the concentration graph a in random variable networks [8]. Classical graphical model selection problem is a particular case of this setting, where associated measure of similarity is a partial correlation. Second, we suggest to use a concentration graph generator based on Cholesky decomposition [9] and show how to obtain a random family of concentration graphs with prescribed expected density. Next we use numerical experiment to compare different algorithms for different setting of parameters. Obtained results show that most of algorithms have a poor quality when recovering concentration graph by observation for the case of higher graph density. It is interesting to note, that naive simultaneous inference algorithm is very competitive with other algorithms including an advanced version of the graphical lasso. Some other new and interesting phenomena are observed and discussed.

The paper is organized as follows. In Sect. 2 we present main definitions and notations. Section 3 is devoted to concentration graph identification algorithms in general setting of random variable networks. In Sect. 4 we develop our approach to measure quality (uncertainty) of identification algorithms. Results of numerical experiences to compare uncertainty of identification algorithms are discussed in the Sect. 6. Finally, in Sect. 7 we discuss obtained results and directions for future research.

2 Basic Definitions and Notations

Random variable network is a pair (X, γ), where X is a random vector $X = (X_1, X_2, \ldots, X_N)$, and $\gamma = \gamma(Y, Z)$ is a measure of pairwise dependence (similarity, association,...) between random variables (Y and Z). Random variable network generates a network model - complete weighted graph Γ with N nodes. Node i is associated with the random variable X_i $(i = 1, 2, \ldots, N)$. Weight of edge (i, j), $(i \neq j)$ is given by $\gamma_{i,j} = \gamma(X_i, X_j)$. *Concentration graph* in random variable network (X, γ) is a subgraph $Conc(\Gamma)$ of the complete weighted graph Γ. Edge (i, j) is included in the concentration graph $Conc(\Gamma)$ if and only if $\gamma_{i,j} \neq 0$. In practice distribution of the vector X is not known and we have only observations of the random vector X, which we model as independent identically distributed random vectors $X(1), X(2), \ldots, X(n)$ (sample of the size n from the distribution X). The main problem in this case is to identify the concentration graph by observations. Identification algorithms can be associated with a map from the sample space to the space of simple undirected, unweighted graphs with N nodes. Quality or uncertainty of identification algorithm therefore is related with estimation of how different can be identified concentration graphs obtained by different samples of observations from the same distribution. Note, that Graphical Model Selection Problem (GMSP) is a particular case of the concentration graph identification problem (CGIP) where the measure of dependence is a partial correlation (conditional dependence).

To measure the quality (uncertainty) of an identification algorithm one can consider the concentration graph identification problem as a binary classification problem for the edges of the complete unweighted graph with N nodes: edge (i, j) is in the class "Yes" if this edge is present in the concentration graph $Conc(\Gamma)$, and edge (i, j) is in the class "No" if this edge is not present in the concentration graph $Conc(\Gamma)$. From this point of view a concentration graph identification algorithm can be considered as classification algorithm. To measure the quality (uncertainty) of identification algorithm one can use the main characteristics of classification algorithms: TP (True Positive) – number of correctly identified edges, TN (True Negative) - number of correctly identified absences of edges, FP (False Positive) - number of incorrectly identified edges, and FN (False Negative) – number of incorrectly identified absences of edges. Various combinations of TP, TN, FP, FN are used to measure the quality of classification algorithms. We will use some of them in the Sect. 4 to measure the quality (uncertainty) of concentration graph identification algorithms.

Uncertainty of a given identification algorithm for the concentration graph identification is related with two aspects of associated random variable network:

- Distribution of the vector $X = (X_1, X_2, \ldots, X_N)$. In a standard situation, vector X has a multivariate Gaussian distribution. In some applications marginal distributions of X have a heavy tails. To model this one can use a large class of elliptical distributions. In particular, multivariate Student distributions are from this class.
- Measure of dependence γ. In standard situation of Graphical Model Selection Problem the Pearson partial correlation is used. This generates a Pearson partial correlation network. Other popular correlation networks are: Pearson correlation networks, Kendall correlation networks, Fechner correlation networks. Important fact is that for elliptical distributions there is a connection between Pearson, Kendall, and Fechner correlations. It is proved in [8] the following relations between these three correlations for elliptical distributions (you can find in this book a detailed description of these correlations and connections between them):

$$\rho^{Kendall} = \rho^{Fechner} = \frac{2}{\pi} \arcsin(\rho^{Pearson}).$$

It implies that concentration graph in Pearson correlation network is the same as concentration graph in Kendall and Fechner correlation networks (for the same distribution). Difference with Pearson correlation network is in a different way to estimate Kendall and Fechner correlations (see details in [8]). We will see that this can provide a substantial improvement of uncertainty of identification algorithms for non Gaussian distributions.

3 Concentration Graph Identification Algorithms

3.1 Multiple Testing Algorithms

In this paper we consider identification algorithms based on multiple hypotheses testing approach. Concentration graph identification problem can be associated with the following multiple hypotheses testing problem:

$$h_{i,j} : \gamma_{i,j} = 0 \ vs \ k_{i,j} : \gamma_{i,j} \neq 0, \ i,j = 1,2,\ldots N, \ i < j.$$

Associated concentration graph identification algorithm can be described as follows: edge (i,j) is included in the concentration graph if the hypothesis $h_{i,j}$ is rejected, edge (i,j) is not included in the concentration graph if the hypothesis $h_{i,j}$ is accepted. Different algorithms were proposed in the literature for multiple hypotheses testing problems. Two measures of quality of multiple hypotheses testing algorithms are investigated in the literature from theoretical point of view: FWER (Family Wise Error Rate) and FDR (False Discovery Rate) [5]. In our case, FWER is the probability to reject at least one true hypothesis $h_{i,j}$, i.e. probability $P(FP > 0)$. FDR is well known in binary classification and it is defined by $FDR = FP/(TP + FP)$. Variety of multiple hypotheses testing algorithms were proposed with FWER and FDR control. General description of this family of algorithms is the following:

1. Fix α - expected level of error.
2. Propose test statistics $T_{i,j}$ for each individual hypothesis $h_{i,j}$ and calculate p-values $p_{i,j}$ of the statistics for a given data (sample of observations).
3. Sort calculated p-values of individual statistics in increasing order.
4. Compare sorted p-values with an appropriate adjustment curve and make a decision to accept or to reject individual hypotheses.

Denote sorted p-values of individual test statistics by

$$p_1 \leq p_2 \leq p_3 \leq \cdots \leq p_M, \; M = \frac{1}{2}N(N-1).$$

We have chosen the following multiple hypotheses testing algorithms for comparison of their quality:

- *Simultaneous inference*: reject all individual hypotheses with p-value less or equal to α. For Pearson partial correlation network (Graphical Model Selection Problem) this algorithms is optimal for the additive loss function (FP+FN) in the class of unbiased algorithms [10]. This optimality can be confirmed for Pearson and Fechner correlation networks too. In what follows we will compare the quality of this optimal algorithm with other multiple hypotheses testing algorithms.
- *Bonferroni correction (B)*: reject all hypotheses with $p_k \leq \frac{\alpha}{M}$. This procedure control FWER at the level α.
- *Holm procedure (H)*: find $R = \min\{k : p_k > \frac{\alpha}{M+1-k}\}$ and reject all hypotheses associated with $k = 1, 2, \ldots, R-1$. This procedure refine the Bonferroni correction and still control FWER at the level α.
- *Hochberg procedure (Hg)*: find $R = \max\{k : p_k \leq \frac{\alpha}{M+1-k}\}$ and reject all hypotheses associated with $k = 1, 2, \ldots, R$. This procedure also refine the Bonferroni correction and control FWER at the level α.
- *Benjamini-Hochberg procedure (BH)*: find $R = \max\{k : p_k \leq \frac{\alpha k}{M}\}$ and reject all hypotheses associated with $k = 1, 2, \ldots, R$. This procedure control FDR at the level α in the case of independent test statistics (this is not the case for concentration graph identification).
- *Benjamini-Yekutieli procedure (BY)*: find $R = \max\{k : p_k \leq \frac{\alpha k}{M c_M}\}$, where $c_M = \sum_{l=1}^{M} \frac{1}{l}$, and reject all hypotheses associated with $k = 1, 2, \ldots, R$. This procedure control FDR at the level α in general case (including dependent test statistics).

All these algorithms are well known in multiple hypotheses testing and they are widely used in applications. Detailed description of the individual test statistics for each network (Pearson partial correlation network, Pearson correlation network, Kendall correlation network, Fechner correlation network) can be found in

Chap. 3 in [8]. For example, in Pearson partial correlation network, the following statistics can be used as individual test statistics:

$$T_{i,j}^{PPC} = r^{i,j}\sqrt{\frac{n-N}{1-(r^{i,j})^2}},$$

where n is the sample size, and $r^{i,j}$ is the sample partial correlation associated with random variables X_i and X_j of the random vector $X = (X_1, X_2, \ldots, X_N)$. It is known that if X has a multivariate Gaussian distribution and hypothesis $h_{i,j}$ is true then $T_{i,j}^{PPC}$ has Student distribution with $(n - N)$ degree of freedom [11]. This gives a way to calculate the p-value of the statistics for a given data (sample of observations).

3.2 Graphical Lasso for Partial Correlation Network

For probabilistic graphical model with Gaussian distribution $X \sim N(\mu, \Sigma)$, conditional independence of X_i and X_j is equivalent to the equation $\rho^{i,j} = 0$, where $\rho^{i,j} = 0$ is the partial correlation between X_i, X_j, all other X_k being fixed. One has [11]:

$$\rho^{i,j} = -\frac{\sigma^{i,j}}{\sqrt{\sigma^{i,i}\sigma^{j,j}}}$$

where $\sigma^{i,j}$ are elements of the matrix Σ^{-1}. Edge (i, j) is included in the concentration graph if $\rho^{i,j} \neq 0$. The concentration graph identification problem in this case is therefore to identify zeros for the inverse of the matrix Σ (covariance matrix of the vector X). Maximum likelihood estimation of the matrix Σ^{-1} with a regularization term is given by the following optimization problem [6]:

$$Trace(S\Omega) - \log\det(\Omega) + \lambda||\Omega||_1 \to \min$$

over all positive semi-definite matrices. Objective function of this optimization problem is convex, and many optimization algorithms can be applied. Zero elements of the obtained solution are identified with absence of edge in the concentration graph. This gives a graphical lasso algorithm. In our experiments we use an advanced version of graphical lasso algorithm with an optimal choice of the regularization parameter λ.

4 Measures of Uncertainty of Concentration Graph Identification

It was already discussed above, that concentration graph identification problem can be considered as a binary classification problem. To measure the quality of concentration graph identification algorithms we propose to use the following characteristics:

- True Negative Rate: $TNR = \frac{TN}{TN+FP}$, proportion of correctly identified absences of edges in concentration graph.

- True Positive Rate: $TPR = \frac{TP}{TP+FN}$, proportion of correctly identified edges in concentration graph.
- False Discovery Rate: $FDR = \frac{FP}{TP+FP}$, proportion of incorrectly identified edges between all identifies edges.
- False Omission Rate: $FOR = \frac{FN}{TN+FN}$, proportion of incorrectly identified absences of edges between all identifies absences of edges.

This four characteristics give in our opinion a complete picture of the quality of any concentration graph identification algorithm. Between various generalized criterion we will chose three:

- Balanced accuracy: $BA = \frac{1}{2}(TPR + TNR)$.
- F1 score: $F1 = 2\frac{(1-FDR)TNR}{(1-FDR)+TNR}$.
- Matthew Correlation Coefficient:

$$MCC = \frac{TN \times TP - FN \times FP}{\sqrt{(NP + FP)(TP + FN)(TN + FP)(TN + FN)}}.$$

Balanced accuracy and F1 score are standard in binary classification. Recently it was shown that MCC has some advantages with respect to BA and F1 in applications [12].

5 Concentration Graph Generator

To study how the quality of a concentration graph identification algorithm is related with graph density we need to generate a family of random concentration graphs with a given expected density. One way to do it is to use Cholesky decomposition. To generate a random symmetric positive definite matrix Ω, a random lower triangular matrix L with a given density of nonzero elements is generated, then the desired matrix is obtained by the formula $\Omega = LL^T$. The connection between density of zeros elements in the matrix Ω and expected density of non zero elements of the matrix L is shown on the Fig. 1. This connection is obtained by numerical experiments. Using this approach we were able to generate a large family of random concentration graphs with a given expected density. More precisely, for a fixed dimension of the random vector X (number of vertex in the concentration graph) and fixed concentration graph density we calculate associated density of non zero elements in the triangular matrix L. Then we generate triangular matrix L and calculate positive definite matrix $\Omega = LL^T$. This matrix is used as inverse $\Sigma^{-1} = \Omega$ to the covariance matrix Σ of X in the case of Pearson partial correlation network, and as the covariance matrix $\Sigma = \Omega$ for the cases of Pearson Kendall, and Fechner correlation networks. For convenience we use diagonal normalization of the matrices under consideration: $diag(\Sigma^{-1}) = (1, 1, \ldots, 1)$ for the partial correlations and $diag(\Sigma) = (1, 1, \ldots, 1)$ for other correlations.

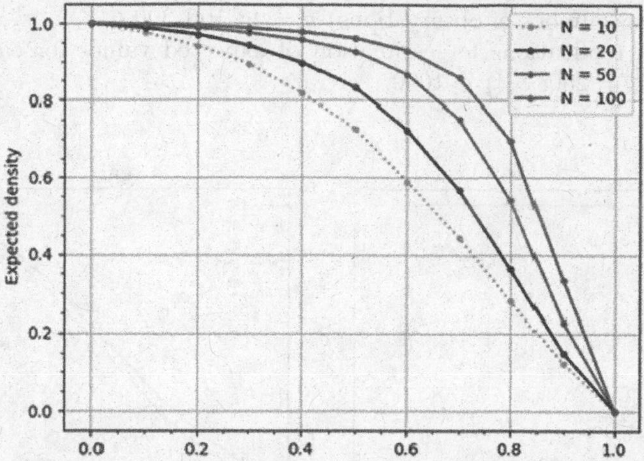

Fig. 1. Expected density of zero elements in positive definite matrix generated by Cholesky decomposition with a given density of zero elements in triangular matrix

6 Comparison of Concentration Graph Identification Algorithms

We investigate the quality of described algorithms for concentration graph identification by a series of numerical experiments. Complete list of results can be found by reference:

One experiment is conducted in 6 steps:
Step 1. Fix value d of the concentration graph density.
Step 2. Generate a concentration graph with d as expected density and fix associated covariance matrix Σ.
Step 3. Generate a sample of the size n from distribution of the random vector X.
Step 4. Apply identification algorithms and identify the concentration graph.
Step 4. Calculate quality characteristics TPR, TNR, FDR, FOR, BA, F1, MCC.
Step 5. Repeat S_{obs} times the Step 3 and calculate average values of quality characteristics.
Step 6. Repeat S_{exp} times the Step 2 and calculate average values of quality characteristics.

Parameters settings are as follows:

- Dimension of the random vector X (number of vertex in the concentration graph), $N = 20$.
- Values of the concentration graph density: $d = 0.1, 0.2, 0.3, \ldots, 0.9$.
- Distribution of the random vector X: multivariate Gaussian and multivariate Student distributions with zero means and covariance matrix Σ.
- Error level $\alpha = 0.05$.

- Sample size (number of observations) $n = 40, 100, 300$.
- Number of replications for estimation of expected values for each value of density $S_{obs} = 200$, $S_{exp} = 1000$.

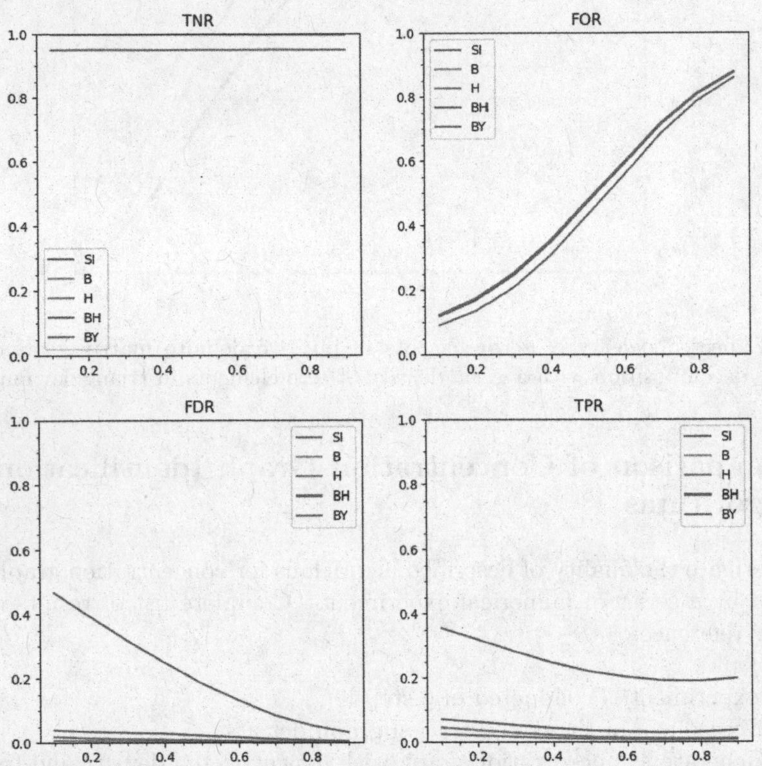

Fig. 2. Dependence of TPR, TNR, FDR, FOR on the concentration graph density. Multiple testing algorithms. Pearson partial correlation network. Gaussian distribution, Sample size $n = 40$.

Obtained results are similar for different networks. Figure 2 shows the dependence of TPR, TNR, FDR, FOR on the concentration graph density for multiple testing algorithms in the case of Gaussian distribution and the sample size $n = 40$. For all algorithms, the TNR (True Negative Rate) for the Simultaneous Inference algorithm is kept at 0.95, which is in good agreement with the selected significance level of individual tests. For the rest of the algorithms, TNR is close to 1, which is explained by the high level of FWER control. The Simultaneous Inference algorithm does not control FDR (False Discovery Rate), whereas other algorithms have such control. As the density of the concentration graph increases, FOR (False Commission Rate) increases to values close to 1, and TPR (True Positive Rate) remains at a low level for all algorithms. This is due to the low power of multiple hypothesis testing algorithms.

Figure 3 shows the dependence of balanced Accuracy, F1 score and Matthew Correlation Coefficient on the concentration graph density for the same case (multiple testing algorithms, Pearson partial correlation network, Gaussian distribution, and the sample size $n = 40$). One can see a poor quality of all algorithms for high density of the concentration graph. In all cases naive Simultaneous Inference algorithm has some advantages with respect to other algorithms.

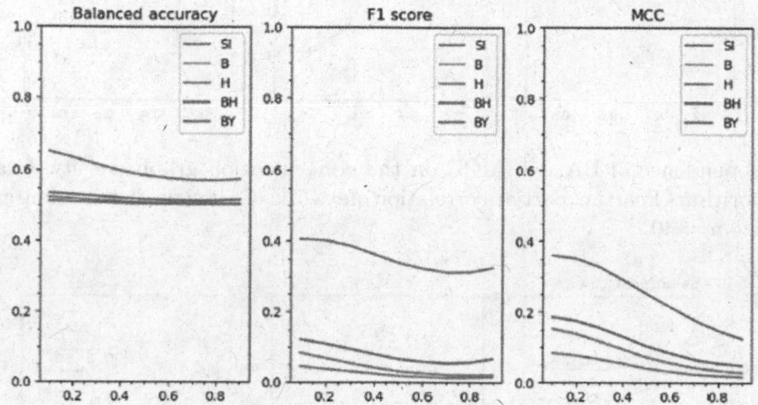

Fig. 3. Dependence of BA, F1, MCC on the concentration graph density. Multiple testing algorithms. Pearson partial correlation network. Gaussian distribution, Sample size $n = 40$.

Figure 4 shows the dependence of balanced Accuracy, F1 score and Matthew Correlation Coefficient on the concentration graph density for graphical lasso algorithm (Pearson partial correlation network, Gaussian distribution and the sample size $n = 40$). One can see a poor quality of graphical lasso for high density of the concentration graph. In this case simultaneous inference algorithm is competitive with graphical lasso too.

The situation does not change essentially in the case of a large sample size. Figure 5 shows the dependence of balanced Accuracy, F1 score and Matthew Correlation Coefficient on the concentration graph density for multiple testing algorithms for the case of Kendall correlation network, Gaussian distribution and the sample size $n = 300$. One can see advantages of the optimal Simultaneous Inference algorithms and poor quality of all algorithms for high density of the concentration graph.

Quality of multiple testing algorithms is decreasing when we go from multivariate Gaussian distribution to multivariate Student distribution in Pearson partial correlation network. This is shown on the Fig. 6. It is interesting to note that this change is much less emphasized in Kendall correlation network

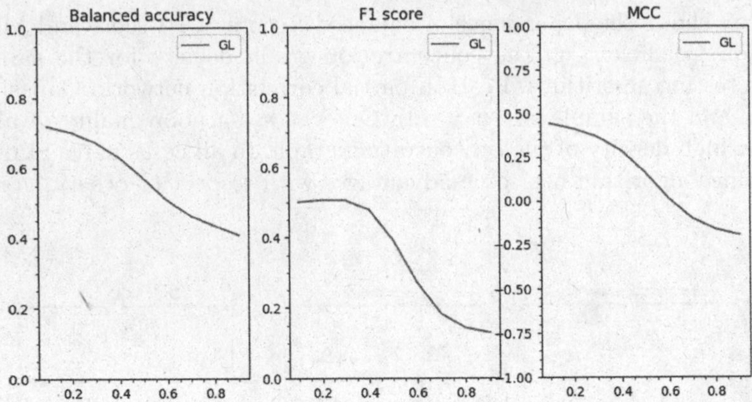

Fig. 4. Dependence of BA, F1, MCC on the concentration graph density. Graphical Lasso algorithm. Pearson partial correlation network, Gaussian distribution and the sample size $n = 40$.

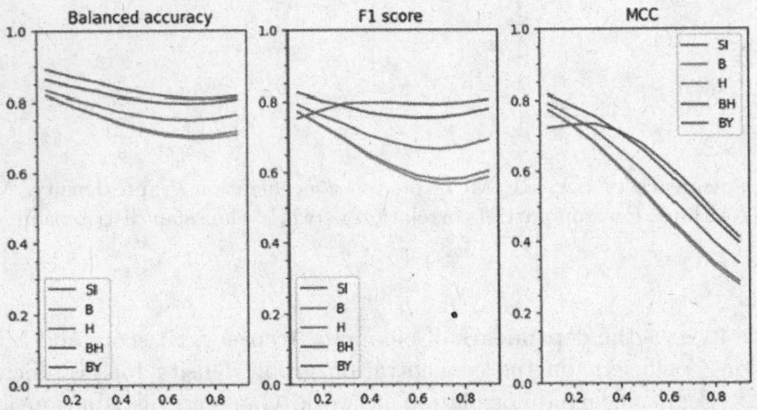

Fig. 5. Dependence of BA, F1, MCC on the concentration graph density. Multiple testing algorithms. Kendall correlation network. Gaussian distribution, Sample size $n = 300$.

and there is no change in quality of algorithms in Fechner correlation network. Figure 7 and Fig. 8, where BA, F1 and MCC characteristics are presented for the sample size $n = 100$ and two distributions show no difference. This phenomena is related with robustness of algorithms with respect to distributions and needs a further investigations.

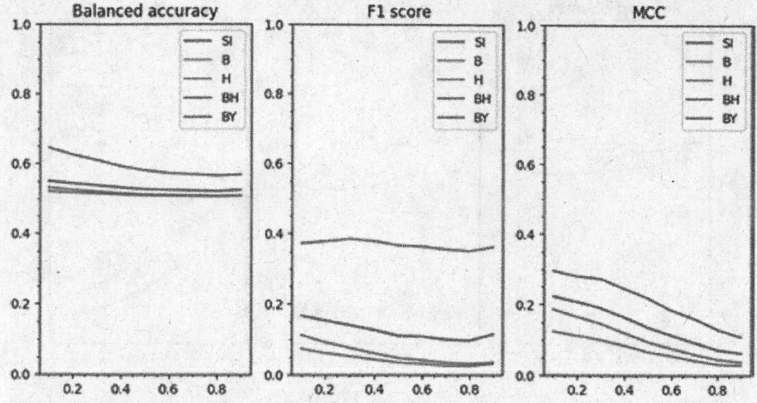

Fig. 6. Dependence of BA, F1, MCC on the concentration graph density. Multiple testing algorithms. Pearson partial correlation network. Student distribution with 3 degree of freedom. Sample size $n = 40$.

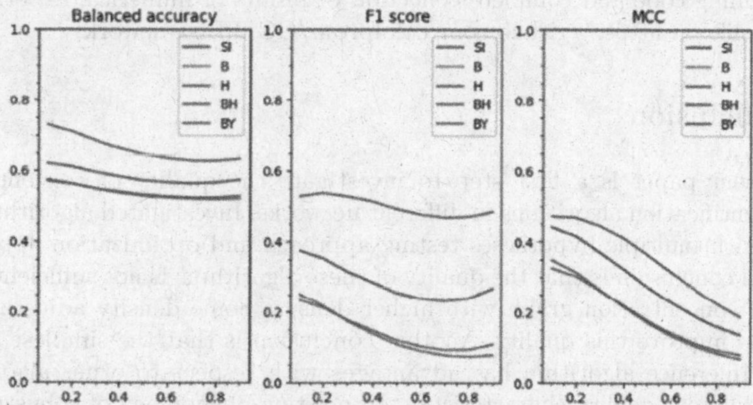

Fig. 7. Dependence of BA, F1, MCC on the concentration graph density. Multiple testing algorithms. Fechner correlation network. Gaussian distribution. Sample size $n = 100$.

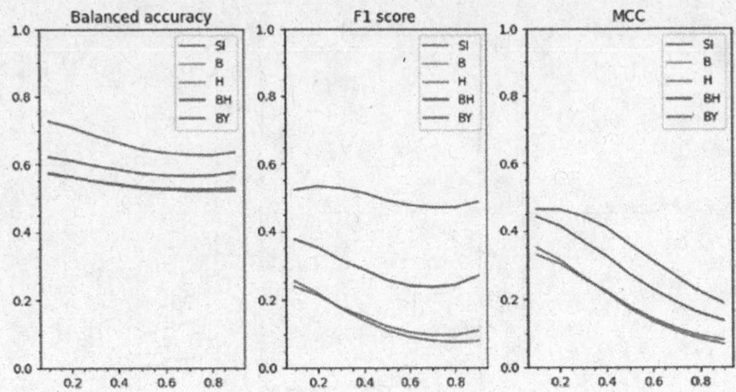

Fig. 8. Dependence of BA, F1, MCC on the concentration graph density. Multiple testing algorithms. Fechner correlation network. Student distribution with 3 degree of freedom. Sample size $n = 100$.

Programming code and complete collection of results of numerical experiments are available on https://github.com/cofofprom/GGMS_framework.

7 Conclusion

The present paper is a first step to investigate the quality of concentration graph identification algorithms in different networks. Investigated algorithms are related with multiple hypotheses testing approach and optimization approach. The main conclusion is that the quality of these algorithms is not sufficient especially for concentration graph with higher density. Some density adjustment is needed to improve this quality. Another conclusion is that the simplest Simultaneous Inference algorithm has advantages with respect to other algorithms and for different quality characteristics. Interesting phenomena of robustness of identification algorithms in Fechner correlation network is observed. This phenomena needs a further investigation. In particular, it can be used to construct robust Graphical Lasso or Graphical Bayesian type algorithms. That will be a topics of further investigation.

Acknowledgements. Chapter 1 was prepared within the framework of the Basic Research Program at the National Research University Higher School of Economics (HSE University), results of the Chapters 3–7 are obtained with a support from RSF grant 22-11-00073. Numerical experiments were conducted using HSE HPC resources [13].

References

1. Liu, J., Peissig, P., Zhang, C., Burnside, E., McCarty, C., Page, D.: Graphical-model based multiple testing under dependence, with applications to genome-wide association studies. In Uncertainty in Artificial Intelligence: Proceedings of the Conference on Uncertainty in Artificial Intelligence, vol. 2012, p. 511. NIH Public Access (2012)
2. Zhou, L., Wang, L., Liu, L., Ogunbona, P., Dinggang, S.: Learning discriminative Bayesian networks from high-dimensional continuous neuroimaging data. IEEE Trans. Pattern Anal. Mach. Intell. **38**(11), 2269–2283 (2016)
3. Drton, M., Maathuis, M.H.: Structure learning in graphical modeling. Annu. Rev. Stat. Appl. **4**, 365–393 (2017)
4. Cordoba, I., Bielza, C., Larranaga, P.: A review of Gaussian Markov models for conditional independence. J. Stat. Plan. Inference **206**, 127–144 (2020)
5. Drton, M., Perlman, M.D.: Multiple testing and error control in Gaussian graphical model selection. Stat. Sci. **22**(3), 430–449 (2007)
6. Friedman, J., Hastie, T., Tibshirani, R.: Sparse inverse covariance estimation with the graphical lasso. Biostatistics **9**(3), 432–441 (2008)
7. Cisneros-Velarde, P., Petersen, A., Oh, S.-Y.: Distributionally robust formulation and model selection for the graphical lasso. In: Proceedings of the Twenty Third International Conference on Artificial Intelligence and Statistics (AISTATS), pp. 108, 756–765. PMLR (2020)
8. Kalyagin, V.A., Koldanov, A.P., Koldanov, P., Pardalos, P.M.: Statistical Analysis of Graph Structures in Random Variable Networks. Springer, Cham (2020). https://doi.org/10.1007/978-3-030-60293-2
9. Cordoba, I., Bielza, C., Larranaga, P., Varando, G.: Sparse Cholesky covariance parametrization for recovering latent structure in ordered data. IEEE Access **8**, 154614–154624 (2020)
10. Kalyagin, V.A., Koldanov, A.P., Koldanov, P., Pardalos, P.M.: Loss function, unbiasedness, and optimality of Gaussian graphical model selection. J. Stat. Plan. Inference **201**, 32–39 (2019)
11. Anderson, T.W.: An Introduction to Multivariate Statistical Analysis, 3rd edn. Wiley, New York (2003)
12. Chicco, D., Jurman, G.: The advantages of the Matthews correlation coefficient (MCC) over F1 score and accuracy in binary classification evaluation. BMC Genom. **21**(1), 1–13 (2020)
13. Kostenetskiy, P., Chulkevich, R., Kozyrev, V.: HPC resources of the higher school of economics. J. Phys. Conf. Ser. **1740**(1), 012050 (2021)

Swarm Intelligence Technique for Capacity Optimization of a Transportation Network

Alexander Krylatov[1,2](✉) and Darya Kuznetsova[1]

[1] Saint Petersburg State University, Saint Petersburg, Russia
a.krylatov@spbu.ru
[2] Institute of Transport Problems RAS, Saint Petersburg, Russia
aykrylatov@yandex.ru

Abstract. Today artificial intelligence systems support efficient management in different fields of social activities. In particular, congestion control in modern networks seems to be impossible without proper mathematical models of traffic flow assignment. Thus, the network design problem can be referred to as Stackelberg game with independent lower-level drivers acting in a non-cooperative manner to minimize individual costs. In turn, upper-level decision-maker seeks to minimize overall travel time in the network by investing in its capacity. Hence, the decision-maker faces the challenge with a hierarchical structure which solution important due to its influence on the sustainable development of modern large cities. However, well-known that a bilevel programming problem is often strongly NP-hard, while the hierarchical optimization structure of a bilevel problem raises such difficulties as non-convexity and disconnectedness. The present paper is devoted to the swarm intelligence technique for capacity optimization of a transportation network. To this end, we develop the bilevel evolutionary algorithm based on swarm intelligence to cope with the continuous transportation network design problem. The findings of the paper give fresh insights to transport engineers and algorithm developers dealing with network design.

Keywords: Bilevel optimization · Network design problem · PSO algorithm

1 Introduction

Efficient traffic management in a modern urban network seems impossible today without the support of artificial intelligence systems that can offer topology optimization strategies based on observed traffic congestions. From mathematical

The work was supported by a grant from the Russian Science Foundation (No. 22-71-10063 Development of intelligent tools for optimization multimodal flow assignment systems in congested networks of heterogeneous products).

N. Olenev et al. (Eds.): OPTIMA 2023, CCIS 1913, pp. 202–213, 2024.
https://doi.org/10.1007/978-3-031-48751-4_15

perspectives, network topology optimization problems can be referred to as the Stackelberg game with independent lower-level decision-makers (drivers) acting in a non-cooperative manner to minimize their individual costs (journey times). In turn, upper-level decision-maker seeks to improve urban network topology by different investment strategies. Generally, the network design problem (NDP) is the bilevel program with the nonlinear upper-level objective function, linear and nonlinear constraints on upper-level variables, and nonlinear lower-level objective function with linear constraints on lower-level variables. Exact solution methods cannot cope with the real-life cases of this problem since even a simpler bi-level network design problem with both linear upper-level and lower-level problems is NP-hard [3,12]. Another issue is the non-convexity of bilevel network design problems, which impedes successful mathematical handling in order to develop efficient bilevel optimization approaches. Even when both the upper and lower level problems are convex, the convexity of the bi-level problem cannot be guaranteed [29].

Due to the intrinsic complexity of these problems, the diversity of solution algorithms is limited, and solution techniques for road network design problems can be categorized into exact methods, heuristic methods, and metaheuristics, which form the vast majority of state-of-the-art methods in the field. Indeed, exact methods such as branch and bound were applied only to small and medium sized networks, the largest being a 40-nodes and 99-links network [6,13,23,28]. Other studies transformed road network design problem into a single level problem and used exact nonlinear programming techniques to solve the resultant problem [17,18,53]. Note that approaches based on single-level reformulations are very popular for coping with bilevel problems [2,20,52]. Another possible way is applying exact me thods to road network design problems reformulated as a mathematical program with equilibrium constraints [26,27,46]. Thus, the bilevel problem can be transformed into a single-level optimization model or a mathematical program with equilibrium constraints, and then the resultant model should be solved [10,24,39]. Worth mentioning before applying heuristics to cope with a bilevel network design problem, a number of researchers used continuous optimization [1,32].

Heuristic approaches have begun developing to solve the bilevel network design problem directly [31,44,50]. Most of such approaches used descent search methods by exploiting the derivative information of the implicit response function obtained from the lower-level problem [7,8,54]. However, a few studies developed single-level models by using the system optimum principle for the lower-level response, which allowed solving the network design problem directly by using heuristic methods [9,43]. More recent studies on bilevel optimum search in network design problem available in the literature rely on metaheuristics or their hybrids [30,33,51]. In almost all of the up-to-date proposed metaheuristic algorithms for a bilevel network design problem, the lower-level problem is solved for each newly generated upper-level solution, while budget constraints are usually handled directly within the algorithmic steps [34,37,38]. Sometimes signal setting constraints are imposed in network design problems, which are often handled also by metaheuristics via getting the lower-level response for each network design scenario [4,16]. Actually, the very first approach to investigate bilevel optimization problems was based on quite a natural idea of replacing the lower-level problem with its optimality Karush-Kuhn-Tucker conditions. Since such a replacement leads to another one nonconvex problem, no wonder most solving algorithms are forced to compute local optimal solutions or stationary points. However, recently it has been shown that local optimal solutions of a mathematical program with complementarity conditions need not be related to locally optimal solutions of the corresponding bilevel optimization problem [11]. Therefore, new ways for the development of solution approaches for bilevel optimization problems are drastically required [19].

In general, the network design problem is a Stackelberg game with the upper-level decision-maker seeking to minimize overall congestion in the network by investing in its capacity, and the lower-level competitive followers (vehicles) influence the final traffic assignment. Assuredly, different traffic assignment models at the lower optimization level lead to significant discrepancies in optimal capacity patterns for the same road network. The major attention in the literature was paid to stochastic and dynamic traffic assignment models at the lower level [15]. However, there is still a gap in studying the network design problem with the traffic assignment modeled by a congestion game [35]. The present paper is devoted to such kind of network design problem. To this end, we give the corresponding formulation of the network design problem. Then we develop the bilevel evolutionary algorithm based on swarm intelligence to cope with capacity optimization of a transportation network. Finally, the computational study completes the paper.

2 The Continuous Network Design Problem

Let us consider an urban network presented by a directed graph $G = (V, E)$, where V represents a set of nodes (intersections), while E represents a set of available arcs between the adjacent intersections. By c_e we denote capacity of the arc $e \in E$, while c is an appropriate column vector of capacities, $c = \{c_e, e \in E\}$. Capacities c_e of all arcs $e \in E$ are assumed to be upper-level decision variables that vary from $l_e \geq 0$ to $u_e > 0$, i.e. $l_e \leq c_e \leq u_e$ for any $e \in E$. Let us introduce differentiable strictly increasing non-negative functions with positive first derivatives on the set of real non-negative numbers $g_e(\cdot)$, $e \in E$, which reflect the investment costs on arcs $e \in E$. Given budget C is sought to be invested in modification of the available arcs: $\sum_{e \in E} g_e(c_e) \leq C$, while l_e and u_e determine the lower and upper bounds for feasible capacity modification. Moreover, we also introduce weight constant $\beta \geq 0$ to set $1/\beta$ as a volume of budget units that may be invested in decreasing overall congestion by a time unit.

By x_e we denote traffic flow on the arc $e \in E$ which is assumed to be lower-level solution variable, while x is an appropriate column vector of all arc-flow variables, $x = \{x_e, e \in E\}$. For any $e \in E$ let us introduce non-negative function $t_e(x_e, c_e)$ which is differentiable strictly increasing (decreasing) function with respect to $x_e \geq 0$ ($x_e \leq 0$) and differentiable strictly decreasing (increasing) function with respect to $c_e \geq 0$ ($c_e \leq 0$). In fact, the function $t_e(x_e, c_e)$ is used to model travel time of a unit of flow on the arc $e \in E$, and it is commonly called arc *delay*, *cost* or *performance* function. Define $W \subseteq V \times V$ as the ordered set of pairs of nodes with non-zero travel demand $F^w > 0$ for any $w \in W$, $F = (F^1, \ldots, F^m)^T$, i.e., $|W| = m$. W is usually called as the set of origin-destination pairs (OD pairs). Demand $F^w > 0$ seeks to be assigned between the available routes R^w: $\sum_{r \in R^w} f_r^w = F^w$, where f_r^w is a variable corresponding to a traffic flow through route $r \in R^w$ between nodes of OD pair $w \in W$. Introduce the vector f associated with the route-flow traffic assignment pattern, such that $f = f_R = \{f_r^w, r \in R^w, w \in W\}$ is actually the column vector of $\{f_r^w\}_{r \in R^w}^{w \in W}$. We also assume that the travel time function of the route $r \in R^w$ between OD pair $w \in W$ is the sum of travel delays on all arcs belonging to this route. Thus, we define travel time through the route $r \in R^w$ between OD pair $w \in W$ as the following separable function

$$t_r^w(f, c) = \sum_{e \in E} t_e(x_e, c_e)\delta_{e,r}^w \quad \forall r \in R^w, w \in W,$$

where, by definition,

$$\delta_{e,r}^w = \begin{cases} 1, \text{ if arc } e \text{ belongs to the route } r \in R^w, \\ 0, \text{ otherwise.} \end{cases} \quad \forall e \in E, w \in W,$$

while, naturally,

$$x_e = \sum_{w \in W} \sum_{r \in R^w} f_r^w \delta_{e,r}^w \quad \forall e \in E,$$

i.e. traffic flow on the arc is the sum of traffic flows through all routes that include this arc.

In this paper, we study the network design problem, which has the following form of the bilevel optimization program [35]:

$$\min_c \sum_{e \in E} t_e(x_e, c_e)x_e + \beta \sum_{e \in E} g_e(c_e), \tag{1}$$

subject to

$$\sum_{e \in E} g_e(c_e) \leq C, \tag{2}$$

$$l_e \leq c_e \leq u_e \quad \forall e \in E, \tag{3}$$

and

$$x = \arg\min_{x \in X} \sum_{e \in E} \int_0^{x_e} t_e(u, c_e)du, \tag{4}$$

where X is given as follows:

$$\sum_{r \in R^w} f_r^w = F^w, \quad \forall w \in W, \tag{5}$$

$$f_r^w \geq 0 \quad \forall r \in R^w, w \in W, \tag{6}$$

while, by definition,

$$x_e = \sum_{w \in W} \sum_{r \in R^w} f_r^w \delta_{e,r}^w \quad \forall e \in E.$$

The network design problem in a form of the bilevel optimization program is motivated by several practical reasons. First of all, the noncooperative manner of drivers' route choosing in real road networks leads to the congestion game at the lower optimization level [22,36]. In fact, problem follows from the assumption every driver tends to minimize personal travel time from origin to destination. Secondly, traffic managers in actual urban areas cannot influence drivers' behavior directly in order to mitigate congestion effects. However, traffic managers can influence network topology via signal control, parking management, or capacity improvements [35]. No doubt, changes in network topology adduct adjustment in route choices, which is reflected in network travel delays. Therefore, traffic managers are assumed to manage network capacities in order to minimize overall travel delays subject to the available budget at the upper optimization level [15]. Let us also mention that the bilevel problem demonstrates a similar structure to well-known interdiction problems [42]. However, there is a significant distinction between these two kinds of tasks. Indeed, the leader in network design problem seeks to minimize the congestion effect by capacity improvement subject to an improvement cost budget. On the contrary, the leader in the interdiction problems, as a rule, maximizes the interdiction effect subject to an interdiction cost budget.

3 Evolutionary Optimization for Network Design

In this section, we provide the bilevel evolutionary algorithm based on swarm intelligence for solving problem (1)–(6). We should note that optimization approaches for the upper and lower levels are different. We find the optimal solution at the upper level using particle swarm optimization (PSO) with selected adaptations, while the Frank-Wolfe (FW) algorithm solves the lower-level problem [25]. In the PSO algorithm, each particle is supposed to be a random upper-level solution from the set of feasible solutions, and a velocity vector is used to move particles around the search space. The approach has three key features:

1. The velocity vector is updated at each iteration.
2. The objective function value is calculated for every particle.
3. The particle remembers its best solution and the best solution of the swarm.

The following equation implements the update of the velocity vector:

$$v_i^q(t+1) = \gamma(t) v_i^q(t) + c_1 r_1 \Big[b_i^q(t) - p_i^q(t) \Big] + c_2 r_2 \Big[b_i(t) - p_i(t) \Big] \quad \forall i \in E, q \in Q,$$

where Q is the set of particles; $p_i^q(t)$ is the i-th component of particle q at iteration t; $v_i^q(t)$ is the i-th component of velocity vector $v^q(t)$ of the particle q at iteration t, $v_{min} \leq v_i^q(t) \leq v_{max}$; $b_i^q(t)$ is the i-th component of the best solution of the particle q at the first t iterations; $b_i(t)$ is the i-th component of the best solution of the swarm (or particle neighborhood topology) at the first t iterations; $\gamma(t)$ is the dynamic coefficient of inertia; r_1 and r_2 are random numbers from the interval $(0, 1)$; c_1 and c_2 are weights. One can see that the equation above has three summands. These summands are called previous velocity, cognitive part, and social part. The cognitive part indicates that a particle learns from its own searching experience, while the social part indicates that a particle can learn from other particles. At every iteration, the position of a particle in the search space is updated according to the following equation:

$$p_i^q(t+1) = p_i^q(t) \ + \ v_i^q(t) \quad \forall i \in E, q \in Q.$$

For every new position, a particle calculates the objective function and checks whether the result is the best one compared to those previously obtained or not.

In this work, we implement PSO with dynamic phases (PSODF) to solve the upper-level problem: the search process is divided into the exploration phase and the exploitation phase. Within our implementation, we use the global star structure for effective exploration and the local ring structure to enhance exploitation. The pseudo-code is given below.

Algorithm 1 PSODP-FW algorithm for capacity optimization

```
initialize a random population of particles
compute the lower-level solutions for all particles
choose the best solution
while counter ¡ stopping criterion do
    for a particle from the swarm do
        if phase change condition met then
            compute a velocity vector for particles from the particle neighbourhood
        else compute a velocity vector for particles from the swarm
        end if
        update the particles
        compute the lower-level solutions
        update the best-found solutions to (1)–(6)
    end for
end while
```

In our implementation of the PSO algorithm, every particle in a swarm is a feasible capacity pattern. We generate an initial population of particles as a random set of upper-level feasible solutions. For each particle, we solve the lower-level problem using a FW algorithm to obtain an equilibrium traffic assignment pattern corresponding to this particle. According to Algorithm 1, at first iterations, we compute velocity vectors for particles from the whole swarm. Then, we compute velocity vectors for particles in a particle neighbourhood. We solve the lower-level problem every time we update particles to obtain corresponding equilibrium traffic assignment patterns. We run Algorithm 1 until the value of the upper-level goal function (1) stops decreasing.

4 Computational Study

Let us consider the results of applying Algorithm 1 to the Sioux Falls city road network. This network is popular among researchers; it consists of 24 nodes and 74 arcs (Fig. 1). The complete data on the Sioux Falls network is available on the web page: https://github.com/bstabler/TransportationNetworks.

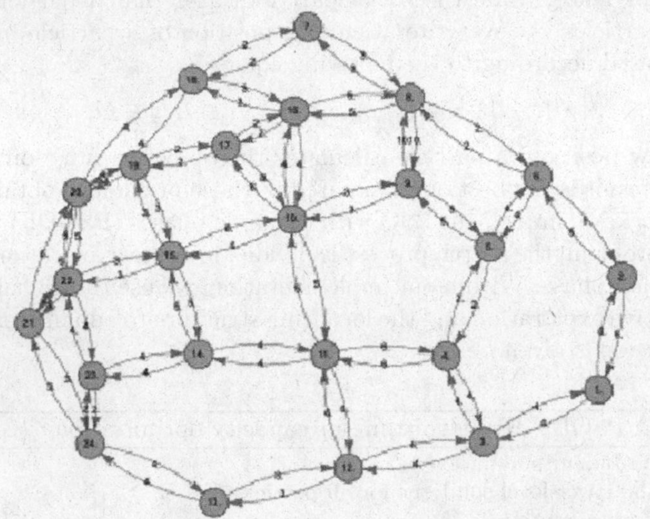

Fig. 1. Sioux Falls road network

In our research, we believe that the lower bound for feasible capacity modification of an arc is the capacity value from the web link above. The upper bound for feasible capacity modification of an arc is the lower bound plus 5000 multiplied by a random scalar from 0 to 1. We define investment cost functions as $g_e(c_e) = (c_e - l_e)^2$, $e \in E$, while $C = 3 \cdot 10^8$ and $\beta = 10^{-7}$. We developed Algorithm 1 in Wolfram Mathematica 13.1 and ran it on the 8-core GPU computer. We used the following adaptation features:

1. *Swarm size.* Algorithm 1 spent significant computational power to solve the lower-level problem for a particle, which made us reduce the size of the population as much as possible. Eventually, we dealt with 20 particles in order to keep the running time within reasonable limits and have a quite diverse population.
2. *Particles neighbourhood topology.* In the beginning, Algorithm 1 used a fully connected neighborhood structure, a global star structure, to move particles around the whole search space. However, after 65% of iterations, the algorithm began to use a local ring structure for a particle to reduce its search space and improve its sensitivity to a local optimum.

3. *Dynamic coefficient of inertia.* Algorithm 1 used the following dynamic coefficient of inertia:

$$\gamma(t) = e^{\frac{-10t}{T}}(\gamma_{max} - \gamma_{min}) + \gamma_{min},$$

where $\gamma_{min} = 0.3$, $\gamma_{max} = 1$, and $T = 130$ is the total number of iterations in the algorithm [5]. However, if the best value of the goal function did not change for 40 iterations, then during subsequent several iterations, the coefficient of inertia increased linearly, as did the random components r_1 and r_2. Such an adaptation strategy made it possible to move a particle from the local minimum to better positions in the search space.

In our implementation, we put $c_1 = c_2 = 0.2$ [14]. Figure 2 demonstrates the dynamic of the best-found values of the goal function with respect to all particles in the swarm (different colors are associated with different particles).

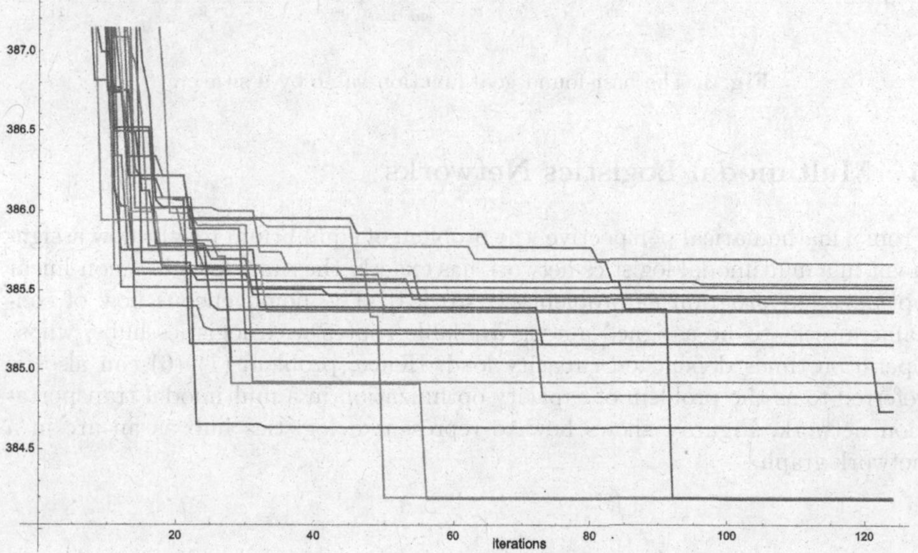

Fig. 2. The best-found goal function value with respect to every particle.

The sequential application of exploration and exploitation phases made it possible to find a more accurate solution. Figure 3 demonstrates the dynamic of the best-found by the swarm values of the goal function. One can see that Algorithm 1 decreased the goal function value from 393.7 to 384.1. Despite $T = 130$, the goal function value decreased for the last time at iteration 116, from 19×10^{-6} to 16.6×10^{-6}, and then the changes stopped.

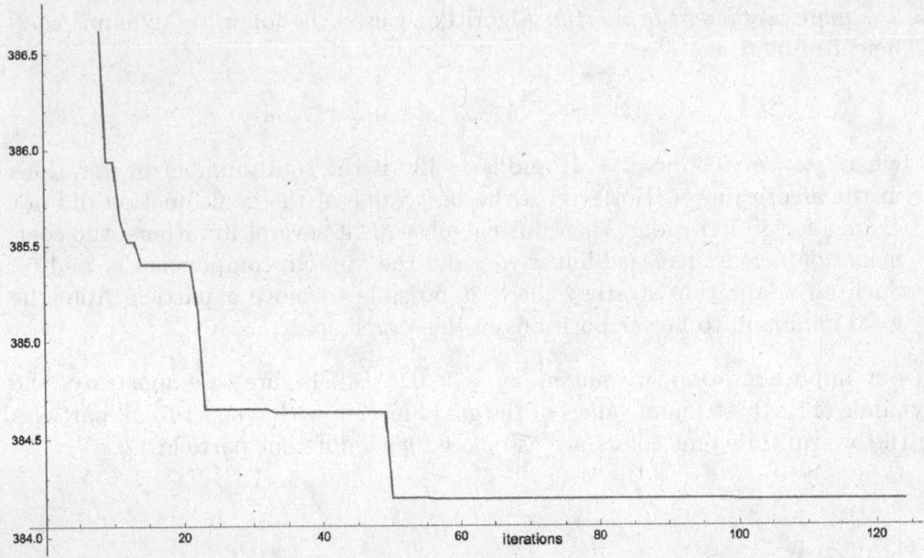

Fig. 3. The best-found goal function value by a swarm.

5 Multimodal Logistics Networks

From a mathematical perspective, the problem of equilibrium freight flow assignment in a multimodal logistics network has exactly the same form of a non-linear optimization program as problem (4)–(6) [21]. The homogeneous flow of containers seeks to be assigned among available routes with logistics hubs, whose operation times depend on capacity load. Hence, problem (1)–(6) can also be referred to as the problem of capacity optimization in a multimodal transportation network. Figure 4 shows how to represent a logistics hub as an arc in a network graph.

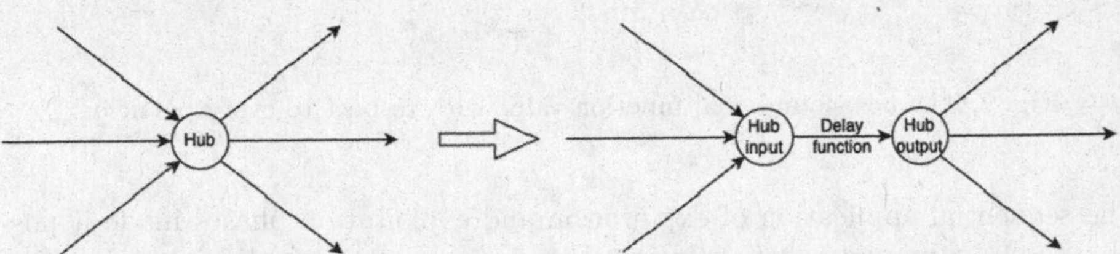

Fig. 4. Representation of a logistics hub as an arc in a network graph.

Once a multimodal logistics network is given in the form of a graph with logistics hubs represented, according to Fig. 4, the developed PSODP-FW algorithm can be directly applied to its capacity optimization. Therefore, the developed swarm intelligence technique can be considered a solver for a quite general capacity optimization problem for networks with homogeneous flows and auction conditions at the lower level.

6 Conclusion

The present paper was devoted to the swarm intelligence technique for capacity optimization of a transportation network. To this end, we developed the bilevel evolutionary algorithm based on swarm intelligence to cope with the continuous transportation network design problem. We showed that the developed technique can be considered a solver for a quite general capacity optimization problem for networks with homogeneous flows and auction conditions at the lower level. The findings of the paper give fresh insights to transport engineers and algorithm developers dealing with network design.

References

1. Abdulaal, M., LeBlanc, L.J.: Continuous equilibrium network design models. Transp. Res. Part B **13**(1), 19–32 (1979)
2. Audet, C., Hansen, P., Jaumard, B., Savard, G.: Links between linear bilevel and mixed 0–1 programming problems. J. Optim. Theory Appl. **93**, 273–300 (1997)
3. Ben-Ayed, O., Boyce, D.E., Blair, C.E., III.: A general bilevel linear programming formulation of the network design problem. Transp. Res. Part B **22**(4), 311–318 (1988)
4. Cantarella, G.E., Pavone, G., Vitetta, A.: Heuristics for urban road network design: lane layout and signal settings. Eur. J. Oper. Res. **175**(3), 1682–1695 (2006)
5. Chen, G., Min, Z., Jia, J., Xinbo, H.: Natural exponential inertia weight strategy in particle swarm optimization. In: Proceedings of the 6th World Congress on Intelligent Control and Automation, pp. 3672–3675 (2006)
6. Chen, M., Alfa, A.S.: A network design algorithm using a stochastic incremental traffic assignment approach. Transp. Sci. **25**(3), 215–224 (1991)
7. Chiou, S.: A hybrid approach for optimal design of signalized road network. Appl. Math. Model. **32**(2), 195–207 (2008)
8. Chiou, S.: Bilevel programming for the continuous transport network design problem. Transp. Res. Part B **39**(4), 361–383 (2005)
9. Dantzig, G.B., Harvey, R.P., Lansdowne, Z.F., Robinson, D.W., Maier, S.F.: Formulating and solving the network design problem by decomposition. Transp. Res. Part B **13**(1), 5–17 (1979)
10. Davis, G.A.: Exact local solution of the continuous network design problem via stochastic user equilibrium assignment. Transp. Res. Part B **28**(1), 61–75 (1994)
11. Dempe, S., Kalashnikov, V., Prez-Valds, G.A., Kalashnykova, N.: Bilevel Programming Problems: Theory. Algorithms and Applications to Energy Networks. Springer, Berlin (2015). https://doi.org/10.1007/978-3-662-45827-3
12. Deng, X.: Complexity issues in bilevel linear programming. In: Migdalas, A., Pardalos, P.M., Värbrand, P. (eds) Multilevel Optimization: Algorithms and Applications. Nonconvex Optimization and Its Applications, vol 20. Springer, Boston, MA (1998). https://doi.org/10.1007/978-1-4613-0307-7_6

13. Drezner, Z., Wesolowsky, G.O.: Selecting an optimum configuration of one-way and two-way routes. Transp. Sci. **31**(4), 386–394 (1997)
14. Engelbrecht, A.P.: Computational Intelligence: An Introduction, Second Edition. John Wiley & Sons Ltd (2007)
15. Farahani, R.Z., Miandoabchi, E., Szeto, W.Y., Rashidi, H.: A review of urban transportation network design problems. Eur. J. Oper. Res. **229**, 281–302 (2013)
16. Gallo, M., D'Acierno, L., Montella, B.: A meta-heuristic approach for solving the urban network design problem. Eur. J. Oper. Res. **201**(1), 144–157 (2010)
17. Gao, Z., Sun, H., Zhang, H.: A globally convergent algorithm for transportation continuous network design problem. Optim. Eng. **8**(3), 241–257 (2007)
18. Gao, Z., Wu, J., Sun, H.: Solution algorithm for the bi-level discrete network design problem. Transp. Res. Part B **39**(6), 479–495 (2005)
19. Kleinert, T., Labbé, M., Ljubić, I., Schmidt, M.: A survey on mixed-integer programming techniques in bilevel optimization. EURO J. Comput. Optim. **9**, 100007 (2021)
20. Kleinert, T., Labbé, M., Plein, F., Martin, S.: Closing the gap in linear bilevel optimization: a new valid primal-dual inequality. Optim. Lett. **15**, 1027–1040 (2021)
21. Krylatov, A.Y., Raevskaya, A.P.: Freight flow assignment in the intermodal logistics network. Transp. Res. Procedia **68**, 492–498 (2023)
22. Krylatov, A., Zakharov, V., Tuovinen, T.: Principles of Wardrop for traffic assignment in a road network. Springer Tracts on Transp. Traffic **15**, 17–43 (2020)
23. LeBlanc, L.J.: An algorithm for the discrete network design problem. Transp. Sci. **9**(3), 183–199 (1975)
24. LeBlanc, L.J., Boyce, D.E.: A bilevel programming algorithm for exact solution of the network design problem with user-optimal flows. Transp. Res. Part B **20**(3), 259–265 (1986)
25. LeBlanc, L.J., Morlok, E.K., Pierskalla, W.P.: An efficient approach to solving the road network equilibrium traffic assignment problem. Transp. Res. **9**(5), 309–318 (1975)
26. Lo, H.K., Szeto, W.Y.: Chapter 9: planning transport network improvements over time. In: Boyce, D., Lee, D.H. (Eds.), Urban and Regional Transportation Modeling: Essays in Honor of David Boyce. E. Elgar, Cop.: Cheltenham (G.B.), Northampton (Mass.), pp. 157–176 (2004)
27. Lo, H.K., Szeto, W.Y.: Time-dependent transport network design under cost-recovery. Transp. Res. Part B **43**(1), 142–158 (2009)
28. Long, J., Gao, Z., Zhang, H., Szeto, W.Y.: A turning restriction design problem in urban road networks. Eur. J. Oper. Res. **206**(3), 569–578 (2010)
29. Luo, Z., Pang, J., Ralph, D.C.: Mathematical Programs with Equilibrium Constraints. Cambridge University Press, Cambridge (1996)
30. Mathew, T.V., Sharma, S.: Capacity expansion problem for large urban transportation networks. J. Transp. Eng. **135**(7), 406–415 (2009)
31. Marcotte, P., Marquis, G.: Efficient implementation of heuristics for the continuous network design problem. Ann. Oper. Res. **34**(1), 163–176 (1992)
32. Marcotte, P.: Network optimization with continuous control parameters. Transp. Sci. **17**(2), 181–197 (1983)
33. Meng, Q., Yang, H.: Benefit distribution and equity in road network design. Transp. Res. Part B **36**(1), 19–35 (2002)
34. Miandoabchi, E., Farahani, R.Z.: Optimizing reserve capacity of urban road networks in a discrete network design problem. Adv. Eng. Softw. **42**(12), 1041–1050 (2010)

35. Migdalas, A.: Bilevel programming in traffic planning: models, methods and challenge. J. Glob. Optim. **7**, 381–405 (1995)
36. Patriksson, M.: The traffic assignment problem: models and methods. VSP, Utrecht, The Netherlands (1994)
37. Poorzahedy, H., Abulghasemi, F.: Application of ant system to network design problem. Transportation **32**(3), 251–273 (2005)
38. Poorzahedy, H., Rouhani, O.M.: Hybrid meta-heuristic algorithms for solving network design problem. Eur. J. Oper. Res. **182**(2), 578–596 (2007)
39. Poorzahedy, H., Turnquist, M.A.: Approximate algorithms for the discrete network design problem. Transp. Res. Part B **16**(1), 45–55 (1982)
40. Sinha, A., Malo, P., Deb, K.: Efficient evolutionary algorithm for single-objective bilevel optimization. ArXiv abs/1303.3901 (2013)
41. Sheffi, Y.: Urban Transportation Networks: Equilibrium Analysis with Mathematical Programming Methods. Prentice-Hall Inc, Englewood Cliffs, N.J. (1985)
42. Smith, J.C., Song, Y.: A survey of network interdiction models and algorithms. Eur. J. Oper. Res. **283**(3), 797–811 (2020)
43. Steenbrink, P.A.: Optimization of Transport Network. Wiley, New York, USA (1974)
44. Suh, S., Kim, T.: Solving nonlinear bilevel programming models of the equilibrium network design problem: a comparative review. Ann. Oper. Res. **34**(1), 203–218 (1992)
45. Szeto, W.Y., Jaber, X., O'Mahony, M.: Time-dependent discrete network design frameworks considering land use. Comput.-Aid. Civil Infrastruct. Eng. **25**(6), 411–426 (2010)
46. Szeto, W.Y., Lo, H.K.: Strategies for road network design over time: robustness under uncertainty. Transportmetrica **1**(1), 47–63 (2005)
47. Szeto, W.Y., Lo, H.K.: Time-dependent transport network improvement and tolling strategies. Transp. Res. Part A **42**(2), 376–391 (2008)
48. Szeto, W.Y., Lo, H.K.: Transportation network improvement and tolling strategies: the issue of intergeneration equity. Transp. Res. Part A **40**(3), 227–243 (2006)
49. Wardrop, J.G.: Some theoretical aspects of road traffic research. Proc. Institut. Civil Eng. **2**, 325–378 (1952)
50. Yang, H.: Sensitivity analysis for the elastic-demand network equilibrium problem with applications. Transp. Res. Part B **31**(1), 55–70 (1997)
51. Yang, H., Wang, J.Y.T.: Travel time minimization versus reserve capacity maximization in the network design problem. Transp. Res. Rec. **1783**, 17–26 (2002)
52. Zare, M.H., Borrero, J.S., Zeng, B., Prokopyev, O.A.: A note on linearized reformulations for a class of bilevel linear integer problems. Ann. Oper. Res. **272**, 99–117 (2019)
53. Zhang, H., Gao, Z.: Bilevel programming model and solution method for mixed transportation network design problem. J. Syst. Sci. Complexity **22**, 446–459 (2009)
54. Ziyou, G., Yifan, S.: A reserve capacity model of optimal signal control with user-equilibrium route choice. Transp. Res. Part B **36**(4), 313–323 (2002)

Improving Background Subtraction Algorithms with Shadow Detection

Oleg Makarov(✉) and Elena Shchennikova

Ogarev Mordovia State University, Saransk, Russia
makaroff297991@gmail.com

Abstract. Motion detection of objects on video is a resource–intensive area of computer vision that requires several stages. The first stage of this task is the background subtraction process. The quality of further recognition fully depends on this stage. Natural or artificial illumination changes could provide a lot of shadows on the video. Such shadows could be falsely interpreted as parts of a moving object or even as the separate objects on the scene. This article proposes an improvement to the background subtraction algorithms by detecting and removing shadows from the images. The proposed approach relies onto spatial correlations between neighboring pixels in combination with the normal distribution of noise. It provides good background subtraction results along with low computational cost.

Keywords: Motion Detection · Background Subtraction · Shadow Detection · Optimization · Foreground Object Mask · Background Model · Normal Distribution · Illumination Changes

1 Introduction

In the field of computer vision, motion detection refers to the process of identifying and tracking movements within a video or image sequence. It involves analyzing consecutive frames to detect changes in pixel values or object positions over time. Motion detection in videos is a challenging task in computer vision, and there is no universal solution that addresses all scenarios. Existing methods have limitations and are not always effective. While some methods work well in specific scenarios, they may not be relevant in others.

Background subtraction is a commonly used real-time method for identifying moving objects in video streams. "Background" refers to the stationary or static elements in a scene or image. It represents the environment or the objects that are not in motion. For example, in a surveillance video, the background would include the walls, furniture, or any other objects that remain still. On the other hand, "subtraction" refers to the process of subtracting the background from the current frame or image. This technique is used to isolate the moving objects or changes in the scene by removing the stationary background. By subtracting the background, we can focus on the elements that are in motion and analyze them

N. Olenev et al. (Eds.): OPTIMA 2023, CCIS 1913, pp. 214–227, 2024.
https://doi.org/10.1007/978-3-031-48751-4_16

separately. Moving objects are also called foreground objects. The background subtraction algorithm takes a sequence of video frames as input and generates binary images, known as foreground object masks, which represent the extracted background. Figure 1 illustrates the background subtraction process.

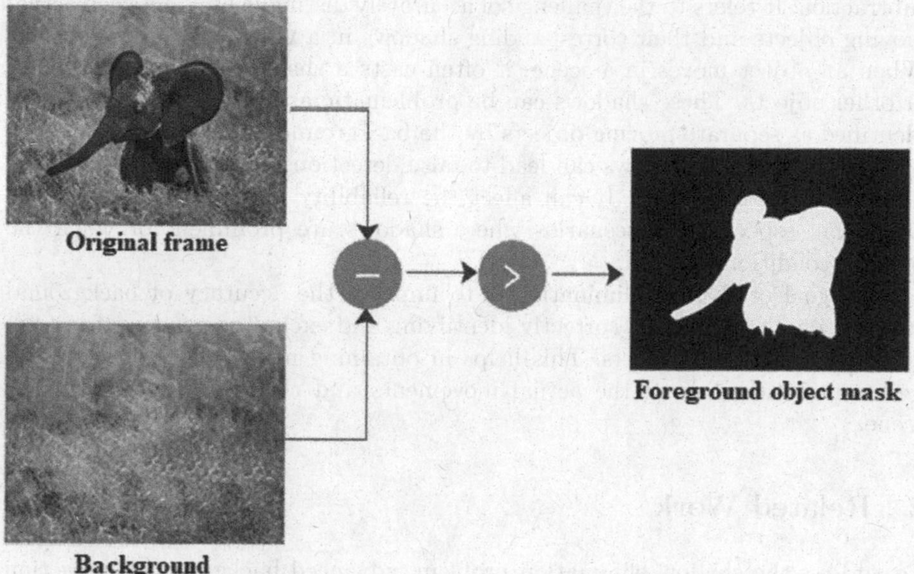

Fig. 1. Background subtraction process

The background subtraction algorithm compares the current frame to the previously generated background model, which does not include foreground objects. During this process, each input pixel is classified as either foreground or background for every video frame. Foreground objects are typically displayed in white while the background is black, allowing the background model to be "subtracted" from the current frame.

However, a technical challenge arises in generating and rebuilding an accurate background model using all available data from previous frames. It can be challenging to determine rules for classifying foreground objects and background areas. In practice, the application layer is usually responsible for classifying foreground objects.

Background subtraction methods are crucial in detecting motion in video sequences, and the quality of the resulting mask from this step is critical for subsequent object recognition by software layers. However, these algorithms can sometimes produce a high number of errors that cannot be corrected in post-processing. An ideal background subtraction approach should be able to adapt to various changes in illumination, camera oscillations, high-frequency background objects and slow changes in background geometry. In some cases, these algorithms need to be integrated into the camera, which can lead to limitations in processing power and RAM. This poses a significant challenge, especially for real-time applications, where the techniques need to be fast and provide accurate results. Unfortunately, updating the background model regularly often consumes a lot of computational resources.

The shadow elimination problem is one of the biggest problems in background subtraction. It refers to the challenge of accurately distinguishing between actual moving objects and their corresponding shadows in a video or image sequence. When an object moves in a scene, it often casts a shadow on the background or other objects. These shadows can be problematic as they can be mistakenly identified as separate moving objects by the background subtraction algorithm.

The presence of shadows can lead to false detections and inaccuracies in the motion detection process. It can affect the reliability and performance of the algorithm, especially in scenarios where shadows are prominent or when the lighting conditions change.

The goal of shadow elimination is to improve the accuracy of background subtraction algorithms by correctly identifying and excluding shadows from the detected foreground objects. This helps in obtaining more reliable results and better understanding of the actual movements and changes occurring in the scene.

2 Related Work

To address the shadow elimination problem, advanced background subtraction algorithms employ various techniques. These techniques may involve modeling and analyzing the color, intensity, or texture differences between the foreground objects and their shadows. Additionally, algorithms may utilize additional information such as motion history or temporal consistency to differentiate between true object movements and shadows.

Chalidabhongse et al. [4] relied on the color constancy of human vision and considered each pixel as a combination of two components: brightness and chromaticity. They examined these components for distortion based on the actual and expected pixel value.

In [2], the authors researched the size and direction of object movement between frames by using normalized cross-correlation to obtain contextual information about moving objects. This allowed them to obtain ground truth results when testing the movement of pedestrians and cars on video.

Other researchers have improved upon the mixed background model proposed by Grimson [5]. They developed a new shadow detection scheme based on the computational color space, which improved object segmentation in the video compared to the original version.

Arun Varghese and Sreelekha G proposed a two-stage shadow detector integrated into the background extraction algorithm [9]. The proposed method significantly reduced the post-processing resources required for the output data.

Bunyak and colleagues [3] considered foreground regions in terms of intensity, chrominance and reflectance, which were extended to three color components for local pixel classification. They applied a compact constraint to the foreground and shadow masks to refine the results. Their algorithm utilized spatial and spectral information. The proposed method used two photometric invariants to ensure reliable detection of the shadow mask. It did not require any prior information about the camera, scene geometry or object model.

Other authors proposed three methods to enhance background extraction algorithms for color images: the utilization of the YCoCg color space, a spherical association volume and a cast shadows management approach [8]. These improvements were based on spectral properties. With this approach, pixels affected by casted shadows were disregarded during the process of updating the background model. The proposed algorithm created new background code words for such pixels to avoid classifying them as background pixels.

Wang et al. also studied colored images [10]. They had two main ideas: when the current pixel point had a shadow, the value of the RGB components of its pixels was smaller than without a shadow; when the current pixel point had a shadow, the reduction of pixel value on its RGB components was the same. Similar to [8], the authors focused on the blue component of the RGB model. They assumed that this component changed more than others when a shadow was applied to the background model.

Agrawal and colleagues [1] proposed a new method for background subtraction using the integrated ABGS segmenter. This segmenter utilizes a two-level adaptive threshold approach to generate a binary mask at the pixel level. The proposed background subtraction model is combined with a chroma-based shadow removal method, which effectively detects and removes shadow pixels without distorting the object's shape. Moreover, this approach can automatically remove multiple shadows present in the current frame.

In their study, Wang et al. [11] focused on grayscale images and explored graphical probabilistic models such as the Bayesian rule and the maximum likelihood method. The authors identified three sources of contextual information that are beneficial for shadow detection. The first source is boundary information, which helps distinguish between different frames. The second source is spatial information, as objects and shadows tend to occupy contiguous regions. The third source is temporal information, where models are updated based on previous segmentation results.

Similarly, other researchers have proposed a background subtraction method integrated with a shadow detection algorithm [12]. This algorithm relies on the position of shadows and their edge attributes. Initially, the shadow is transformed into the HSV color space. Then, the algorithm determines the direction of shadow movement based on edge information and its positions according to horizontal and vertical projections. Subsequently, the position of the shadow is further refined using the aspect ratio method and the shadow is ultimately removed from the background model. Experimental results have demonstrated that this method yields satisfactory outcomes, particularly when the color invariance principle is ineffective.

3 The Proposed Solution

In this paper, the Gaussian Mixture Model (GMM) algorithm serves as the foundation for our proposed method. The GMM algorithm is a popular choice for background subtraction due to its widespread usage and familiarity. However, it is important to note that our proposed improvement in shadow detection is not limited to the GMM algorithm. It can be applied to other algorithms as well, as it does not depend on the specific implementation details of background models.

The proposed algorithm consists of several stages, each contributing to its overall functionality. To provide a better understanding of our technique, we present a short overview described in Algorithm 1:

Input : Image frame, computed background, foreground object mask
Output: Foreground object mask with detected shadows
Stage 1: Preprocessing;
Apply morphological operations to improve foreground object mask;
Stage 2: Shadow Detection;
foreach *foreground pixel in foreground object mask* **do**
  if *pixel = background pixel + normal noise* ***and*** *some neighboring pixels are classified as shadowed* **then**
    Classify this pixel as shadowed pixel
  else
    Classify this pixel as foreground pixel
  end
end
Stage 3: Post-processing;
Apply morphological operators to clean output foreground object mask

Algorithm 1: The proposed shadow detection algorithm

Then consider a more detailed and comprehensive description of the algorithm. The first stage involves preprocessing the foreground object masks to improve the shapes of foreground objects. Some types of illumination changes and weather conditions can introduce "grainy" noise on video frames after background subtraction, particularly in outdoor scenes. For example, leaves of trees sway in windy weather, and the water surface glistens in the sun. These effects often result in a large number of point-based changes in the background model, which are undesirable. To suppress these effects, morphological operations [7] can be used. There are currently many approaches for solving such problems. To achieve an optimal balance between performance and results, it is recommended to use a complex morphological opening operation, which is defined as the dilation of the erosion. This operation allows us to filter out small areas that are falsely classified as foreground objects and then restore the connectivity of object boundaries with minimal computational resources. Furthermore, morphological processing operators can help improve the contrast of darkened images and expand the range of pixel intensities used in the frame.

Additionally, at this stage, it is advisable to utilize the information that shadow intensity is always less than the background intensity. In other words, shadows cannot be lighter than the surface on which they are cast. Therefore, after applying morphological operators, it is necessary to ignore pixels that are lighter than the background model.

At the second stage, we assume that the shadow can be represented as a linear combination of the background pixel intensity and a normally distributed noise value. We also expect this noise value to be constant in a small neighborhood of the frame. In other words, the intensity value of the current pixel can be defined as:

$$I_s(x) = aI(x) + b(x), \tag{1}$$

Here, $0 < a < 1$, $b \sim \mathcal{N}(0, \sigma_b^2(x))$. The coefficient a is the normalizing constant for the considered area, $I(x)$ is the pixel intensity value in the background model, and b represents the normally distributed noise value. It is important to note that the shadow is always darker than the background, so $I_s(x) \leq I(x)$. The coefficient a reflects the influence of the light source on the background model.

Based on formula (1), we can determine the following normally distributed random variable:

$$G(x) = \frac{I_s(x)}{I(x)} = a + \frac{b(x)}{I(x)} \sim \mathcal{N}(a, \left(\frac{\sigma_b(x)}{I(x)}\right)^2). \tag{2}$$

Let $S(x)$ be a small neighborhood around the pixel x, and let n be the number of pixels lying inside this neighborhood. We can empirically determine the mean and standard deviation of this random variable using a sample of pixels in the neighborhood $S(x)$:

$$\mu_G(x) = \frac{1}{n} \sum_{y \in S(x)} G(y), \tag{3}$$

$$\sigma_G(x) = \sqrt{\frac{1}{n} \sum_{y \in S(x)} (G(y) - \mu_G(x))^2}. \tag{4}$$

Let also introduce another difference random variable:

$$\Delta(x) = G(x) - \mu_G(x) \sim \mathcal{N}(0, \sigma_\Delta(x)). \tag{5}$$

where

$$\sigma_\Delta(x) = \frac{1}{n^2} \left((n-1)^2 \sigma_G^2(x) + \sum_{\substack{y \in S(x), \\ y \neq x}} \sigma_G^2(y) \right). \tag{6}$$

Now, we can evaluate the function $\Delta(x)$ and determine how close the pixels are to their expected values within the considered neighborhood. For each foreground pixel x, we need to estimate the probability of the value $\Delta(x)$ falling within the interval determined by the pre-selected confidence level β:

$$P\{|\Delta(x)| < l(\beta)\} > \beta. \tag{7}$$

In the simplest case, we can use the three-sigma rule of thumb. Let $\beta = 0.9973$ and $l(\beta) = 3\sigma_{\Delta}(x)$.

If the pixel truly belongs to the foreground object, then in most cases condition (7) will not be met because there will be no correlation between the background model and the pixel being considered. If this condition is met, then it is likely that the pixel belongs to the shadow of a foreground object, but further checks are also required. It is better to check this condition for neighboring pixels since shadows are usually displayed in connected regions on the frame. In the simplest case, we can take a circle with a radius of r centered on the pixel being considered. Note that in general, this area does not coincide with the neighborhood used in formulas (3–6). Then we can define the following criterion: if condition (7) is met for at least half of the pixels inside the circle, the proposed algorithm should classify the pixel being considered as belonging to the shadow. We recommend using radius values of 2, 3, or 4. It is not recommended to use large values because it increases CPU usage and also affects the results for edge pixels, as the algorithm can capture an area outside the shadow. Additionally, if the radius value is greater than the width of the shadow, it will negatively impact the results. This is especially true for low-resolution videos.

In the third stage, the proposed algorithm performs post-processing of the detected shadows by applying morphological operators similar to the first stage in order to combine unrelated shadow regions and exclude falsely detected shadows.

4 Results

The proposed approach was tested using datasets from the publicly available CDNET database [13]. This resource provides a set of unique input frames from various video recordings, along with the processed ground truth recognition results, which can be used to test motion detection algorithms. The proposed algorithm was tested on six datasets from the "Shadow" section in the "dataset 2014" video collection: "Backdoor", "Bungalows", "Bus stop", "Cubicle", "Copy machine" and "People in shade". Due to the low resolution of most videos in this set (360 × 240 pixels), we experimentally determined the following optimal parameters for the algorithm: $a = 0.3$, S is an area of 3×3, $n = 9$, $\beta = 0.9973$, $r = 2$.

Figures 2, 3, 4, 5, 6, 7 show the results of the shadow detection algorithm. Foreground objects are represented in white, while the background model is represented in black. Shadows detected by the algorithm are shown in gray.

To test the algorithm, we will be using the following metrics: true positives (TP), which refers to the number of correctly detected foreground pixels; false positives (FP), which refers to the number of background pixels incorrectly classified as foreground; true negatives (TN), which refers to the number of correctly

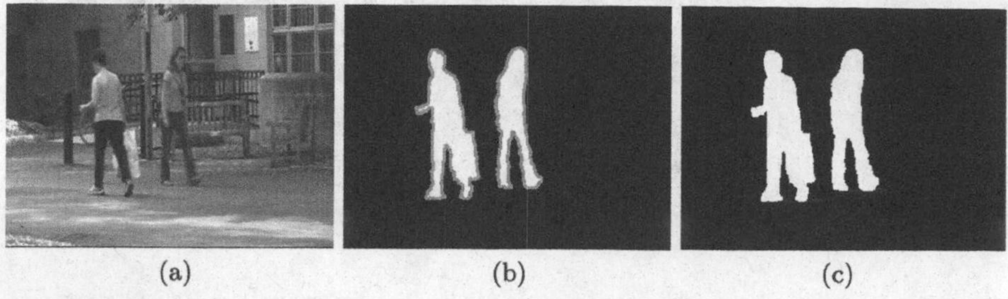

(a) (b) (c)

Fig. 2. Example of shadow detection on the "Backdoor" dataset: the input frame (a), the ground truth mask (b), the result of background subtraction (c).

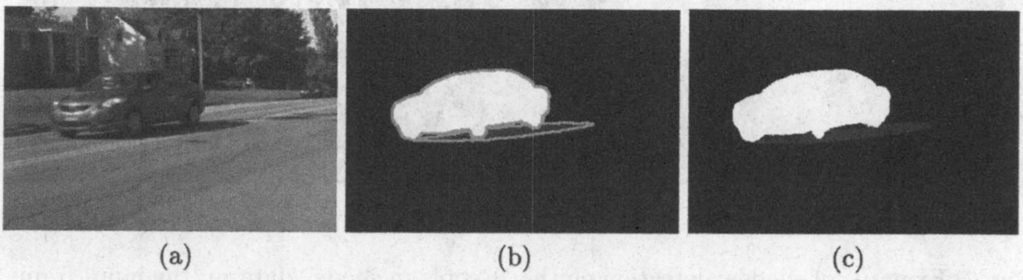

(a) (b) (c)

Fig. 3. Example of shadow detection on the "Bungalows" dataset: the input frame (a), the ground truth mask (b), the result of background subtraction (c).

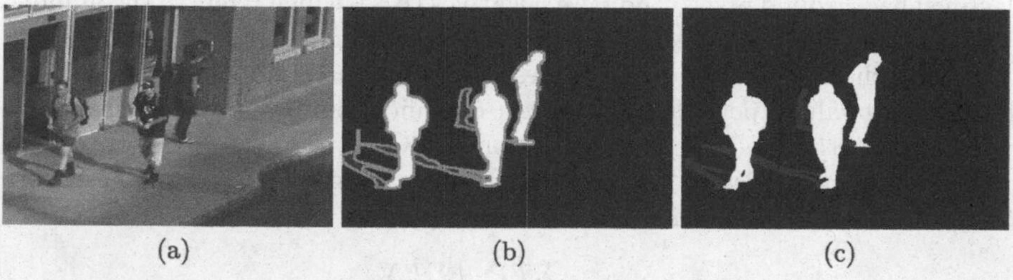

(a) (b) (c)

Fig. 4. Example of shadow detection on the "Bus stop" dataset: the input frame (a), the ground truth mask (b), the result of background subtraction (c).

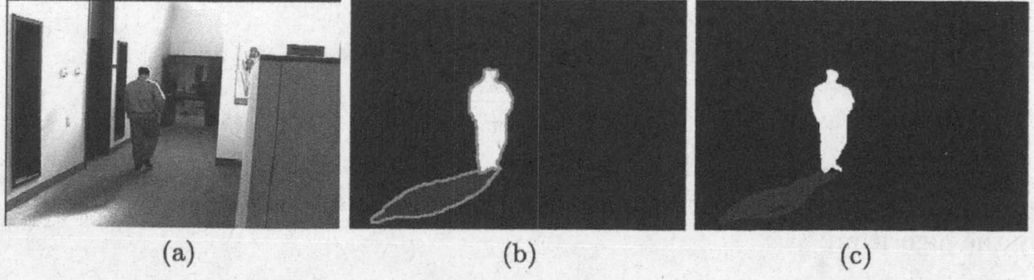

(a) (b) (c)

Fig. 5. Example of shadow detection on the "Cubicle" dataset: the input frame (a), the ground truth mask (b), the result of background subtraction (c).

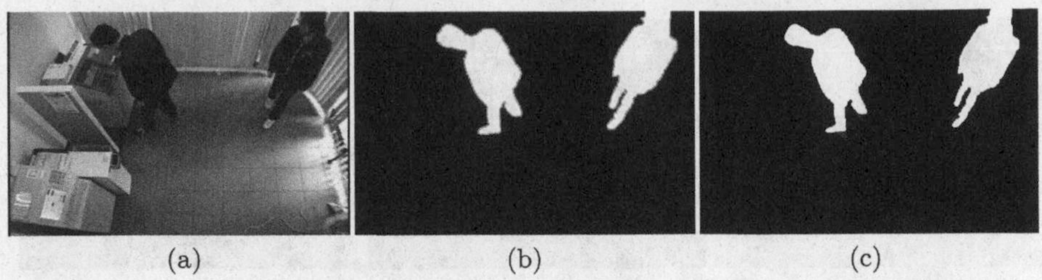

(a) (b) (c)

Fig. 6. Example of shadow detection on the "Copy machine" dataset: the input frame (a), the ground truth mask (b), the result of background subtraction (c).

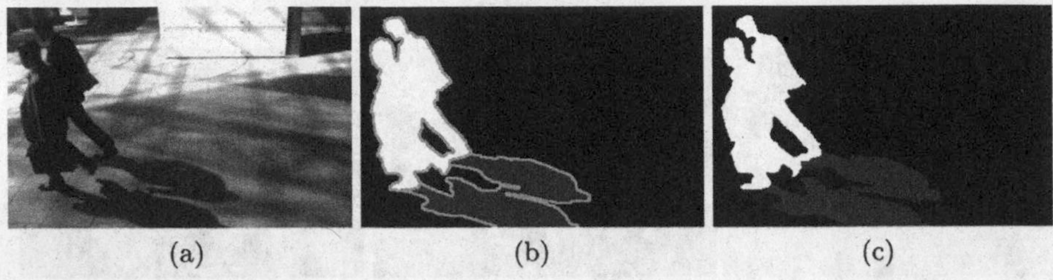

(a) (b) (c)

Fig. 7. Example of shadow detection on the "People in shade" dataset: the input frame (a), the ground truth mask (b), the result of background subtraction (c).

classified background pixels; and false negatives (FN), which refers to the number of foreground pixels incorrectly classified as background.

We will then evaluate the effectiveness of the proposed method by calculating the following values: percentage of correct classification, recall and precision [6].

The percentage of correct classification (PCC) is the most commonly used metric in computer vision. It is calculated by dividing the total number of correctly classified pixels by the total number of pixels:

$$PCC = \frac{\Sigma TP + \Sigma TN}{\Sigma TP + \Sigma FP + \Sigma TN + \Sigma FN}. \tag{8}$$

Recall (Rcl) is calculated by dividing the total number of true positive pixels detected by the algorithm by the total number of true positive pixels in the ground truth frame:

$$Rcl = \frac{\Sigma TP}{\Sigma TP + \Sigma FN}. \tag{9}$$

Precision (Prc) is calculated by dividing the total number of true positive pixels detected by the algorithm by the total number of positive pixels detected by the algorithm:

$$Prc = \frac{\Sigma TP}{\Sigma TP + \Sigma FP}. \tag{10}$$

Table 1 displays the results of testing the proposed algorithm on six video sequences from the CDNET database.

Table 1. Performed metrics of the proposed algorithm

Dataset	Metrics*						
	TP	TN	FP	FN	Rcl	Prc	PCC
Backdoor (320 × 240)	8683	60197	4886	3034	74.1%	64.0%	89.7%
Bungalows (360 × 240)	35856	49094	929	521	98.6%	97.5%	98.3%
Bus stop (360 × 240)	21828	58788	3345	2439	89.9%	86.7%	93.3%
Cubicle (352 × 240)	17497	64735	1413	835	95.4%	92.5%	97.3%
Copy machine (720 × 480)	58739	274343	7437	5081	92.0%	88.8%	96.4%
People in shade (380 × 244)	8691	79877	2469	1683	83.8%	77.9%	95.5%
Total:							**95.1%**

* TP - true positive pixels, TN - true negative pixels, FP - false positive pixels, FN - false negative pixels, Rcl - Recall metric, Prc - Precision metric, PCC - percentage of correct classifications

Additionally, the proposed method was tested on the same datasets as other shadow detection methods described in [6]. These include the statistical parametric approach (SP), the statistical nonparametric approach (SNP), and two deterministic non-model algorithms (DNM1 and DNM2). The results of the comparison are shown in Table 2. The results of the usual GMM algorithm are also included in Table 2 for comparison.

Table 2. Comparison of motion detection algorithm results based on the PCC metric

Methods	Datasets						Total
	Backdoor	Bungalows	Bus stop	Cubicle	Copy machine	People in shade	
GMM	64.4%	86.2%	83.3%	87.8%	76.9%	70.3%	78.2%
SP	78.3%	90.9%	87.6%	92.3%	89.4%	90.7%	88.2%
SNP	85.6%	94.5%	**93.4%**	90.1%	94.1%	91.3%	91.5%
DNM1	86.6%	89.3%	90.4%	90.8%	92.4%	93.1%	90.4%
DNM2	**90.1%**	93.5%	91.2%	**98.1%**	90.2%	91.9%	92.5%
Proposed	89.7%	**98.3%**	93.3%	97.3%	**96.4%**	**95.5%**	**95.1%**

5 Discussion

Based on the results in Tables 1 and 2, the proposed algorithm successfully detects motion in videos.

The proposed method exhibited the best results when processing videos from the following sections: "Bungalows", "Cubicle" and "Copy machine". There are several reasons for this. Firstly, these videos have good lighting conditions with no rapid illumination changes. The scenes are consistently illuminated, resulting in more uniform shadows created by the light sources and clearer boundaries. Secondly, these videos have monochrome background surfaces, such as the road in the "Bungalows" video or the floor and walls in the "Cubicle" and "Copy machine" videos. When homogeneous shadows fall on these homogeneous backgrounds, they uniformly change neighboring groups of pixels, which is ideal for the proposed algorithm. Thirdly, these videos do not contain any other negative factors that could affect object recognition, such as wind effects or camera oscillations.

It should be noted that the overall accuracy of processing the "Copy machine" dataset is slightly lower than the other datasets. This is because this video sequence has the highest resolution (720 × 480). To achieve better performance, it is recommended to adjust some parameters of the algorithm, such as setting $S = 5x5$ pixel area, $n = 25$, and $r = 3$. However, for the sake of experimental objectivity, the algorithm parameters were not changed during testing for all video recordings.

The algorithm yielded the worst results for the data sets "Backdoor", "Bus stop" and "People in shade". The processing of the first video, "Backdoor", showed poor results due to the presence of wind in the scene, causing the leaves on the trees to sway. The algorithm mistakenly identified some of these small movements as foreground objects. On the other hand, it should be noted that the proposed algorithm produced acceptable results in other similar cases by suppressing small background fluctuations, although it was not specifically designed to handle wind in video recordings. This is because the algorithm attempts to establish probabilistic relationships between changes in neighboring groups of pixels but cannot accurately determine it and eliminates pixels that were mistakenly classified by the original GMM algorithm as foreground objects.

When analyzing the results of testing the "Bus Stop" dataset, we discovered multiple overlapped shadows, such as the shadow of a person within the shadow of a building or another object. This reduced the overall accuracy of object recognition in some frames. The video "People in shade" contained many narrow penumbra zones created by the ceiling of the hall. The cloudy weather resulted in periodic uneven changes in illumination in the background model. For instance, light pixels quickly increased their intensity when the sun emerged from behind clouds, while dark shadows from opaque elements of the ceiling remained unchanged. As a consequence, the shadows of foreground objects fell on the unevenly distributed background (the floor of the hall). Consequently, the algorithm struggled with certain areas of penumbra. It is worth mentioning that reducing the update time of the background model could provide acceptable results in such situations.

A comparison of recall and precision metrics reveals that the algorithm tends to mistakenly classify detected foreground pixels as background more frequently than identifying background pixels as parts of foreground objects.

In comparison to other common methods for detecting shadows in video sequences, the proposed algorithm effectively classifies foreground objects and their shadows. On average, the proposed method outperforms other approaches across all sets. However, when considering certain test sets individually, the proposed algorithm performs slightly worse in some scenarios. This is due to the specific characteristics of the test sets and the fact that the algorithm parameters remained unchanged for all six video sequences during testing.

One of the main performance criteria for background subtraction algorithms is the average speed of sequential processing of video frames during recognition. Our testing was conducted on an Intel Core2 Duo E7300 processor (2.66 GHz) with 6 GB of RAM. During testing, each algorithm produced no more than 10–15 frames per second for low-resolution frames, and at least 5–7 frames per second for the "Copy machine" dataset. Other algorithms provided similar performance results. Therefore, we can conclude that the proposed algorithm, with its average set of parameters, has a good processing speed comparable to other approaches, with minimal consumption of computing resources. However, if we increase certain parameters for processing high-resolution videos (such as the area S or the radius r of the neighboring circle of pixels), the processing time will also increase proportionally. This trend is true for other parametric background subtraction algorithms as well. However, with the rapid growth of the computing electronics market, the performance indicator of algorithms is becoming less important due to cheaper microprocessor manufacturing technology. The fps metric only plays a significant role in real-time applications where the speed of video processing needs to be maximized.

The main advantage of the proposed algorithm is its high precision in detecting motion. When compared to other commonly used approaches, this algorithm has shown the best results. Another advantage is that it consumes fewer computational resources. The algorithm provides an acceptable speed with average parameter settings, which is comparable to other approaches. Additionally, the proposed algorithm can handle overlapped shadows caused by multiple lighting sources, as it is not dependent on the shape and number of shadows.

However, the main disadvantage of the proposed algorithm is its inability to effectively handle well-known problems in motion detection, such as changes in illumination, small camera displacements, various weather conditions and combinations of these factors. This especially affects the updating of the background model for surfaces with uneven colors where shadows are present. Future research should focus on addressing these issues to minimize their impact on the motion detection process. Nevertheless, in some cases, the algorithm successfully suppresses falsely detected foreground objects in the initial stages.

It is important to note that the proposed algorithm is parametric. On one hand, this can be considered an advantage as accurate parameter selection can significantly improve recognition quality for specific video recordings and allow for a balance in computing resource consumption. On the other hand, this can also be seen as a disadvantage since correct parameter selection requires the presence of an expert during the initial configuration stage or access to training datasets, which may not be feasible in certain scenarios. However, parametric algorithms are still in demand today because most consumers prioritize acceptable recognition results over ease of use and configuration, even if the algorithm is less efficient.

6 Conclusion

This article proposes an improvement for background subtraction algorithms related to shadow detection. The presented approach can be applied as a complement to the post-processing stages of any motion detection algorithm. The idea behind the developed method is to assume that the effect of applying a shadow to the background model is the same for a group of neighboring pixels in a small neighborhood. The proposed algorithm classifies the part of an object moving in the frame as a shadow if the majority of pixels in the specified area satisfy the condition for matching the sample to the normal distribution. Additionally, it is suggested to use different sets of morphological operators before and after the main stage of shadow processing in order to reduce the influence of "grainy" noise on the motion detection result.

The presented method is parametric, meaning it depends on a set of initially configured values. It is used to balance recognition quality and resource consumption. However, this article also proposes a universal set of parameters that is suitable for most cases. Using these parameters provides acceptable recognition results with low computational costs.

The proposed technique was tested on six publicly available sets of video data, all of which contain shadows of different intensity and shape in the scene. The proposed approach provides acceptable recognition results and performs better than other similar background subtraction algorithms described in the article [6].

The developed method has the following advantages: high-quality recognition of moving objects and shadow detection along with low consumption of computing resources. The main disadvantage of the proposed technique is its inability to adapt to external negative factors on the scene, such as changes in illumination, the presence of wind, small camera oscillations, etc. These problems are planned to be addressed in future research.

References

1. Agrawal, S., Natu, P.: ABGS segmenter: pixel wise adaptive background subtraction and intensity ratio based shadow removal approach for moving object detection. J. Supercomput. **79**, 7937–7969 (2022). https://doi.org/10.1007/s11227-022-04972-9

2. Barbuzza, R., Dominguez, L., Perez, A., Esteberena, L.F., Rubiales, A., D'amato, J.P.: A shadow removal approach for a background subtraction algorithm. In: De Giusti, A.E. (ed.) CACIC 2017. CCIS, vol. 790, pp. 101–110. Springer, Cham (2018). https://doi.org/10.1007/978-3-319-75214-3_10
3. Bunyak, F., Ersoy, I., Subramanya, S.: Shadow detection by combined photometric invariants for improved foreground segmentation, pp. 510–515 (2005). https://doi.org/10.1109/ACVMOT.2005.108
4. Chalidabhongse, T., Harwood, D., Davis, L.: A robust background subtraction and shadow detection. In: Proceedings on Asian Conference on Computer Vision (2000)
5. Kaewtrakulpong, P., Bowden, R.: An improved adaptive background mixture model for realtime tracking with shadow detection. In: Remagnino, P., Jones, G.A., Paragios, N., Regazzoni, C.S. (eds.) Advanced Video-Based Surveillance Systems, pp. 135–144. Springer, Boston (2002). https://doi.org/10.1007/978-1-4615-0913-4_11
6. Prati, A., Mikic, I., Trivedi, M., Cucchiara, R.: Detecting moving shadows: algorithms and evaluation. IEEE Trans. Pattern Anal. Mach. Intell. **25**(7), 918–923 (2003). https://doi.org/10.1109/TPAMI.2003.1206520
7. Srisha, R., Khan, A.: Morphological operations for image processing: understanding and its applications (2013)
8. St-Laurent, L., Prévost, D., Maldague, X.: Optimization of color-based foreground/background segmentation for outdoor scenes (2012). https://doi.org/10.2316/P.2012.778-042
9. Varghese, A., Sreelekha, G.: Sample-based integrated background subtraction and shadow detection. IPSJ Trans. Comput. Vis. Appl. **9**, 1–12 (2017). https://doi.org/10.1186/s41074-017-0036-1
10. Wang, M., Yang, J.: Moving object detection and shadow removal algorithm. In: Proceedings of the 2016 International Forum on Management, Education and Information Technology Application, pp. 792–796. Atlantis Press (2016). https://doi.org/10.2991/ifmeita-16.2016.145
11. Wang, Y., Tan, T.: Adaptive foreground and shadow detection inimage sequences (2012). https://doi.org/10.48550/arXiv.1301.0612
12. Zhu, S., Guo, Z., Ma, L.: Shadow removal with background difference method based on shadow position and edges attributes. EURASIP J. Image Video Process. **2012** (2012). https://doi.org/10.1186/1687-5281-2012-22
13. Changedetection.net (cdnet). http://changedetection.net/. Accessed 29 June 2023

Optimal Route for Drone for Monitoring of Crop Yields

Tatiana Makarovskikh(✉), Anatoly Panyukov, Mostafa Abotaleb, Valentina Maksimova, Olga Dernova, and Eugeny Raschupkin

South Ural State University, Chelyabinsk, Russia
{Makarovskikh.T.A,paniukovav,abotalebm,maksimovavn,dernovaoa}@susu.ru
http://www.susu.ru

Abstract. The article deals with the solution of the routing problem with restrictions. The task is to find the optimal drone flight trajectory for monitoring agricultural land. The main restrictions on building a route are the drone flight time before recharging and the location of the operator performing the survey. The initial data are the coordinates of the contours of agricultural land and the permissible location points of the drone operator. In formal terms it is generalized travelling salesman problem. The article presents an algorithm for solving the problem in the presence of a number of assumptions related to the specifics of the application area.

Keywords: Routing · Plane graph · Algorithm · Optimization · Constraints · Drone and truck problem · Precision farming

1 Introduction

All kinds of trajectory problems are universal mathematical models of optimization and control tasks. The routing tasks with different types of constraints are devoted to finding the routes satisfying certain constraints have arisen from specific practical situations. For example, in [1] the authors consider some algorithms solving routing tasks with constraints for control of cutting machines in the industry; the paper [2] is devoted to the routing problem for cutting blanks from sheet material. There are a great number of papers like [3] in which the problem of optimal routing of movements with additional constraints such as precedence conditions and cost functions depending on the list of jobs is considered. In this paper we consider defining the optimal route for so called drone and truck problem [4] satisfying the set of constraints. The solution of this task can be included as a separate function in the geoportal of the region and used when planning an order for monitoring of agricultural land.

A number the researches are devoted to precision farming technologies like [5], where the online adaptive path-planning solution based on the fusion of rapidly exploring random trees and deep reinforcement learning algorithms applied to the generation and control of drone trajectory during an inspection task. Here

N. Olenev et al. (Eds.): OPTIMA 2023, CCIS 1913, pp. 228–240, 2024.
https://doi.org/10.1007/978-3-031-48751-4_17

and later we consider a drone as it is commonly used in the research community [4]. In [6] the authors consider the algorithm for efficient, low consumption and accurate plant protection route considering the flight characteristics of drones and orchard planting characteristics. This study proposes a plant protection route planning algorithm to solve the waypoint planning problem of drone multi-objective tasks. By improving the heuristic function in Ant Colony Optimization (ACO), the algorithm combines corner cost and distance cost for multi-objective node optimization. Paper [7] is devoted to the problem of flying over restricted areas by a drone. The optimal control synthesis problem is solved as an analytical definition of the optimal control of a linear non-stationary system based on the quadratic quality functional. A mathematical model of drone motion in the horizontal plane is proposed, in the form of a system of ordinary differential equations in the Cauchy form. The article [8] discusses the use of a genetic algorithm for calculating the optimal closed routes on the Euler-Hamiltonian graph of reference points on the ground for organizing various monitoring schemes. The mathematical model of finding the routes is based on minimizing a non-negative objective function of integer variables using a genetic algorithm. One more interesting research [9] considers a simplified new modification of the mathematical model of the flight of a drone to a given target and back in a given time, in the form of Newton-Euler differential equations. Time is determined by the energy resource of its design. Particular attention is paid to some original way of guaranteed landing of the drone in the "nest", a place for recharging the battery that ensures the movement of the device.

Summarizing the above, it can be noted that the authors of works on drone routing use either the solution of systems of differential equations (when searching for the exact flight trajectory) or various heuristic approaches to solving optimization problems. In this article, we consider the problem of planning the drone path, thus, the algorithms for controlling the vehicle and maintaining course stability will not be considered. Our aim, unlike other published works, is to start developing the exact algorithm for solving the task with given constraints and to try get as best as possible polynomial algorithm for the given statement. We are interested in estimating the drone flight time for monitoring the objects of interest to the customer and determining the number and place for recharging. Speaking on recharging we use another term "truck" which is used for drone delivery and charging. To solve these problems, it is sufficient to apply optimization methods and represent the considered objects as a planar graph. The existing conditions for organizing the survey of objects will be considered as a set of constraints for the task.

2 Statement of the Problem and Constraints

Let's consider a farm that owns N fields, each field is inscribed in a rectangle $F_i = \{l_i \times w_i\}$; $i = \overline{1, N}$, the coordinates of the upper left corner of each such rectangle are saved in WGS84[1] system and its dimensions l_i and w_i, $i = \overline{1, N}$ are given.

The company that monitors this territory using drone provides the customer with the following conditions:

- one rising of a drone cost C_{call} (includes the working time of an operator and one battery charge, the number of charges for one battery is limited and the cost of a new battery is proportionally divided between all its charges);
- the cost of transfer the brigade truck to the place of shooting C_{way};
- characteristics of the selected drone for research: average flight time before recharging T_{av}, average velocities of ascent to a given height V_{up} and flight V_{flight};
- the size of an area taken from above S_a;
- camera aspect ratio a_x, a_y;
- the height of monitoring h;
- drone restart cost $C_{recharge}$;
- image overlay percentage $P \in [0; 1)$;
- drone returns for recharging to the truck that is placed in a starting point or moves to any allowed point at the boundary of a field, and continues monitoring after recharging.

The task is to search a drone flight path for the given map of areas that satisfies the imposed constraints and parameters, determine the optimal starting point, minimize the number of drone recharges and minimize the distances between the launch points of the device. The solving of this task may be subdivided into the following stages.

1. Let flight time is not limited and explored area is connected. We consider this task while building the graph model for the task.
2. Let the flight time is not limited but explored area is not connected. Here we connect the areas for which the time for drone relocation is higher than continuing flight and catching some additional pictures.
3. Let the flight time is limited by the drone battery capacity, but the location of start is any point. The field is not connected. Here we have the task close to previous one.
4. The general case when the flight time is limited by battery capacity, and location of possible starting points is fixed due to some restrictions (for example, the proximity of trees, swampy areas, hard-to-reach points etc.)
5. The general case as the previous one, but the investigated area has any shape, not only rectangle.
6. To take into account the influence the external factors to improve the drone flight trajectory.

In [4] the author discusses several mathematical models and problems based on the concept of DTCO (drone-truck combined operations), which can be

[1] World Geodetic Parameters of the Earth 1984, which includes the system of geocentric coordinates. Unlike local systems, it is a single system for the entire planet.

roughly divided into travelling salesman problem with drone (TSPD), and vehicle routing problem with drones (VRPD). The most of the task considered in [4] are devoted to the delivery task when the aim is to determine the order of drone flyby or the territory served (FSTSP, PDSTSP [10], PDSTSP+DP [11], TSP-D [12] etc.). The problem describes the optimization of single drone and single truck scenario where the objective is to minimize the tour completion time. In our paper we discuss the drone trajectory used for monitoring of crop yields, i.e. the minimal sequence of rectangles covering the investigated area. For this purpose we also need one drone and one truck. For solving this task we can use the approach [12] where the nodes are divided into three categories: 1) drone node: a node visited by a drone only, 2) truck node: a node visited by a truck only, and 3) combined node: a node visited by a truck and a drone (mounted on the truck). In this case a solution to TSP is the combination of a truck route (composed of the truck and combined nodes) and a drone route (composed of the drone and combined nodes). But the considered case of yields monitoring is to be a particular case allowing solving our task by polynomial time. Let us discuss it.

3 Graph Model and the Case of Routing with Unlimited Time for Connected Area

To solve the task let us consider the plane graph $G_{F_i} = (V_{F_i}, E_{F_i})$ corresponding to one investigated area (field) F_i, the set of vertices V_{F_i} of which be the points of shooting by the drone camera, and E_{F_i} be the connections between the nearest neighbours. As soon as each $S_a = (a_x \cdot K) \cdot (a_y \cdot K) = K^2 \cdot a_x a_y$, then the lengths of each photo size are equal to

$$X = a_x \cdot K = a_x \cdot \sqrt{\frac{S_a}{a_x a_y}}, \quad Y = a_y \cdot K = a_y \cdot \sqrt{\frac{S_a}{a_x a_y}}.$$

The point of shooting be the center of the picture. So, the point of the next shooting for current shot (x_{cur}, y_{cur}) with overlay P can be defined as following:

$$x_{next} = x_{cur} + X - X \cdot P, \quad y_{next} = y_{cur}$$

for moving along the rectangle width or

$$x_{next} = x_{cur}, \quad y_{next} = y_{cur} + Y - Y \cdot P$$

for moving along the rectangle height.

To obtain the optimal path $P(F_i)$ for the rectangular area F_i we need moving only horizontally or vertically, so the other trajectories are not considered. After counting all the coordinates of shots we obtain graph G_{F_i} vertices $v_{ij} \in V_{F_i}$, $i = \overline{1, M}$, $j = \overline{1, L}$, $|V_{F_i}| = M \cdot L$. The edges $e \in E_{F_i}$ correspond to possible moves from each vertex, $|E_{F_i}| = (M-1) \cdot (L-1)$. Hence, such a graph has only vertices of degree 2 (4 vertices in the corners), 3 (vertices incident to outer face), and 4 (all the other vertices). The structure of graph is shown in Fig. 1. One half of edges $e \in E_{F_i}$ has length $l_{vert} = Y - Y \cdot P$ and another one has length $l_{hor} = X - X \cdot P$. The task is to define the Hamiltonian cycle[2] of the shortest length in this graph. Since $a_x \neq a_y$, without loss of generality, let's consider that $a_x > a_y$, and $l_{vert} < l_{hor}$.

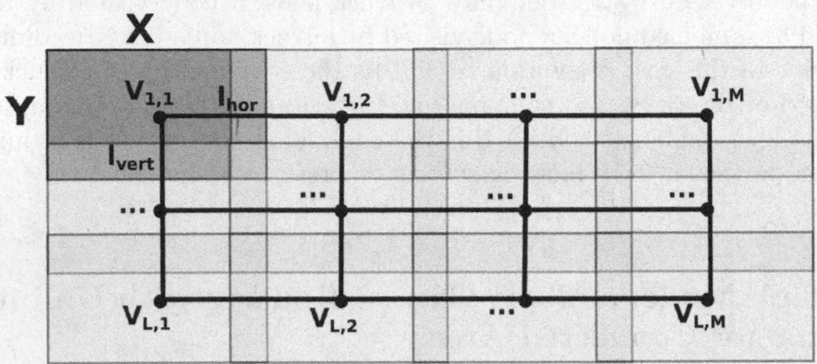

Fig. 1. The structure of graph with possible ways and points of shots for drone

So, half of edges is shorter then the others, then to get the shortest path we need include as much as possible these short edges of length l_{vert} and as less as possible of long edges of length l_{hor}. In any case the minimal number of long edges in a route is equal to $2 \cdot (M-1)$. This value is explained by the fact that it is necessary to fly around the object of interest twice (from the starting point to the far boundary and back).

Despite the $\mathcal{NP}$-hardness of the problem of finding a Hamiltonian cycle in a graph, for the case under consideration, one can find a general approach to determining the path of the shortest length in polynomial time. It is very easy to define the shortest path of flyby a single F_i because only the following cases are available:

- M is even for any L. The length of the route (see Fig. 2) is equal to

$$R = 2(M-1)l_{hor} + (2(L-1) + (M-2)(L-2))l_{vert};$$

[2] The cycle $C = v_0 v_1 v_2 \dots v_n v_0$ in graph $G_{F_i} = (V_{F_i}, E_{F_i})$ is called Hamiltonian if it passes each vertex v_i, $i = \overline{0, n}$ only once (starting vertex v_0 is visited twice), and for each v_i, $i = \overline{1, n}$ there exists $e = v_{i-1} v_i$, $e \in E_{F_i}$.

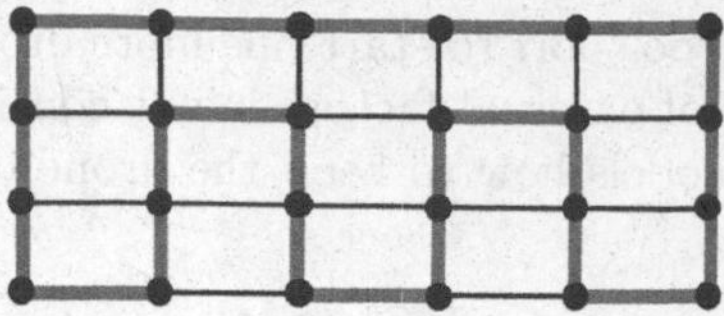

Fig. 2. The example of shortest drone route $P(F_i)$ for graph with even $M = 6$

- M is odd, L is even. The length of the route (see Fig. 3) is equal to

$$R = (2(M-1) + (L-2))l_{hor} + (2(L-1) + (M-2)(L-2))l_{vert}$$

  if we avoid idle passes or

$$R = 2(M-1)l_{hor} + ((L-1) + (M-1)(L-2))l_{vert} + L_{idle}, \quad L_{idle} = 2l_{hor} + (L-1)l_{vert}$$

  if we allow idle passes.
- both M and L are even. The length of the route (see Fig. 3(b)) is calculated the same way as for the previous case with idle pass.

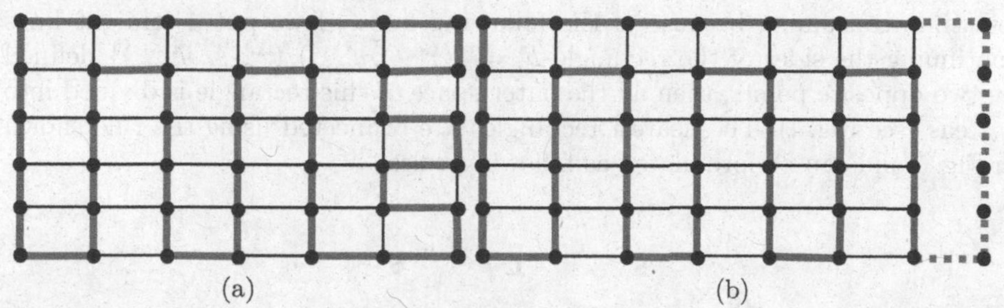

Fig. 3. The example of shortest drone route $P(F_i)$ for graph with odd $M = 7$ and even $L = 4$: (a) without idle shots; (b) with idle shots

The idle shooting is used to fly and monitor some outer area (is the research company organizing monitoring can sell these shots to the owner of the area that is to be observed) to avoid the double shooting the vertices of G_{F_i} and to return to the starting vertex. If we have some time limitations then L_{idle} may be shortened but its minimal length is $(L-1)l_{vert}$ which is the minimal way to reach the place not shoot yet. The sequence of edges in a route may vary from the example if possible.

The path $P(F_i)$ of flying and shooting for a single object F_i, $i = 1, \overline{N}$ may be obtained by using schemes in Figs. 2 and 3. Nevertheless there are some tasks that need additional consideration. In particular, the obtained value may exceed the maximal time of drone flight, we have more than one object under consideration, sometimes it is useful to unite two or more objects to one since the distance between them is too short to start one more drone flight. Moreover, the wind speed and influence of external factors is not considered. If we need them me may use the known models how to keep the drone on its way, for example, [13].

4 Algorithm for Shortest Connecting of Areas

Let the farmer needs monitoring of more than one field. If the boundaries of the fields are combined or are at a distance of several tens of meters these fields can be united to one investigated area. In common fields may lie in distance of several kilometers from each other, and there should exist an approach for building of drone route to investigate these fields. The approach used for connecting the different areas under research may be similar to [3].

Let's consider the peculiarities of the applied task.

The main constraint used for uniting of areas to one route is the following: if the distance between boundaries of two fields takes time less than drone landing and takeoff time then these fields are connected to one object. To solve this task we need define the shortest distance between boundaries of fields F_i. We use only vertices of graph G belonging to outer face.

Since all fields F_i are considered to be rectangles then the task becomes less time consuming because of the following facts. If we put 4 straight lines continuing the sides of the rectangle $F_i = \{(x^i_{min}, y^i_{min}), (x^i_{max}, y^i_{max})\}$ defined by two opposite points, then all the outer space of this rectangle is divided into 8 areas (see Fig. 4). The nearest rectangles are connected using the rule shown in Fig. 5 and can be formulated as following rules.

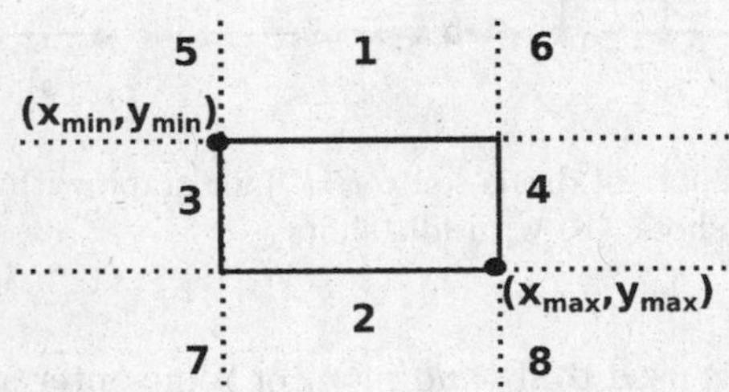

Fig. 4. Dividing the outer space of a rectangle to 8 areas

1. If the straight line l passing along the boundary of rectangle F_i, $i = \overline{1, N}$ intersects the boundary of rectangle F_j, $j \neq i$, $j = \overline{1, N}$ then the segment of l between F_i and F_j be the shortest way between them.
2. For each pair of rectangles F_i and F_j, $j \neq i$, $i, j = \overline{1, N}$ there exists the segment connecting them.

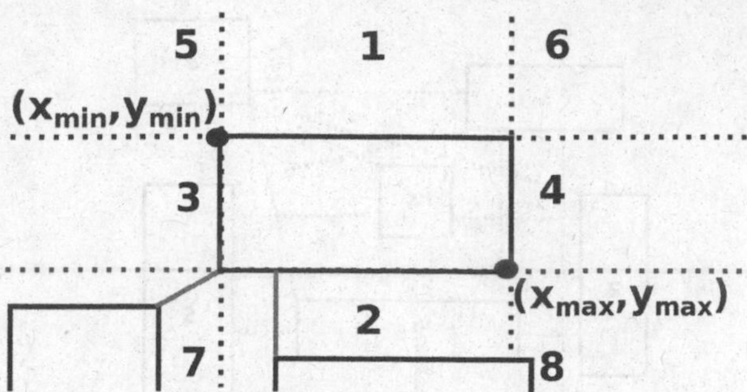

Fig. 5. The shortest connections between rectangles belonging to different areas

3. If there are 2 segments A and B between rectangles F_i and F_j, $j \neq i$, $i, j = \overline{1, N}$, and their lengths $L(A) = L(B)$ then for any segment C between A and B the equality $L(A) = L(B) = L(C)$ holds.
4. If F_j belongs to areas 5–8 of rectangle F_i outer space (see Fig. 5), $j \neq i$, $i, j = \overline{1, N}$, then the shortest segment connects one of the following pairs of points:
   - (x^i_{min}, y^i_{min}) and (x^j_{max}, y^j_{max}) (area 5 of F_i) or (x^j_{min}, y^j_{min}) and (x^i_{max}, y^i_{max}) (area 8 of F_i);
   - (x^i_{min}, y^i_{max}) and (x^j_{max}, y^j_{min}) (area 6 of F_i) or (x^j_{max}, y^j_{min}) and (x^i_{min}, y^i_{max}) (area 7 of F_i).

Using these rules we can calculate the shortest connections between all pairs of rectangles F_i, $i = \overline{1, N}$. If we consider F_i be the vertices of the connections graph G_c, then its edges $e \in E(G_c)$ be the connections, and weight $w(e)$ of each $e \in E(G_c)$ be the length of the segment. Then G_c be a full weighted graph $K_{|F_i|}$. The example of connecting the rectangles is shown in Fig. 6. The dotted lines show the edges of $K_{|F_i|}$, the rectangles of F_i are enumerated from 1 to 7. Hence, to get the shortest connection between F_i we need define the minimal spanning tree $T(G_c)$ (see Fig. 7) using any known algorithm best suitable for the given initial data and given computational resources.

Theorem 1. *The obtaining the minimal length connections between all F_i, $i = \overline{1, N}$ can be run in polynomial time.*

Proof. There are $C^2_N = \frac{N(N-1)}{2}$ connections between rectangles F_i. Each connection can be defined by linear time. It is known that obtaining the minimal spanning tree $T(G_c)$ needs time not less than $O(|V(G_c)| \cdot |E(G_c)|)$.

After obtaining $T(G_c)$ we have a connected graph $G_p = (V_p, E_p)$ (Fig. 8), $E_p = E(P(F_i)) \cup E(T(G_c))$ corresponding to a drone path. As we see in Fig. 8(a) in common the degree of $v_i \in V_p$ may be either odd or even. So, to fly by the defined path with the defined connections and return to initial point (drone returns to truck) we need pass edges $E(T(G_c))$ connecting the odd vertices twice or define the shortest matching between them.

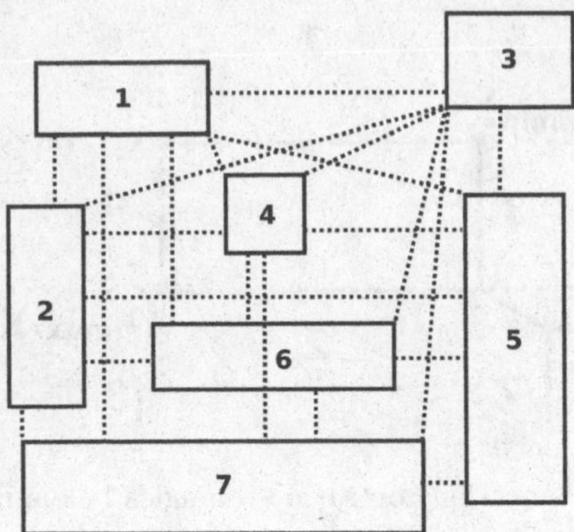

Fig. 6. The shortest connections between rectangles belonging to different areas

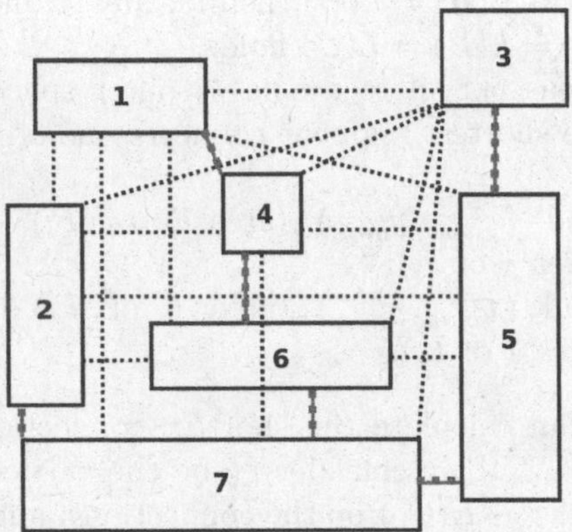

Fig. 7. The minimal spanning tree for the given flight plan

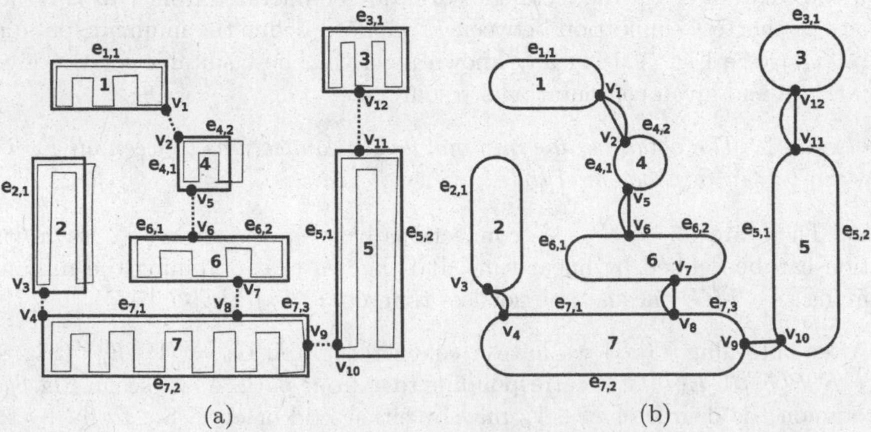

Fig. 8. Graph $G_p = (V_p, E_p)$: (a) With paths of flyby and connections; (b) Only graph structure

One of the solutions of this task is shown in Fig. 8(b) where the odd vertices are connected twice. In this case we have connected 4-regular graph with cut vertices. Hence, we need get Euler cycle for G_p. One of algorithms [1] allowing to get Euler path for G_p gets an AOE-chain for any plane 4-regular graph. This algorithm was developed for routing of a cutter with several technological constraints, and it gets a closed chain $C(G_p)$ satisfying some additional constraints that are not principal for the task under consideration. But the obtained path $C(G_p)$ is one of the possible solutions of the considered graph:

$$C(G_p) = v_1 e_{1,1} v_2 e_{4,1} v_5 v_6 e_{6,1} v_7 v_8 e_{7,1} v_4 v_3 e_{2,1} v_3 v_4 e_{7,2} v_9 v_{10} e_{5,2} v_{11} v_{12} e_{3,1} v_{12} v_{11} e_{5,1} v_{10} v_9 e_{7,3} v_8 v_7 e_{6,2} v_6 v_5 e_{4,2} v_2 v_1.$$

5 Drone Path for the Limited Time of Flight

As we mentioned before for solving this task we can use the approach [12] where the nodes are divided into three categories: 1) drone node: a node visited by a drone only, 2) truck node: a node visited by a truck only, and 3) combined node: a node visited by a truck and a drone (mounted on the truck). If we consider the task when a drone needs recharging by truck then we have only two types of vertices:

1. drone node: a node visited by a drone only, they are the points of a field F_i not belonging boundary or nodes marked as unreachable by truck;
2. combined node: a node visited both by a truck and a drone (for both monitoring and recharging), it is a node belonging boundary of the field reachable by truck.

So, if a drone needs recharging during monitoring the fields, we need define the combined nodes to send the truck there so drone can recharge. Hence, for graph $G_p = (V_p, E_p)$, $V_p = V_{drone} \cup V_{truck}$, $V_{comb} = V_{drone} \cap V_{truck}$. To define the recharging points we calculate $T_{flight} = T_{up} + T_{explore} + T_{down}$, where T_{up} be the time for rising the drone, T_{down} be the time for drone descent, and $T_{explore}$ be the time for shooting. If we do not take into account any external factors such as air temperature, wind speed, cloudiness etc., then $S_{flight} = V_{flight} \cdot T_{explore}$ be the maximal length of drone path. For safe drone descent we need pass the way shorter than S_{flight}, so drone goes down in the nearest combined node $v^* \in V_{comb}$ before length S_{flight} is reached. We need put the truck to node v^*. Knowing the way for a drone $C(G_p)$ we can define the nodes $v_1^*, v_2^*, \dots v_k^*$ where the truck waits for drone for recharging.

To calculate the drone optimal path we can use algorithm suggested in [1] where all the vertices of graph (nodes) are defined as V_{in} (be the starting points of chains) and V_{out} (be the ending nodes of chains) and algorithm defines the path in a graph with optimal length connections between odd vertices. Moreover, the defined path satisfies some technological restrictions having no value in the considered task of drone-truck routing.

The open tasks for drone path for the limited time of flight are taking into account different external factors such as temperature, humidity, wind speed, and others to estimate T_{flight} for them.

6 Testing

To implement and test the algorithms proposed in the article, we developed graphical application providing the user the opportunity to select equipment, the explored area and calculate the drone flight path taking into account the selected equipment. All the data on equipment is saved to a database containing the tables of available drones, trucks and cameras. This application runs for any shape of the explored area. The discretisation of this area runs in polynomial time. The example of running discretisation algorithm is presented in Fig. 9. Here, the rectangles included in the survey area are highlighted in purple, the survey points are marked in red. For clarity, a zero overlap coefficient was used. To construct the drone flight path, the first version of our application uses the nearest neighbour algorithm, which results in a construction similar to that considered above for a rectangular area. An example of the result of the algorithm is shown in Fig. 10. The starting point is marked in red, the points of work (shooting) are marked in green. Let us note that the developed software product takes into account only connected areas and drone needing no recharging during exploring and monitoring (and thus does not involve the truck as an option). All this and taking into account the influence of external factors are directions for further research and development.

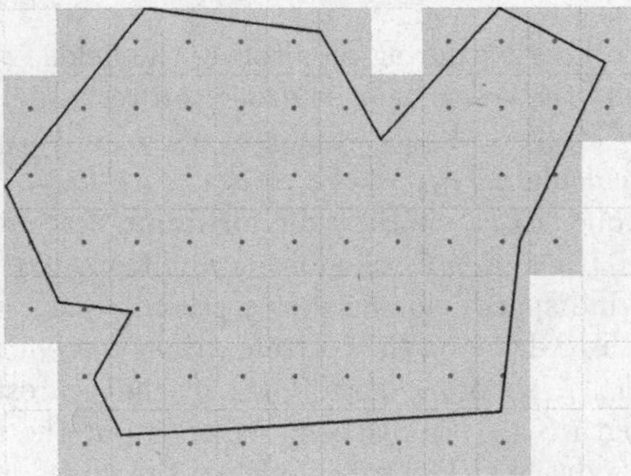

Fig. 9. The example of running the discretisation algorithm for polygonal area

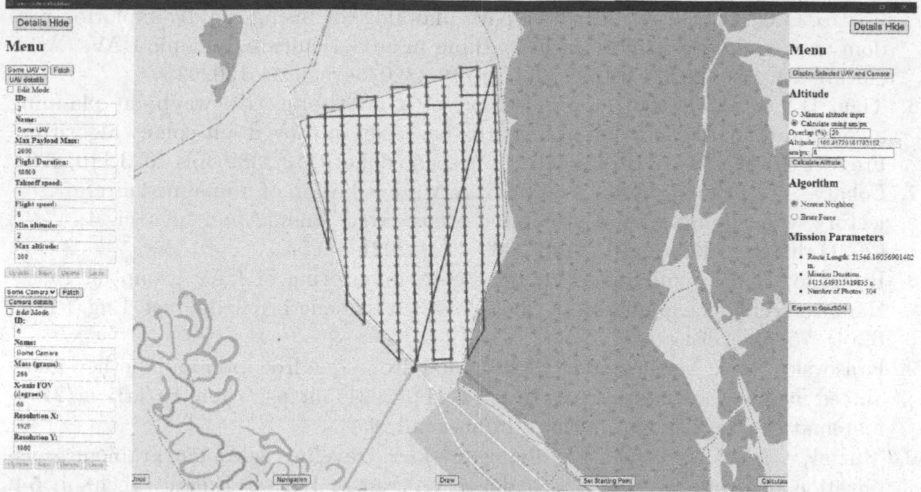

Fig. 10. Building a trajectory along which the drone performs its work

7 Conclusion

The considered type of drone and truck problem has lot of unsolved tasks concerning different technological constraints. Restrictions on flight time and height, on placement of truck, on properties of camera, and weather conditions are among them. In our article, we considered the problem of planning the drone path with several easy constraints: we run our drone and truck system in the ideal weather (no wind, positive temperature), and suppose that truck can parking in almost all outer border points. We defined the way of obtaining the drone flight trajectory for monitoring the objects of interest to the customer and determining the number and place for recharging. For this purpose we used the developed earlier polynomial routing algorithm. We see there are lot of unsolved tasks for different constraints that are the topic of our future research. After solving these tasks the proposed algorithm can be implemented to practice.

References

1. Makarovskikh, T., Panyukov, A.: Special type routing problems in plane graphs. Mathematics **10**, 795 (2022). https://doi.org/10.3390/math10050795
2. Petunin, A.A., Polishchuk, E.G., Ukolov, S.S.: On the new algorithm for solving continuous cutting problem. IFAC-PapersOnLine **52**, 2320–2325 (2020)
3. Petunin, A.A., Chentsov, A.G., Chentsov, P.A.: Optimal routing in problems of sequential bypass of megacities in the presence of restrictions. Chelyab. Phys.-Math. J. **7**(2), 209–233 (2022)
4. Chung, S.H., Sah, B., Lee, J.: Optimization for drone and drone-truck combined operations: a review of the state of the art and future directions. Comput. Oper. Res. **123**, ID 105004 (2020). https://doi.org/10.1016/j.cor.2020.105004

5. Castro, G.G.R.D., et al.: Adaptive path planning for fusing rapidly exploring random trees and deep reinforcement learning in an agriculture dynamic UAVs. Agriculture **13**, 354 (2023). https://doi.org/10.3390/agriculture13020354
6. Tian, H., et al.: Design and validation of a multi-objective waypoint planning algorithm for UAV spraying in orchards based on improved ant colony algorithm. Front. Plant Sci. **14**, 1101828 (2023). https://doi.org/10.3389/fpls.2023.1101828
7. Lobaty, A.A., Bumai, A.Y., Avsievich, A.M.: Formation of unmanned aircraft trajectory when flying around prohibited areas. Syst. Anal. Appl. Inf. Sci. **4**, 47–53 (2021). https://doi.org/10.21122/2309-4923-2021-4-47-53
8. Rudenko, E.M., Semikina, E.V.: Intelligent monitoring of UAV group on Eyler-Hamilton reference graphs on the local. Institut inzhenernoj fiziki [Inst. Eng. Phys.] **62**(4), 75–80 (2021)
9. Krasovsky, A.N.: Mathematical flight model drone-quadrocopter and method guaranteed fit into the "nest". Curr. Res. **14**(41) (2021). https://apni.ru/article/2164-matematicheskaya-model-polyota-drona-kvadroko
10. Murray, C.C., Chu, A.G.: The flying sidekick traveling salesman problem: optimization of drone-assisted parcel delivery. Transp. Res. C Emerg. Technol. **54**, 86–109 (2015)
11. Ham, A.M.: Integrated scheduling of m-truck, m-drone, and m-depot constrained by time-window, drop-pickup, and m-visit using constraint programming. Transp. Res. Part C Emerg. Technol. **91**, 1–14 (2018). https://doi.org/10.1016/j.trc.2018.03.025
12. Agatz, N., Bouman, P., Schmidt, M.: Optimization approaches for the traveling salesman problem with drone. Transp. Sci. **52**(4), 965–981 (2018)
13. Yu, X., Luo, Y., Liu, Y.: A novel adaptive two-stage approach to dynamic optimal path planning of UAV in 3-D unknown environments. Multimed. Tools Appl. **82**, 18761–18779 (2023). https://doi.org/10.1007/s11042-022-14254-4

Algorithm for Multi-criteria Optimization of Robot Parameters for Fruit Harvesting Based on Evolutionary Methods, Taking into Account the Importance of Criteria

Larisa Rybak, Dmitry Malyshev(✉), and Artem Voloshkin

Belgorod State Technological University named after V.G. Shukhov, Belgorod, Russia

rlbgtu@gmail.com

Abstract. The article develops a multi-criteria optimization method based on parallel modifications of single-criteria evolutionary algorithms, including Genetic, Grey Wolf Optimization, Particle Swarm Optimization. It is proposed to use two-component relative importance coefficients. An algorithm for normalizing the first component of coefficients based on solving a system of equations using Interval contractors has been developed. A multicomponent software package has been developed that allows implementing the developed optimization method and algorithm using parallel computing. Using the developed software package, the method was tested to solve the problem of optimizing the geometric parameters of a 3-DOF parallel robot.

Keywords: Multi-criteria optimization · Harvesting Robot · Parallel robot · Interval contractor

1 Introduction

The tasks of multi-criteria optimization arise at all stages of the creation of new devices, from the design of the structure to control methods. These problems have several conflicting objective functions, so it is impossible to obtain a single optimal solution during optimization, and to obtain a compromise set of optimal solutions, several target values must be taken into account. Solving complex multi-criteria problems with nonlinearity and high dimensionality involves the use of highly efficient methods, including those based on population algorithms.

Evolutionary algorithms are one of the subclasses of population algorithms and simulate the natural process of evolution and optimization using group

The investigation was supported by a grant from the Russian Science Foundation No. 22-19-20153, https://rscf.ru/project/22-19-20153/ and the Government of the Belgorod Region, Agreement No. 4, using equipment of High Technology Center at BSTU named after V.G. Shukhov.

N. Olenev et al. (Eds.): OPTIMA 2023, CCIS 1913, pp. 241–255, 2024.
https://doi.org/10.1007/978-3-031-48751-4_18

search technology. In comparison with traditional approaches to optimization, evolutionary algorithms have a number of advantages, such as the; applicability to problems that do not have an accurate mathematical representation; high noise immunity; suitability for lifelong learning; [4]. However, this approach also has disadvantages, which include a large number of computational operations, as well as premature convergence ("premature convergence" means that the population for the optimization problem converged too early, resulting in sub-optimality). Genetic optimization also does not guarantee obtaining the global minimum or maximum, however, with proper application, it is possible to find a compromise between the time spent and the result obtained. The evolutionary algorithm is applicable to a wide subclass of optimization problems and produces highly interpretable results.

One of the fundamental works devoted to the possibility of using genetic search to solve many problems is the work of Rosenberg in the late 60s [1]. This area of research (now called evolutionary multi-objective optimization) has a significant number of papers. A typical representative of evolutionary algorithms is Holland's genetic algorithm [2], which is a minimal computer model of natural selection, which made it possible to study the effect of manipulating specific parameters on the evolutionary process. The work [3] presents an overview of the most important multicriteria optimization methods based on evolution, emphasizing the importance of analyzing their roots and developing new approaches that use the search capabilities of evolutionary algorithms. Each method is briefly described, indicating its advantages and disadvantages, degree of applicability, and some known applications.

In [5], the solution to a multicriteria optimization problem for decision support using an evolutionary algorithm that finds a Pareto-optimal front in order to minimize two objective functions is considered. Multicriteria optimization using the developed evolutionary algorithm is implemented to determine the values of the fuel supply parameters of the main ship engine at full load modes in order to obtain minimum values for two objective functions: the content of nitrogen oxides in the exhaust gases and specific effective fuel consumption. In this work, a modification of the SPEA2 algorithm is implemented, a Pareto-optimal front is obtained, containing solutions to support the operator in choosing the operating mode of the main ship engine.

In [6], the authors developed a method for investigations the parameter space for setting and solving applied optimization problems with many quality criteria. This method is based on the construction of an acceptable Pareto-optimal set approximation of solutions. Three practically significant classes of multi-criteria tasks are considered, such as: design, identification, design with control. In [7], an analytical method is proposed for assessing the stability of a multi-criteria choice using the theory of the importance of criteria when changing the boundaries of the interval limiting the possible values of the degree of superiority in importance of one of the criteria over the other.

Nowadays, the majority of Models evolve decision variables in the same way without taking into account their various characteristics, and they struggle to strike a good balance between population diversity and convergence [8]. Recently, there have been related article based on decision variable analysis (DVA), which the decision variables are divided into different types by determining the dominant relationship between perturbed individuals [9]. This problem was solved in [10] using an optimization algorithm based on the relationship of the decision variables (DVR). In [11], a multi-criteria optimization algorithm with two alternative optimization methods (LSMOEA-TM) is proposed, which use two grouping strategies to separate solution variables in order to efficiently solve large-scale multi-criteria optimization problems. A Bayesian-based parameter tuning strategy to reduce computational costs is also presented. In [12], advanced strategies of a multi-purpose particle swarm optimization algorithm are summarized, the basic theories of multicriteria optimization and particle swarm optimization are considered, and complex problems associated with multicriteria optimization are analyzed. In [13], a two-stage fast convergent search algorithm for MDOvS is proposed. First, the multi-criteria optimization problem under consideration is decomposed into several subtasks of single-criteria optimization, and for each subtask, a Pareto retrospective approximation method is used to obtain an approximate optimal solution. After that, from the solutions generated at the first stage, a multi-criteria local stochastic search with a revised modeling distribution rule is used to investigation the entire Pareto front approximation. In [14], a dynamic multicriteria evolutionary algorithm (DMOEA) is used, which is an means of solving problems of dynamic multicriteria optimization (DMOP).

To solve optimization problems in the field of agriculture, evolutionary algorithms are also used, for example, in [15] the optimal design of the suspension system of an agricultural mobile robot, developed on the basis of the suspension design on double wishbones, is investigated. To carry out multi-criteria optimization of the suspension design, a genetic algorithm of non-dominant sorting III (NSGA-III) was chosen in combination with the technique of order preference by similarity to the ideal solution (TOPSIS).

Earlier, the authors developed parallel modifications of evolutionary algorithms and successfully tested them to solve the problem of planning the trajectory of a Delta robot [16] and single-criteria optimization of the geometric parameters of the Gough-Stewart platform [17], taking into account minimizing the size of the structure and providing the required workspace.

The current work proposes the application of parallel modifications of evolutionary algorithms for the problem of multi-criteria optimization with enumeration of the importance of criteria and preliminary normalization of criteria based on an interval contractor.

2 Problem Statement and Development of a Multi-criteria Optimization Algorithm

Complex practical decision-making tasks, as a rule, turn out to be multi-criteria: the consequences of decisions taken have to be assessed using not one, but several criteria. [18] Due to the of using evolutionary methods to solve single-criteria optimization problems with criterion functions that cannot be specified analytically, the task arises of developing algorithms that allow solving the problem with these methods in a multi-criteria formulation. In the case of solving the problem by iterative single-criteria optimization using criteria convolution, it becomes necessary to obtain relative importance coefficients that provide a weighted sum of equivalent criteria. To do this, we divide the relative importance coefficients α_i into two components. The first component β_i, which ensures the equivalence of the criteria, is determined by solving a system of equations of the form

$$\begin{cases} \beta_1 K_1 = \beta_2 K_2 = \ldots = \beta_n K_n \\ \beta_1 + \beta_2 + \ldots + \beta_n = 1 \end{cases} \tag{1}$$

where K_i is the average value of the i criterion.

To solve the system of equations, we use interval analysis methods, in particular the Interval Contractor [7]. The second component γ_i of the relative importance coefficients will change during optimization. We synthesize a multi-criteria optimization algorithm to solve the problem (Algorithm 1).

Algorithm 1. Multi-objective Optimization

Input: $\delta, p = \{p_1, p_2, ..p_m\}$ ▷ δ is accuracy, p is parameters
Output: Pareto Set approximation P

1: **for** j:=1 **to** l **do** ▷ l is number of iterations for average K_i calculation
2:     **for** k:=1 **to** m **do** ▷ m is number of parameters
3:         $p_k^{(j)} = \text{random}(\underline{p_k}, \overline{p_k})$
4:     **end for**
5:     **for** i:=1 **to** n **do** ▷ n is number of criteria
6:         $K_i^{(j)} = f(p^{(j)})$ ▷ Criteria calculation
7:     **end for**
8: **end for**
9: **for** i:=1 **to** n **do**
10:     $K_i = \sum_{j=1}^{l} K_i^{(j)}/n$ ▷ Average criteria calculation
11: **end for**
12: $\beta = \{\beta_1, \beta_2, ..., \beta_n\} =$
13: `Normalization`(δ, K) ▷ β part of criteria weights
14: **for** each node j of grid **do**
15:     $s_\gamma = 0$
16:     **for** i:=1 **to** n **do**
17:         $\gamma_i = f(i)$ ▷ γ part of criteria weights. Calculated based on i value.
18:         $s_\gamma = s_\gamma + \gamma_i$
19:     **end for**
20:     **for** i:=1 **to** n **do**
21:         $\alpha_i = \beta_i \gamma_i / s_\gamma$ ▷ α is criteria weights
22:     **end for**
23:     `Single-objective optimization`
24:     `Update the Pareto Set approximation` P
25: **end for**

In steps 1–11 (Algorithm 1), the average values of the K_i criteria are calculated. At step 12 (Algorithm 1), the criteria are normalized by selecting the coefficients β_i, at which the equality of the average values of the criteria K_i is achieved.To perform normalization based on the solution of system (1) using interval contractors, a procedure is used, the pseudocode of which is given in Algorithm 2. Let's denote the intervals of possible values of the coefficients β_i as B_i. Normalization includes the following steps. B_i equal at initialization [0;1] (Algorithm 2, step 2). Step 3 calls the Contractor procedure (Algorithm 3). The following iterative narrowing can be performed

$$B_i = B_i \cap \left(\frac{B_j K_j}{K_i}\right), i \neq j, i, j \in 1, n \tag{2}$$

$$B_i = B_i \cap \left(1 - \sum_{j=1}^{i-1} B_j - \sum_{j=i+1}^{n} B_j\right) \tag{3}$$

The procedure is performed until the intervals B_i are narrowed to a length less than the approximation accuracy δ or the number of iterations `iterCounter` will reach the specified limit `iterLimit` (Algorithm 3, step 3). In the general case, the condition for the approximation accuracy δ (Algorithm 3, step 13) is triggered, but there are special cases for which the procedure does not allow obtaining the result, for example, in the case when the average values of the criteria $K1$ are equal. Because of this the intervals are not reduced at each iteration. In these cases, the condition on step 4 (Algorithm 2) is triggered and in steps 5–18 an iterative bisection (Algorithm 2, step 8, 10) of intervals is performed, the contractor procedure is repeated for the intervals obtained after bisection (Algorithm 2, step 9, 11), and then the union of the resulting intervals is performed (Algorithm 2, step 13). After achieving the required approximation accuracy (Algorithm 2, step 16,17), the cycle stops (Algorithm 2, step 5). As a result, the coefficients β_i are defined as the middle of the intervals B_i(Algorithm 2, step 22)

$$\beta_i = \frac{(\overline{B_i} + \underline{B_i})}{2} \tag{4}$$

After the criteria normalization is performed, the multi-criteria optimization algorithm performs iterative single-criteria optimization (Algorithm 1, step 23) and updating of the Pareto set approximation using evolutionary algorithms with different relative importance coefficients a_i, which are defined as

$$\alpha_i = \frac{\beta_i \gamma_i}{\sum_{i=1}^{n} \gamma_i} \tag{5}$$

where $\gamma_i \in [0,1]$ are the components of the relative importance coefficients, which are changed after each iteration of single–criteria optimization in accordance with an n-dimensional grid having ranges for each of the dimensions [0;1] and a step of δ_y.

Thus, the developed algorithm can be applied to solve the problem of multi-criteria optimization with any number of criteria. The number of parameters,

Algorithm 2. Normalization

```
function NORMALIZATION(δ, K = {K_1, K_2, ..., K_n})
    B_i = [0; 1], i ∈ 1, .., n                ▷ Initial intervals of β part of criteria weights
    {B,stopFlagGlobal} = Contractor(B)
    if stopFlagGlobal=false then
        while stopFlagGlobal=false do
            for i_1 := 1 to n do
                B^* = B, B^{**} = B
                B^*_{i1} = [\underline{B_{i1}}; (\overline{B_{i1}} + \underline{B_{i1}})/2]
                B^* = Contractor(B^*)
                B^{**}_{i1} = [(\overline{B_{i1}} + \underline{B_{i1}})/2; \overline{B_{i1}};]
                B^{**} = Contractor(B^{**})
                for i_2 := 1 to n do
                    B_{i2} = B^*_{i2} ∪ B^{**}_{i2}
                end for
            end for
            if ∀B_i, \overline{B_i} − \underline{B_i} ≤ δ then
                stopFlagGlobal=true
            end if
        end while
    end if
    for i:=1 to n do
        β_i = (\overline{B_i} + \underline{B_i})/2
    end for
    return {β_1, β_2, ..., β_n}
end function
```

Algorithm 3. Contractor

```
function CONTRACTOR(B = {B_1, B_2, ..., B_n})
    iterCounter = 0, stopFlagLocal = false
    while stopFlagLocal=false ∧ iterCounter < iterLimit do    ▷ Contractor loop
        for i:=1 to n do
            s_B = [0; 0]
            for j:=1 to n do
                if i ≠ j then
                    s_B = s_B + B_j
                    B_i = B_i ∩ (B_j K_j / K_i)
                end if
            end for
            B_i = B_i ∩ (1 − s_B)
            if ∀B_i, \overline{B_i} − \underline{B_i} ≤ δ then
                stopFlagLocal=true
            end if
        end for
    end while
    return B,stopFlagLocal
end function
```

constraints and the nature of the objective functions also do not affect the applicability of the algorithm, but should be taken into account when choosing the algorithm of single-criteria optimization performed at step 22 (Algorithm 1). If there are objective functions that are not defined analytically, evolutionary algorithms can be successfully used for this task. Let's consider the application of the synthesized algorithm on the example of solving the problem of optimizing the geometric parameters of the robot.

3 Application of the Algorithm to Optimize the Geometric Parameters of the Robot

Consider the design of a fruit harvesting robot, the 3D model of which is shown in Fig. 1a. The robot includes a mobile platform with a fruit basket, on which a parallel structure mechanism is installed [20], in the center of the mobile platform of which a telescopic link is installed for access to fruits at high altitude. The mechanism of the parallel structure consists of three drive kinematic chains of the RR<u>P</u>S type and a central kinematic chain with a spherical hinge rigidly connected to fixed and moving platforms. Thus, by changing the lengths l_i of the rods A_iB_i, the required rotation of the movable platform is ensured along all axes relative to the center of the joint C. The fixed platform $A_1A_2A_3$ and the moving platform $B_1B_2B_3$ of the mechanism are regular triangles with radii R_1 and R_2, respectively.

a)

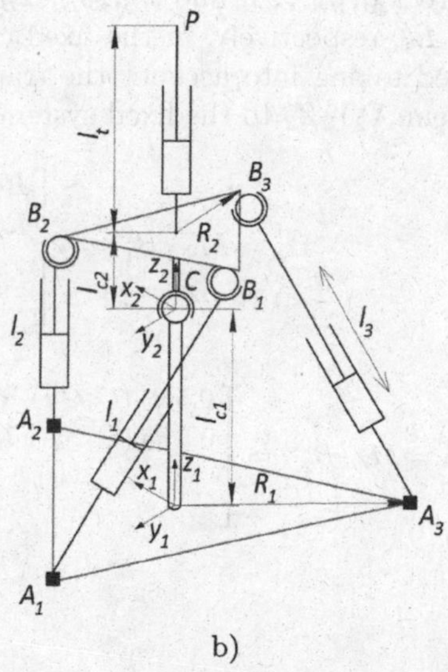

b)

Fig. 1. Conceptual design work for fruit harvesting: a) 3D model; b) calculation scheme of the platform

The input coordinates are the lengths of the drive links l_1, l_2, l_3, the output coordinates are the point P of the output link x_P, y_P, z_P. A detailed solution of the inverse kinematics problem and the task of determining the workspace were previously considered by the authors in [21]. The amount of workspace is one of the most important characteristics of the robot and, based on this, is used as an optimization criterion [22]. For the considered robotic system, the workspace is of particular importance, determining the possibility of collecting fruits with different positions on the tree. The shape and dimensions of the robot's working area depend on the range of length variation l_t of the telescopic link, as well as on the orientation ranges of the mobile platform $B_1B_2B_3$. The orientation of the output link can be set using the Euler angles α, β, γ, and the orientation reachability check is performed using the solution of the inverse kinematics problem.

We introduce the following notation: $C_a = \cos\alpha, S_a = \sin\alpha, C_\beta = \cos\beta, S_\beta = \sin\beta, C_\gamma = \cos\gamma, S_\gamma = \sin\gamma$.

The reachability of the platform orientation is determined based on the limitation on the length of the drive rods l_i

$$l_{\min} \le l_i \le l_{\max}, \tag{6}$$

where l_{min}, l_{max} are determined by the design parameters of the mechanism; l_i is the length of the i-th rod, which is defined as

$$l_i = \sqrt{\left(x_{Bi} - x_{Ai}\right)^2 + \left(y_{Bi} - y_{Ai}\right)^2 + \left(z_{Bi} - z_{Ai}\right)^2} \tag{7}$$

where x_{Ai}, y_{Ai}, z_{Ai} and x_{Bi}, y_{Bi}, z_{Bi} are the coordinates of the joint centers A_i and B_i, respectively, in the fixed coordinate system $X_1Y_1Z_1$, which are determined taking into account the transition matrix from the movable coordinate system $X_2Y_2Z_2$ to the fixed system $X_1Y_1Z_1$ [21] as

$$\mathrm{B}_1 = M_{2\_1}B_1^{(2)} = \begin{bmatrix} R_2\left(C_aC_\gamma - C_\beta S_a S_\gamma\right) + l_{c2}S_aS_\beta \\ R_2\left(C_\gamma S_a + C_aC_\beta S_\gamma\right) - l_{c2}C_aS_\beta \\ l_{c1} + R_2S_\beta S_\gamma + l_{c2}C_\beta \\ 1 \end{bmatrix} \tag{8}$$

$$\mathrm{B}_2 = M_{2\_1}B_2^{(2)} = \begin{bmatrix} 0.5R_2\left(C_aC_\gamma - C_\beta S_aS_\gamma - \sqrt{3}C_aS_\gamma - \sqrt{3}C_\beta C_\gamma S_a\right) + l_{c2}S_aS_\beta \\ 0.5R_2\left(C_\gamma S_a + C_aC_\beta S_\gamma + \sqrt{3}C_aC_\beta C_\gamma - \sqrt{3}S_aS_\gamma\right) - l_{c2}C_aS_\beta \\ l_{c1} + 0.5R_2\left(S_\beta S_\gamma + \sqrt{3}C_\gamma S_\beta\right) + l_{c2}C_\beta \\ 1 \end{bmatrix} \tag{9}$$

$$\mathrm{B}_3 = M_{2\_1} B_3^{(2)} = \begin{bmatrix} 0.5R_2\left(C_\beta S_a S_\gamma + \sqrt{3} C_a S_\gamma + \sqrt{3} C_\beta C_\gamma S_a - C_a C_\gamma\right) + l_{c2} S_a S_\beta \\ 0.5R_2\left(\sqrt{3} S_a S_\gamma - C_\gamma S_a - C_a C_\beta S_\gamma - \sqrt{3} C_a C_\beta C_\gamma\right) - l_{c2} C_a S_\beta \\ l_{c1} - 0.5R_2\left(S_\beta S_\gamma + \sqrt{3} C_\gamma S_\beta\right) + l_{c2} C_\beta \\ 1 \end{bmatrix} \tag{10}$$

$$A_1 = \left[R_2\ 0\ 0\ 1\right]^T \tag{11}$$

$$A_2 = \left[0.5R_2\ 0.5\sqrt{3}R_2\ 0\ 1\right]^T \tag{12}$$

$$A_3 = \left[-0.5R_2\ -0.5\sqrt{3}R_2\ 0\ 1\right]^T \tag{13}$$

Thus, substituting (7)–(13) into (7), we obtain an analytical dependence

$$l_i = f(\alpha, \beta, \gamma) \tag{14}$$

In addition to constraints on the length of the drive rods, it is necessary to take into account the constraints associated with the intersections of links and special positions of the robot, when hit in which the dynamic loads on the links increase significantly, and the robot loses controllability, as well as the intersections of links. Earlier, during the analysis of the workspace for determining singularities, the method proposed in [23], based on the analysis of the Jacobi matrix, was considered. It was found that the workspace with different signs of the determinant of the Jacobi matrix is identical in shape and size. Based on this, within the framework of the current work, constraints on singularities are not introduced. The exclusion of intersections of links can be taken into account using the method described earlier by the authors in [24].

The first optimization criterion was chosen based on ensuring the compactness of the design.

$$K_1 = R_1 + R_2 + l_{\min} + l_{\max} + p_1\left(\left(\frac{\mathrm{k}_R R_1}{R_2}\right)^{2\vartheta-1} - 1\right) \rightarrow \min \tag{15}$$

where R_1 is the radius of the fixed platform, R_2 is the radius of the mobile platform, l_{min} and l_{max} are the limit values for changing the lengths of the rods, p_1 is the specified penalty coefficient, $k_R = 0.7$ is the recommended ratio of the radii of the platforms according to [25] R_2/R_1, ϑ is the Heaviside function

$$\vartheta = \begin{cases} 1, & \text{if} \frac{R_2}{R_1} < 0.7 \\ 0 - \text{else} \end{cases} \tag{16}$$

To ensure sufficient space inside the rod to accommodate ball-screw pairs, it is also necessary to ensure the minimum recommended ratio of the minimum and maximum lengths of the rods, on the basis of which the following optimization constraint is formed

$$\frac{l_{\min}}{l_{\max}} \geq 0.6 \tag{17}$$

Taking into account the dependence of the workspace volume on the range of changes in the length l_t of the telescopic link and the orientation ranges of the mobile platform, as the second criterion for optimizing geometric parameters, which has a direct relationship with the volume of the workspace, we use the value N_w^-, which characterizes the orientation ranges

$$K_2 = N_w^- = |S| - |W|, W \subseteq S \tag{18}$$

where S is the set of possible platform orientations with a discrete step, W is the set of achievable platform orientations, N_w^- is the number of unreachable platform orientations.

Thus, the optimization problem looks like this.

- parameters: radii of platforms R_1 and R_2, angles of mutual arrangement of joints ψ and ω, limit values of lengths of rods L_{min} and L_{max}, as well as lengths l_{c1} and l_{c2} of links of the central kinematic chain;
- criteria: functions K_1 and K_2 calculated by formulas (15), (18);
- constraint: condition (17).

Let's perform optimization using synthesized algorithms.

4 Numerical Simulation

For this optimization, a software package in the C++ programming language has been developed. As algorithms for single-criteria optimization, modifications of algorithms [19] with parallel calculation of fitness functions are used: Genetic Algorithm (GA) [26,27], Grey Wolf Optimization algorithm (GWO) [28,29], Particle Swarm Optimization algorithm (PSO). [30,31]. Parallel computing is implemented using the OpenMP library. Visualization of the results was performed using the developed Python script (Matplotlib and JSON libraries). Let's set the initial data for a computational experiment. Ranges of optimization parameters: $R_1 \in [200; 1000]$, $R_2 \in [200; 1000]$, $L_{min} \in [200; 1200]$, $L_{max} \in [400; 1400]$, $l_{c1} \in [200; 1400]$, $l_{c2} \in [0; 800]$, the penalty coefficient $p_1 = 1000$, the number of grid nodes for each dimensions to enumerate the γ_i component of the relative importance coefficients is 20.

GA algorithm parameters: the number of individuals in the initial population $H = 1000$, the number of generations $W = 20$, the number of crosses at each iteration $S_{GA} = 500$, the number of possible values of each of the parameters $g = 2^{25}$, the probability of mutation $p_m = 80\%$. The total number of generations as a result is 20 (sampling the weights of the first criterion) * 20 (sampling the weights of the second criterion) * 20 (number of generations for one weight) = 8000.

GWO algorithm parameters: $H = 1000$, $W = 20$, the number of new individuals at each iteration $S_{GWO} = 500$.

PSO algorithm parameters: $H = 1000$, $W = 20$, the number of groups $G = 2$, the values of the free parameters $\alpha = 0.7$, $\beta = 1.4$, $\gamma = 1.4$.

Each iteration of single-criteria optimization is performed in two stages. At the first stage, the range of parameters was changed in ranges corresponding to the overall dimensions of the workspace for each of the coordinates. The ranges of parameters at each subsequent stage decreased by 10 times. At the same time, the center of the ranges corresponded to the best result obtained at the previous stage. The Table 1 shows the Pareto set approximation at different iterations of evolutionary algorithms. By iterations we mean the number of processed grid nodes (loop at step 14, Algorithm 1).

Figure 2 shows the Pareto set approximation obtained using each of the algorithms. The able to work of algorithms is the proportion of points of the Pareto front approximation obtained by each of the algorithms. Figure 3 shows the Pareto set approximation obtained using the Pareto sets approximation of each of the algorithms. It consists of 96 points. Of these, only 2 were obtained using the GA algorithm, 13 were obtained using the GWO algorithm. The remaining 81 points were obtained using the PSO algorithm, which shows its higher efficiency for solving the problem.

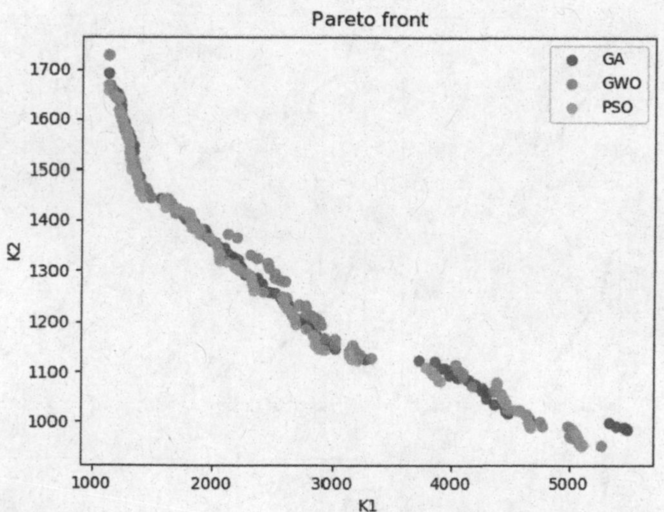

Fig. 2. Pareto approximation multiplicity: (a) GA; (b) GWO; (c) PSO.

The minimum and maximum values of parameters and criteria are shown in Table 2.

Table 1. Iterations of evolutionary algorithms

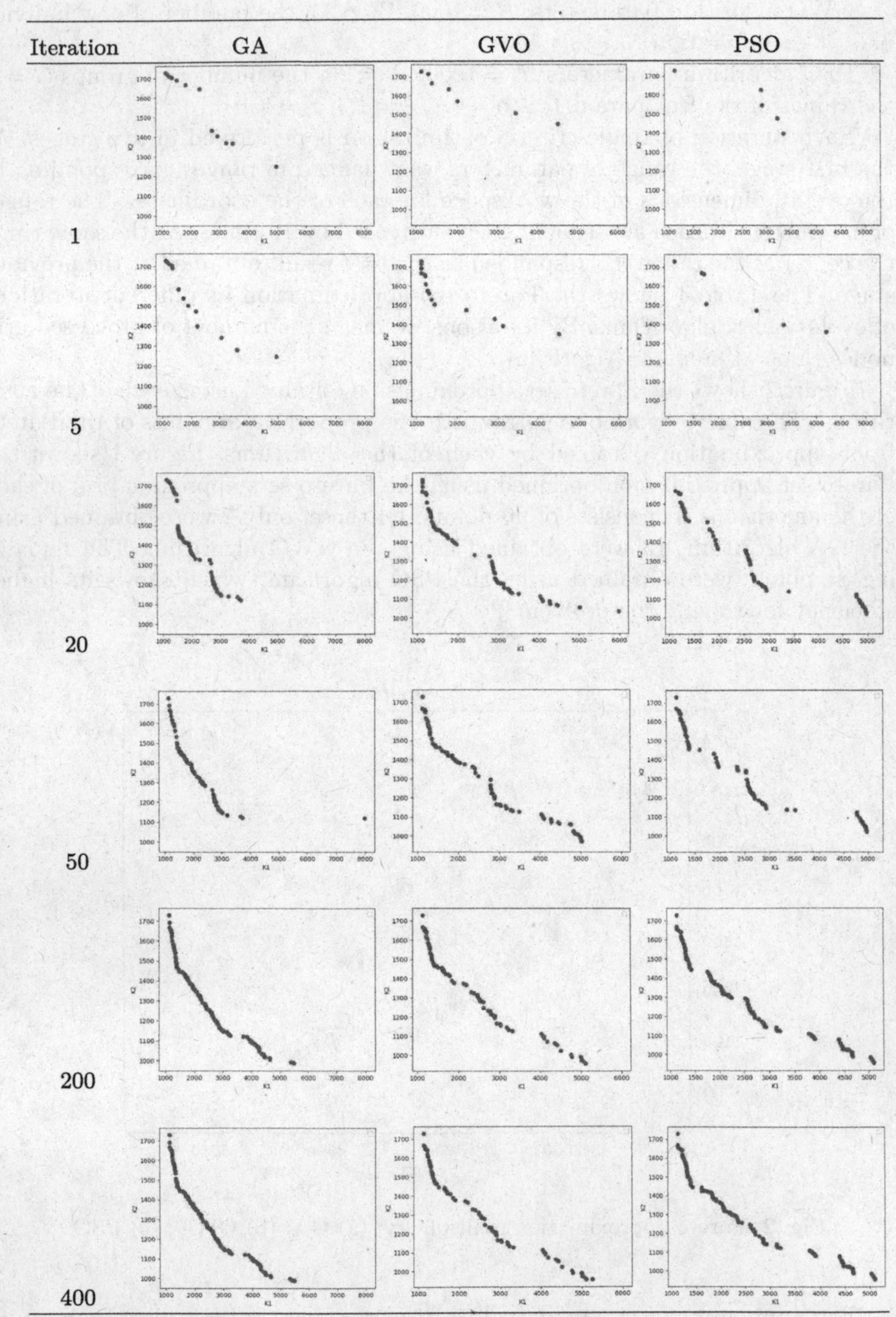

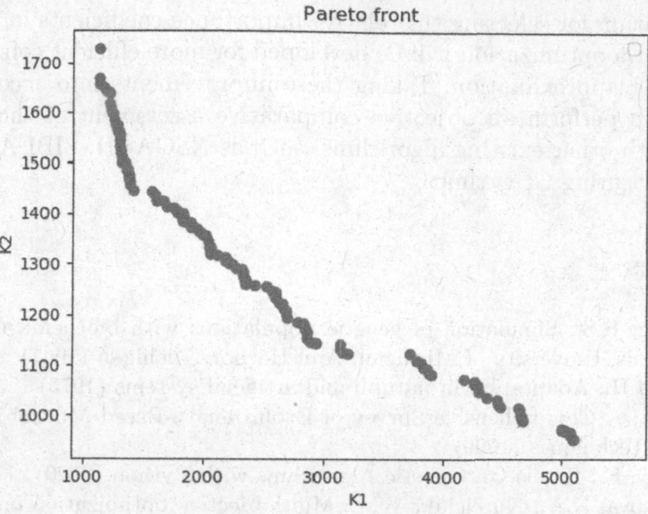

Fig. 3. The final Pareto set approximation.

Table 2. Optimization results.

Parameter	Minimum value	Maximum value	Average value
R_1	277,474	1000	481,077
R_2	200	281	225,354
L_{min}	240,011	847,629	603,981
L_{max}	400	1400	999,803
l_{c_1}	200	1066,07	580,506
l_{c_2}	80,0839	800	109,996
K_1	1125,74	5152,26	2769,685
K_2	954	1728	1282,74

The Pareto set approximation of optimal parameters obtained can be used in the future to select the dimensions of the structural elements of the robot, taking into account dynamic loads.

5 Conclusion

The use of evolutionary algorithms made it possible to solve the problem of multi-criteria optimization of the geometric parameters of the robot. The PSO algorithm showed the best performance indicators and allowed to obtain more than 86.4% of the points of the Pareto set approximation, while GA and GWO allowed to obtain only 2.1% and 13.5%, respectively. As part of the future work, various modifications of evolutionary algorithms, including hybrid ones, will be used, and an algorithm for selecting the relative importance coefficients in the process of multi-criteria optimization will be developed for more efficient construction of the Pareto set approximation. Taking these improvements into account, it will be possible to perform an objective comparative assessment of the developed algorithm with other existing algorithms, such as NSGA-III, SIBEA or SEMO, including measuring CPU time.

References

1. Rosenberg, R.S.: Simulation of genetic populations with biochemical properties. Ph.D. thesis, University of Michigan, Ann Harbor, Michigan (1967)
2. Holland, J.H.: Adaptation in natural and artificial systems (1975)
3. Coello, C.: A Comprehensive Survey of Evolutionary-Based Multiobjective Optimization Techniques (1999)
4. Wirsansky, E.: Hands-On Genetic Algorithms with Python (2020)
5. Polkovnikova, N.A., Kureichik, V.M.: Multiobjective optimization on the base of evolutionary algorithms. Izvestiya SFedU. Eng. Sci. 149–162 (2015)
6. Sobol, I.M., Statnikov, R.B.: Choice of Optimal Parameters in Problems with Many Criteria. Nauka, Moscow (1981)
7. Jaulin, L., Kieffer, M., Didrit, O., Walter, E.: Applied Interval Analysis, with Examples in Parameter and State Estimation, Robust Control and Robotics. Springer, London (2001). https://doi.org/10.1007/978-1-4471-0249-6
8. Jiang, M., Qiu, L., Huang, Z., Yen, G.G.: Dynamic multi-objective estimation of distribution algorithm based on domain adaptation and nonparametric estimation. Inf. Sci. **435**, 203–223 (2018)
9. Zheng, J., Zhou, Y., Zou, J., Yang, S., Junwei, O., Yaru, H.: A prediction strategy based on decision variable analysis for dynamic multi-objective optimization. Swarm Evolut. Comput. **60**, 100786 (2021)
10. Nelyubin, A.P.: Criteria importance theory: sensitivity analysis of multicriterial choice using interval importance information. Am. J. Control Syst. Inf. Technol. **1**(1), 13–17 (2013)
11. Ziyu, H., Zihan, L., Lixin, W., Hao, S., Xuemin, M.: A dynamic multiobjective optimization algorithm based on decision variable relationship. Neural Comput. Appl. (2023)
12. Tianyu, L., Junjie, Z., Lei, C.: A stable large-scale multiobjective optimization algorithm with two alternative optimization methods. Entropy **25**, 561 (2023)
13. Feng, Q., Li, Q., Quan, W., Pei, X.-M.: Overview of multiobjective particle swarm optimization algorithm. Chin. J. Eng. **43**, 745–753 (2021)
14. Fei, L., Qingfu, Z.: A Two-Stage Algorithm for Integer Multiobjective Simulation Optimization (2023)
15. Peidi, W., Yongjie, M.: A dynamic multiobjective evolutionary algorithm based on fine prediction strategy and nondominated solutions-guided evolution. Appl. Intell. (2023)
16. Qu, Z., et al.: Optimal design of agricultural mobile robot suspension system based on NSGA-III and TOPSIS. Agriculture **13**, 207 (2023)
17. Pisarenko, A., Malyshev, D., Rybak, L., Cherkasov, V., Skitova, V.: Application of evolutionary PSO algorithms to the problem of optimization of 6-6 UPU mobility platform geometric parameters. Procedia Comput. Sci. **213**, 643–650 (2022)

18. Podinovsky, V.V.: Ideas and methods of the theory of the importance of criteria. Nauka (2019)
19. Malyshev, D., Cherkasov, V., Rybak, L., Diveev, A.: Synthesis of trajectory planning algorithms using evolutionary optimization algorithms. In: Advances in Optimization and Applications: 13th International Conference (2022)
20. Merlet, J.-P.: Parallel Robots, vol. 28, 2nd edn. Springer, Heidelberg (2007)
21. Malyshev, D., Rybak, L., Gaponenko, E., Voloshkin, A.: Algorithm for determining the singularity-free and interference-FreeWorkspace of a robotic platform for fruit harvesting. Eng. Proc. **33**, 33 (2023)
22. Maminov, A.D., Posypkin, M.A.: Research and developing methods of solving engineering optimization problems for parallel structure robots. Int. J. Open Inf. Technol. **7**(11), 1–7 (2019)
23. Gosselin, C., Angeles, J.: Singularity analysis of closed-loop kinematic chains. IEEE Trans. Robot. Autom. **6**(3), 281–290 (1990)
24. Behera, L., Rybak, L., Malyshev, D., Gaponenko, E.: Determination of workspaces and intersections of robot links in a multi-robotic system for trajectory planning. Appl. Sci. **11**, 4961 (2021)
25. Stocco, L., Salcudean, S., Sassani, F.: Fast constrained global minimax optimization of robot parameters. Robotica **16**(06), 595–605 (1998)
26. Holland, J.H.: Adaptation in Natural and Artificial Systems. MIT Press, Cambridge (1975)
27. Diveev, A.I., Konstantinov, S.V.: Evolutionary algorithms for the problem of optimal control. RUDN J. Eng. Res. **18**(2), 254–265 (2017)
28. Mirjalili, S., Mirjalili, S.M., Lewis, A.: Grey wolf optimizer. Adv. Eng. Softw. **69**, 46–61 (2014)
29. Diveev, A.I., Konstantinov, S.V.: Optimal control problem and its solution by grey wolf optimizer algorithm. RUDN J. Eng. Res. **19**(1), 67–79 (2018)
30. Kennedy, J., Eberhart, R.: Particle swarm optimization. In: Proceedings of IEEE International Conference on Neural Networks, vol. 4, pp. 1942–1948 (1995)
31. Shi, Y., Eberhart, R.C.: A modified particle swarm optimizer. In: Proceedings of IEEE International Conference on Evolutionary Computation, vol. 7, no. 1, pp. 69–73 (1998)

Application of Recursive Algorithms for Optimization and Approximation of Workspace of Parallel Robots

Anton Pisarenko, Dmitry Malyshev(✉), Larisa Rybak, and Victoria Perevuznik

Belgorod State Technological University named after V.G. Shukhov, Belgorod 308012,, Russian Federation
rlbgtu@gmail.com

Abstract. The article discusses the application of various methods for transforming the workspace of a parallel 3-PRRR robot to reduce data storage and simplify the visualization process. Two approaches to transforming the workspace have been proposed and described. Transformation is a re-decomposition of the workspace into geometric shapes without loss of approximation quality. The first approach is finding the minimum covering a set of rectangles for each face of the outside of the workspace. The basis of the approach is the application of the depth-first search algorithm for a bipartite graph. The second approach is based on a constrained Delaunay triangulation algorithm to each outer face of the workspace. A comparative assessment of the results of the proposed approaches is given.

Keywords: Robot workspace · Depth-first search · Delaunay triangulation

1 Introduction

In the process of developing modern productive robotic mechanisms and platforms, several problems arise, such as: solving the direct problem of kinematics, increasing the accuracy of determining the workspace, designing an effective control system, etc. [1]. The introduction and improvement of the spatial parallel mechanisms requires the solution of these problems. Compared with serial mechanisms, they have several advantages, such as: high positioning accuracy, effective lifting capacity, speed and smooth running [2]. One of the most important tasks in designing robots is workspace determination. Workspace determination is based on various methods, including geometric, discrete, and interval analysis. Geometric and discrete methods have several disadvantages. In particular, geometric methods need to be adapted individually for each spatial robot structure.

This work was supported by the Russian Science Foundation, the agreement number 19-19-00692. The work was realized using equipment of High Technology Center at BSTU named after V.G. Shukhov.

N. Olenev et al. (Eds.): OPTIMA 2023, CCIS 1913, pp. 256–267, 2024.
https://doi.org/10.1007/978-3-031-48751-4_19

It is difficult to consider all the limitations, and therefore, it is difficult to use for complex spatial robots. The discrete methods do not allow for a workspace with a high approximation accuracy. The representation of the workspace boundary can include a large number of nodes, which requires a lot of resources for computing and storing data. The universal STL format is widely used for representing three-dimensional models, in which information about an object is stored in the form of a list of triangular faces and their normals. The number of faces determines the volume of the model and the speed of processing. For this reason, the problem arises of optimizing data for export to a specified format while visualizing a three-dimensional model of the robot's workspace. Often the workspace of the robot has a complex shape with internal cavities. Based on this, the task of reducing the number of triangular faces (polygons) describing the covering set of the workspace appears. Let's consider the solution of this problem for the workspace of the 3-PRRR mechanism [4], which can be used as part of a robotic system for rehabilitation of the lower limbs [5]. Section 2 will present the robot model and the robot's workspace before transformation. Section 3 will describe approaches to transforming the workspace (re-decomposition) without losing approximation accuracy. Section 4 will present the simulation results.

2 Determination of the Workspace

The 3-PRRR parallel mechanism (Fig. 1) has three degrees of freedom namely movement along the axes. The mechanism consists of three serial kinematic chains and has three degrees of freedom – for translational movements along each of the axes. The position of the rehabilitation platform, which is an equilateral triangle $D_1D_2D_3$ centered at point P and the radius of the circumscribed circle R, is determined by linear displacements $q_1q_2q_3$. Let us introduce the following notation: a_i is the distance between points A_i and B_i, b_i is between B_i and C_i, c_i is between C_i and D_i.d_i is between B_i and D_i.

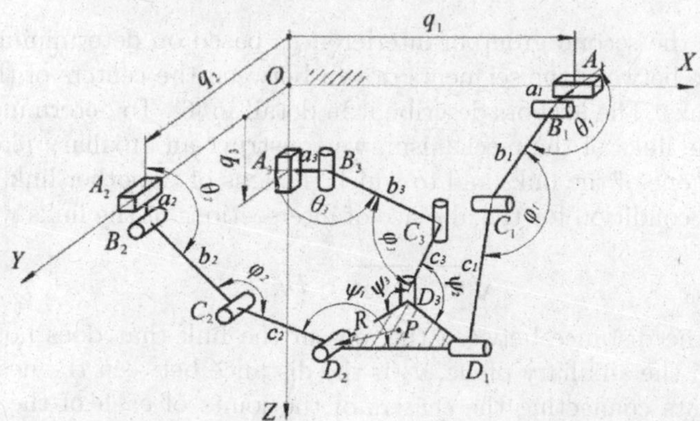

Fig. 1. Structural diagram of a parallel 3-PRRR robot

We will take into account the following design constraints on the parameters of the mechanism:

- on the drive coordinate q:

$$q_i \in [q_{min}, q_{max}] \tag{1}$$

- the distance d between the centers of the joints B_i and D_i

$$d_i \in [0; d_{max}], \tag{2}$$

where $d_{max} = b + c$.

The distances d_i can be defined as the distance between B_i and D_i:

$$d_1 = \sqrt{\left(y_P + \frac{R}{2} - a_1\right)^2 + (z_P)^2}, \tag{3}$$

$$d_2 = \sqrt{\left(x_P - \frac{\sqrt{3}}{2}R\right)^2 + (z_P - a_2)^2}, \tag{4}$$

$$d_3 = \sqrt{(x_P - a_3)^2 + (y_P - R)^2} \tag{5}$$

- the first group of link intersections are intersections at small angles between links connected by rotary joints.

- the second group of link intersections are the intersection of links that are not connected.

The first group can be determined, taking into account the restrictions on the angles of rotation in the joints B_i,C_i and D_i:

$$\begin{cases} \theta_i \in [\theta_{min}, \theta_{max}], \\ \phi_i \in [\phi_{min}, \phi_{max}], \\ \psi_i \in [\psi_{min}, \psi_{max}] \end{cases} \tag{6}$$

We define the second group of interferences based on determining the minimum distance between the segments drawn between the centers of the joints of each of the links. The authors describe it in detail in [6]. To determine the intersections of the links of the mechanism, we construct an auxiliary plane parallel to the axis of one of the links and to which the axis of the other link belongs. In this case, the condition for the absence of intersections of the links will take the form

$$\sqrt{{u_1}^2 + {u_2}^2} > D_{link}, \tag{7}$$

where u_1 is the distance between the axis of the link that does not belong to the plane and the auxiliary plane, u_2 is the distance between the nearest points of the segments connecting the centers of the joints of each of the links when projecting a segment that does not belong to the auxiliary plane onto this plane, D_{link} is the diameter of the links.

The algorithm for determining the workspace, using the above restrictions, as well as its analysis was considered in detail by the authors earlier in [7]. The covering set of the resulting workspace (Fig. 2) is presented as a partially ordered set of integers.

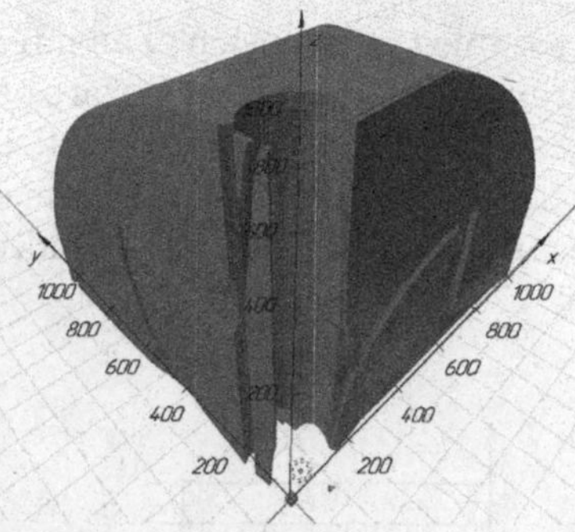

Fig. 2. Workspace of the 3-PRRR mechanism, taking into account the intersections of the links.

The workspace consists of many elongated boxes (Fig. 3). Each of the boxes is described in one ("long") dimension by two integers – the lower and upper boundaries of the interval. For each of the other dimensions, the box has one number describing the coordinate. For processing and rendering, auxiliary coefficients are used for each of the dimensions, allowing the coordinates to be recalculated in real number space. Rendering is done by dividing the rectangular faces of the boxes diagonally into two triangles and forming an STL file from these triangular faces.

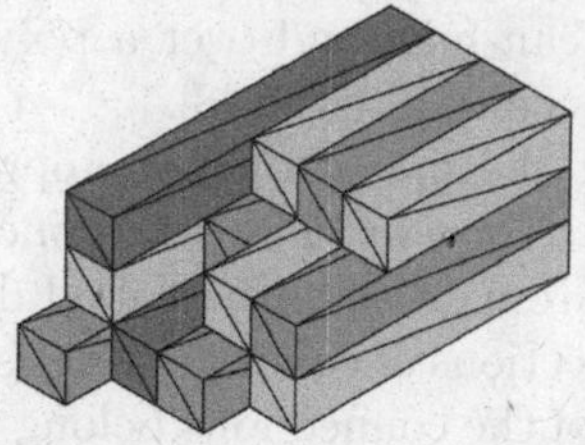

Fig. 3. Sketching data representation using boxes.

Consider approaches to transforming the resulting covering set to reduce the number of triangular faces in a three-dimensional model of the workspace. Reducing the number of triangles is important to reduce the amount of model data and speed up rendering. Transformation is a re-decomposition of the workspace into geometric shapes without loss of approximation quality. As a result of the two types of transformation, a set of triangles will be obtained, covering the outer edges of the workspace in different ways.

3 Transforming the Covering Set for Rendering

The proposed approaches assume the exclusion of all internal surfaces that do not need to be used for rendering and for external surfaces to minimize the number of triangular faces. Let us consider two approaches to finding the minimum covering by triangles of external two-dimensional surfaces. As an example, let's take a random set of polygons (Fig. 4). The faces of individual boxes are marked with numbers (1 - 18).

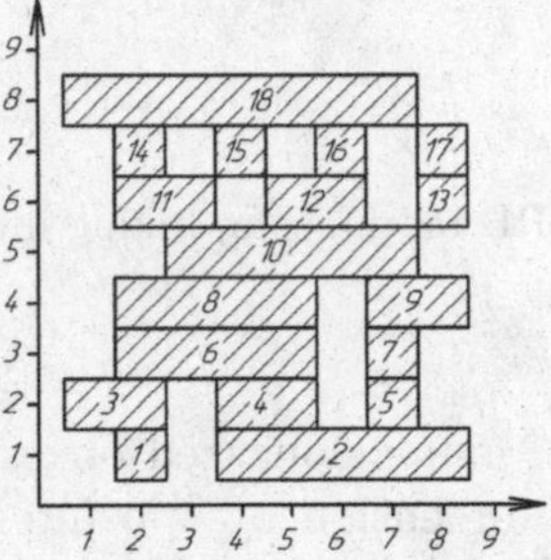

Fig. 4. Sketch of the original data representation.

The first approach is to find the maximum number of non-intersecting line segments connecting the concave vertices of the source area under consideration. We take a bipartite graph as the form of representation of the intersection of segments, therefore it is necessary to find the maximum independent set of a bipartite graph. Such a set can be found over a polynomial time interval. The solution to this problem was considered in [8].

Consider the features of the first proposed approach. First, we define the boundary points of each box. Then we define the concave vertices of the polygon (Fig. 5a). Concave vertices are marked with shaded dots.

We find all possible connections of concave vertices, horizontal or vertical segments, so that all segments of the connections belong to the workspace (Fig. 5b). If for a concave vertex there is no other concave vertex that could be connected vertically or horizontally, without leaving the workspace, then we skip this point.

We compose a bipartite graph of intersections of segments of concave vertices. The vertices of the graph correspond to segments. An intersection is considered to be the case when two random segments have at least one common point [9]. Thus, the edges of a bipartite graph correspond to the intersection of two independent segments (Fig. 6).

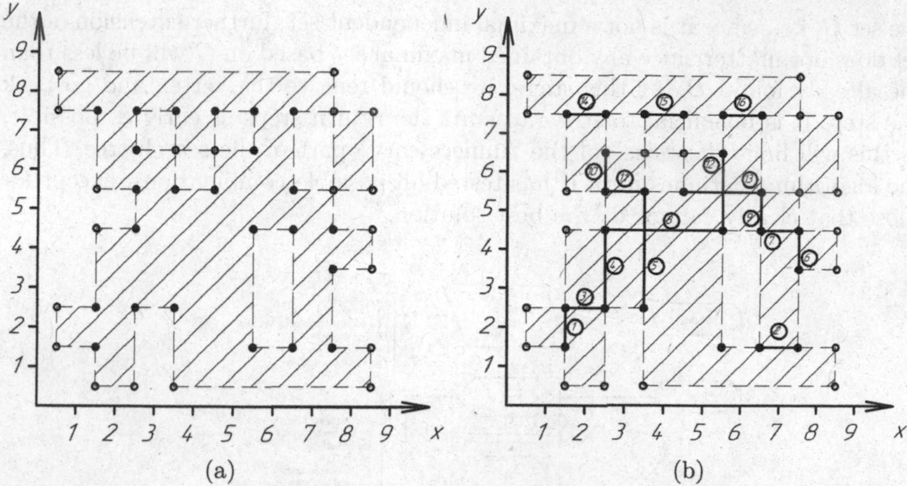

Fig. 5. Sketches of the sequence of actions: (a) sketch of vertices: black dots - concave vertices (b) sketch of connected concave vertices.

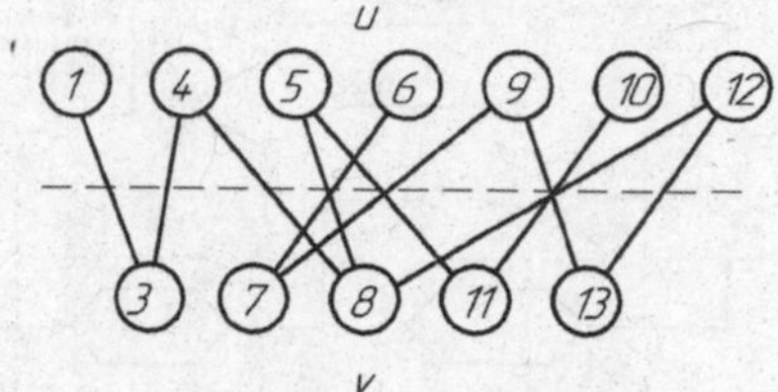

Fig. 6. Bipartite graph.

After compiling a bipartite graph, we can start looking for the maximum independent set. As a search algorithm, we take Depth-first search (DFS). Depth-first search consists in sequentially expanding an independent set, moving along only one branch of the search tree. Let's consider the proposed search algorithm (Fig. 7). We take a bipartite graph $G\{V, E\}$, where V is the set of vertices of the graph, E is the courage of its edges. We will store the list of visited vertices in the set q. We start traversing the graph from any available vertex V. For any vertex V adjacent to only one other vertex ($e = 1$), only one traversal is possible namely from the opposite adjacent vertex. In this case, at the initial stage, it makes sense to exclude the vertices that have only one connection with another vertex, thereby reducing the total number of vertices that need to be visited ($G \setminus V_j$), and, consequently, the number of iterations will also decrease. With an increase in the number of vertices a problem arises such as limited computer resources. The solution of this problem can be facilitated by comparing the new obtained set G with the best found B. At the next extension of the independent set G, one should check its size with the size of the best found maximum set. If the size of G set is equal to the size of B set and there are extension paths for the set G, i.e., since it is not a maximal independent set, further extension of the set does not matter since any obtained maximal set based on G will be less than the already found B. At this stage, we should remove the vertex and go back one step. It is beneficial to strive to start the return steps as early as possible, as this will limit the "size" of the "unnecessary" part of the search tree. Thus, the algorithm continues until it has tested all possible combinations, except for those that clearly cannot be the best solution.

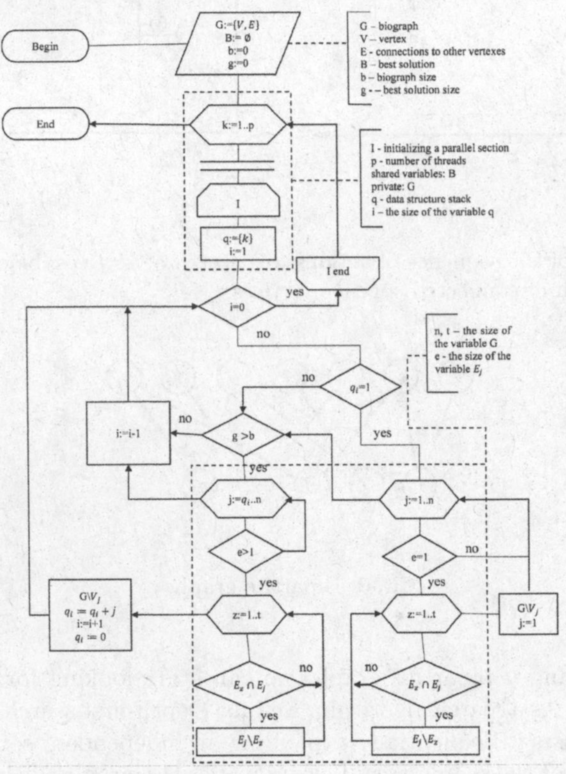

Fig. 7. Algorithm for finding the maximum independent set

After finding the maximum independent set, we can start transforming the original workspace. First, distribute the original boxes into new polygons or rectangles. The restrictions are: the maximum independent set of a bipartite graph found, corresponding to the segments connecting the curved vertices, and also to these segments, we add segments that did not initially intersect. After the transformations, the new workspace will take the form shown in Fig. 8a. Divide the resulting polygons from each remaining concave vertex. We do it until all the shapes are transformed into rectangles (Fig. 8b). The final step will be to split the rectangles along the diagonal (Fig. 8c) and export the resulting data to STL format.

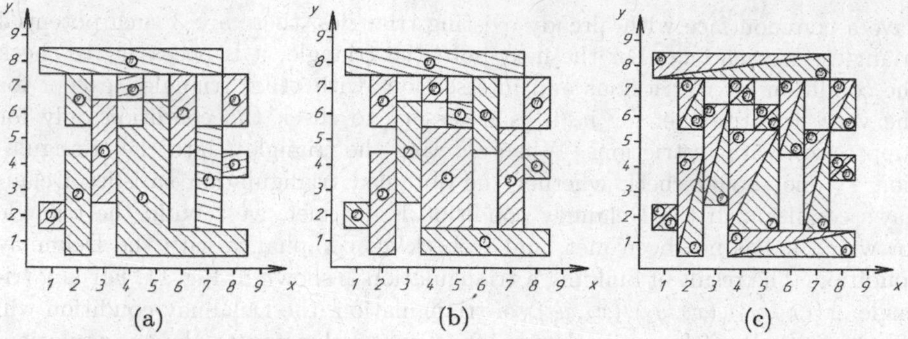

Fig. 8. Transform results (a) the result after the first transformation, (b) the result after the second transformation, (c) the final result.

The second approach is to perform Delaunay triangulation [10] of two-dimensional contours of the outer surfaces of the workspace with constraints, which are the boundary of the contour of the workspace faces. Various types of triangulations and methods of implementation are discussed in detail in [11]. Delaunay triangulation is such a triangulation in which for any constructed triangle it is true that inside the circumscribed circle there are no vertices from the original set (Fig. 9a). For nonconvex quadrangles, the Delaunay condition is always satisfied, and for convex quadrangles (whose vertices do not lie on the same circle) there are exactly 2 possible triangulations (one of which is a Delaunay triangulation) (Fig. 9b).

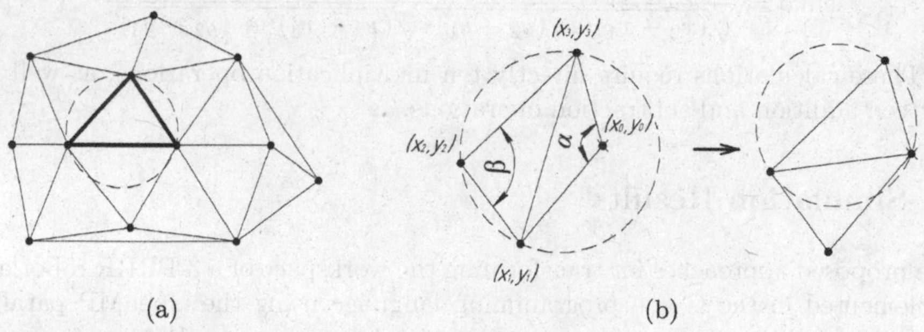

Fig. 9. (a) verification of the Delaunay condition, (b) rebuilding algorithm.

We start by defining the boundary points of the box (Fig. 5a) using the coordinates of all the vertices A and the constraints B, which are the boundaries of the workspace. Let us consider the algorithm for constructing a triangulation (Fig. 10). We build the first triangle, one face of this triangle must necessarily belong to the border of the workspace, in this case these are points A_1 and A_2. From each vertex of the original set A, we construct new triangles t that have a common face with already existing triangles, there are 3 such potential triangles for each face. For the next potential triangle, it is necessary to check the condition for restrictions and intersections with other triangles, except for the very first triangle, for it, it is necessary to check the condition only for suppression with restrictions. When entering the triangle t into the triangulation T, one should check whether the obtained triangulation satisfies Delaunay's condition. If the Delaunay condition is not met, we rebuild the triangle for which it has not been met, and recheck for compliance with the Delaunay condition. The result of building a triangulation is shown in Fig. 11. For any triangle $\Delta((x_1, y_1), (x_2, y_2), (x_3, y_3))$ of triangulation, the Delaunay condition will be satisfied only if for any other point (x_0, y_0) belonging to the triangulation, the condition $\alpha + \beta \leq \pi$ (Fig. 9, b), this condition is equivalent to:

$$\sin\alpha\cos\beta + \cos\alpha\sin\beta \geq 0 \tag{8}$$

The values of the sines and cosines of the angles can be calculated using the formulas:

$$\cos\alpha = \frac{(x_0 - x_1)(x_0 - x_3) + (y_0 - y_1)(y_0 - y_3)}{\sqrt{(x_0 - x_1)^2 + (y_0 - y_1)^2}\sqrt{(x_0 - x_3)^2 + (y_0 - y_3)^2}} \tag{9}$$

$$\cos\beta = \frac{(x_2 - x_3)(x_2 - x_1) + (y_2 - y_3)(y_2 - y_1)}{\sqrt{(x_2 - x_3)^2 + (y_2 - y_3)^2}\sqrt{(x_2 - x_1)^2 + (y_2 - x_1)^2}} \tag{10}$$

$$\sin\alpha = \frac{(x_0 - x_1)(y_0 - y_3) - (x_0 - x_3)(y_0 - y_1)}{\sqrt{(x_0 - x_1)^2 + (y_0 - y_1)^2}\sqrt{(x_0 - x_3)^2 + (y_0 - y_3)^2}} \tag{11}$$

$$\sin\beta = \frac{(x_2 - x_3)(y_2 - y_1) - (x_2 - x_1)(y_2 - y_3)}{\sqrt{(x_2 - x_1)^2 + (y_2 - y_1)^2}\sqrt{(x_2 - x_3)^2 + (y_2 - y_3)^2}} \tag{12}$$

These calculations require directly ten multiplication operations, as well as thirteen addition and subtraction operations.

4 Simulation Results

The proposed approaches for transforming the workspace of a 3-PRRR robot are implemented in the C ++ programming language using the OpenMP parallel computation library. The calculations were carried out on a 16-core processor with 32 threads. In the process of calculations, the inner surfaces were excluded from the coverage of the workspace, and for each of the outer surfaces, represented as a two-dimensional partially ordered set of integers [3], the above transformation approaches were implemented. In the first approach, the search section for the maximum independent set of a bipartite graph was parallelized. Each thread was assigned one "branch" of the search tree, and an individual variable of the "stack" type was created, which stores information about the visited graph vertices. Simulation was performed for the following initial data: b = 800 mm, c = 550 mm, a = 50 mm, R = 50 mm q_{min} = 0 mm, q_{max} =

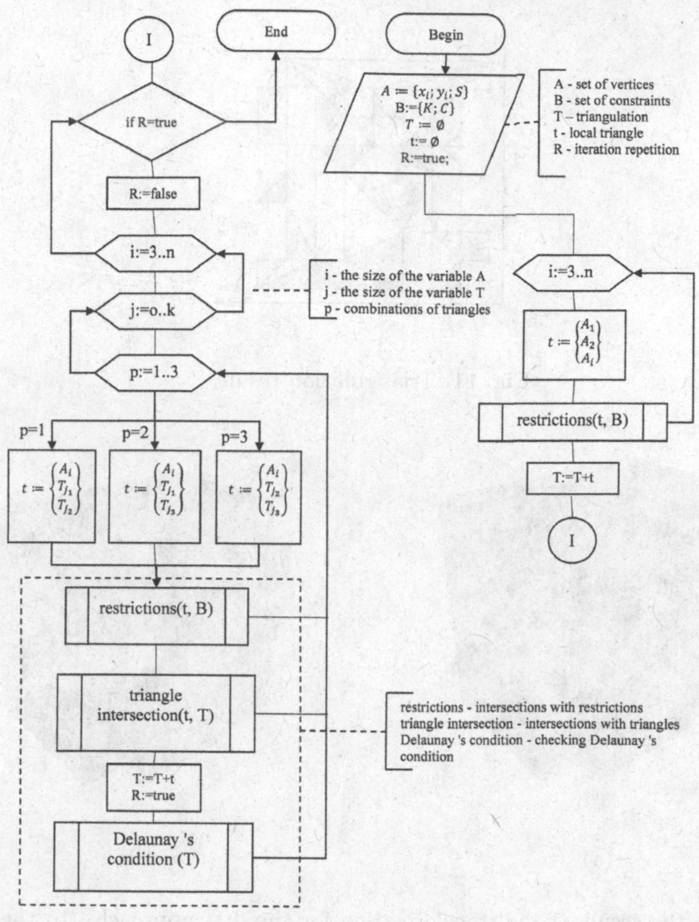

Fig. 10. Constrained Delaunay triangulation algorithm

1000 mm, $\phi_{min} = \theta_{min} = \psi_{min} = 10°$, $\phi_{max} = \theta_{max} = \psi_{max} = 170°$, $D_{link} = 20$ mm. The simulation results with an approximation accuracy of $\delta = 16$ mm for the first approach are shown in Fig. 12a, for the second in Fig. 12b. The exclusion time for inner surfaces was 4 s. The transformation time of the outer surfaces for both approaches was less than 1 s.

The quantitative characteristics of the simulation results are shown in Table 1.

The minimum number of triangular faces and the size of the 3D model of the workspace have been achieved for the second approach. Compared to the original workspace, the number of triangles has decreased by 96.12%. Separately, it should be noted that the original workspace is represented as an ordered set of integers [3]. Therefore, representing the resulting workspace as a widely used

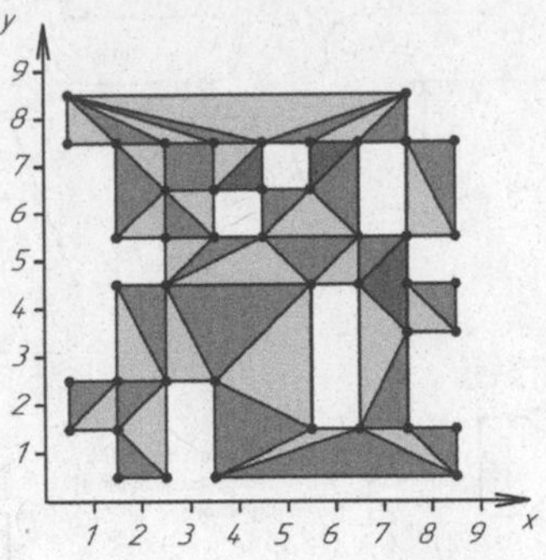

Fig. 11. Triangulation result.

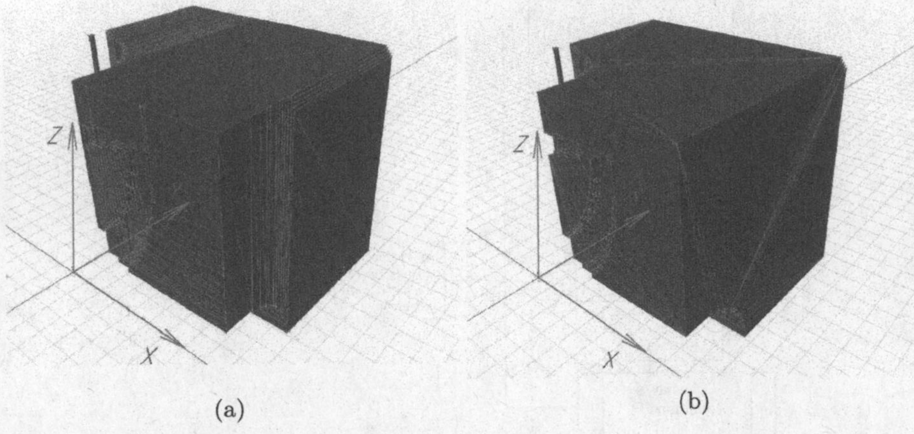

Fig. 12. (a) the result of the transformation for the first approach, (b) the result of the transformations for the second approach.

set of points with a discrete step would include 223 641 points, which is significantly more than the proposed approaches. Conclusion. The simulation result confirmed the efficiency of workspace transformations obtained using interval analysis methods. The size of the STL file of the 3D model for both the first approach and the second approach is significantly reduced compared to the original representation of the workspace. Compared to the original workspace, the number of triangles has decreased by 96.12%. The resulting workspace model can be easily exported to any CAD system with STL format support, for subsequent processing and analysis. In this case, the boundary of the transformed covering set fully corresponds to the boundary of the original covering set obtained using interval analysis methods, which means maintaining the specified approximation accuracy. As part of future work, the number of obtained triangular faces will be reduced by combining them and reshaping the contours of the surfaces. It

Table 1. Transformation results.

	The number of triangular faces	STL file size, bytes
Original workspace	50 844	2 542 284
Without internal surfaces	6 534	326 784
First approach	1 994	99 784
Second approach	1 972	98 684

is planned to optimize the algorithm for eliminating internal surfaces, and the algorithms will be modified using parallel computations.

References

1. Chablat, D., Wenger, P.: Architecture optimization of a 3-DOF translational parallel mechanism for machining applications, the orthoglide. IEEE Trans. Robot. Autom. **1733**, 403–410 (2003). https://doi.org/10.1109/2003.810242
2. Merlet, J.P.: Parallel Robots. Springer, Dordrecht (2006). https://doi.org/10.1007/1-4020-4133-0
3. Rybak, L., Malyshev, D., Gaponenko, E.: Optimization algorithm for approximating the solutions set of nonlinear inequalities systems in the problem of determining the robot workspace. In: Olenev, N., Evtushenko, Y., Khachay, M., Malkova, V. (eds.) OPTIMA 2020. CCIS, vol. 1340, pp. 27–37. Springer, Cham (2020). https://doi.org/10.1007/978-3-030-65739-0_3
4. Kong, X., Gosselin, C.: Kinematics and singularity analysis of a novel type of 3-CRR 3 DOF translational parallel manipulator. Int. J. Robot. Res., 791–798 (2002)
5. Mohan, S., Sunilkumar, P., Rybak, L., Malyshev, D., Khalapyan, S., Nozdracheva, A.: Conceptual design and control of a sitting-type lower-limb rehabilitation system established on a spatial 3-PRRR parallel manipulator. In: Zeghloul, S., Laribi, M.A., Sandoval Arevalo, J.S. (eds.) RAAD 2020. MMS, vol. 84, pp. 345–355. Springer, Cham (2020). https://doi.org/10.1007/978-3-030-48989-2_37
6. Behera, L., Rybak, L., Malyshev, D.: Determination of workspaces and intersections of robot links in a multi-robotic system for trajectory planning. Appl. Sci., 4961 (2021). https://doi.org/10.3390/app11114961
7. Ahmetzhanov, M., Rybak, L, Malyshev, D., Mohan, S.: Determination of the workspace of the system based on the 3-PRRR mechanism for the lower limb rehabilitation. in: Advances in Mechanism Design III. TMM (2020). Mechanisms and Machine Science, vol. 85, pp. 193–203. Springer, Cham, (2020). https://doi.org/10.1007/978-3-030-83594-1_20
8. Wu, S., Sahni, S.: Fast algorithms to partition simple rectilinear polygons. VLSI Design **1**(3), 193–215 (1994)
9. Kristofides, N.: Graph theory. An Algorithmic Approach. Mir, Moscow (1978)
10. Delaunay, B.: On the empty sphere. Iz-vo. AN USSR. OMEN, pp. 793–800 (1934)
11. Skvortsov, A., Mirza, N.: Algorithms for constructing and analyzing triangulation. Tomsk: Izd-vo vol. un-ta (2006)

Discrete Tomography Problems with Paired Projections and Complexity Characteristics

Hasmik Sahakyan[1(✉)] and Levon Aslanyan[2]

[1] Institute for Informatics and Automation Problems of NAS RA, 1 P.Sevak, 0014 Yerevan, Armenia
hsahakyan@sci.am

[2] National Academy of Sciences of Republic of 24, Bagramyan avenue, 0019 Yerevan, Armenia
lasl@sci.am

Abstract. This paper presents a study of discrete tomography problems with horizontal, vertical, and paired projections, where paired projections reflect the conditions of generalized degree sequences of hypergraphs. Paired projections, as far as we know, are considered for the first time in the problems of discrete tomography. The general purpose of introducing these projections is to involve more information about the object under consideration in order to narrow the class of all possible solutions when a solution exists. On the other hand, the consideration of pairwise projections successfully complements our previous research, which is devoted to the study of structures with different classes. In this paper, we consider different types of pairwise projections, including the set of all pairs of columns, the set of all pairs with one fixed column, and the set of all pairs of neighboring columns; and investigate the complexity characteristics of the corresponding problems.

Keywords: Discrete tomography · Pair-based projections

1 Introduction

Discrete Tomography [9] deals with the reconstruction of discrete objects and structures [1,6,23] from their projections [2,16] composed along a given set of directions. Discrete tomography has various applications in different fields, including image processing, material science, medical problems, and others.
In the simplest case, discrete object T is a subset of vertices of the integer lattice Z^m. Due to the complexity of the problem (the consistency and reconstruction problems are NP-hard for $m \geq 2$ directions and for $n \geq 3$ non-parallel projections [11], special attention has been given to the 2-dimensional case where

Partially supported by grants No. 21T-1B314 and No. 21SC-BRFFR-1B029 of the Science Committee of MESCS RA.

N. Olenev et al. (Eds.): OPTIMA 2023, CCIS 1913, pp. 268–280, 2024.
https://doi.org/10.1007/978-3-031-48751-4_20

discrete objects can be presented as binary images or binary matrices. Numerous researches are devoted to the case of horizontal and vertical projections. In terms of binary matrices, these are the row and column sums of the matrix, respectively. The existence and reconstruction problems of binary matrices by their row and column sums is considered and solved in 1957 [10,16]. But in this case, the number of solutions can be exponentially large [15]. An idea to reduce the set of possible solutions is to use a priori information/property of the object to be recovered, if such property exists [24]. Two commonly used in this context geometrical properties are convexity and connectivity. The existence problem of a binary matrix is NP-complete for horizontal or vertical convex, as well as for horizontal or vertical convex and connected matrices [4]. NP-completeness of the case of 4-connected matrices, as well as of horizontal and vertical convex matrices is proved in [11,22]. The case of horizontal and vertical convex and connected matrices is solvable in polynomial time [7,14]. Another idea to reduce the number of possible solutions is to take further projections along different lattice directions. Reconstruction problem for the case of horizontal, vertical and diagonal projections is considered and NP-completeness is proved in [5]. For some cases (horizontal, vertical, diagonally connected and convex matrices) the problem is solvable in polynomial time [4]. The uniqueness and reconstruction problems for the case of diagonal and anti-diagonal projections are considered in [21]. The case of diagonal and horizontal projections is investigated in [20]. A special series of investigations is devoted to the structures with the different rows [3,19].

In this work, we follow the concept of extending the information about the reconstructed object, to achieve a narrowing of the set of possible solutions. To do this, we introduce pairwise projections and study the reconstruction of discrete sets in Z^2 with respect to horizontal, vertical, and pairwise projections.

It is worth noting that pair-based projections reflect the conditions of generalized degree sequences of hypergraphs. On the other hand, pair-based projections can be advantageous in specific situations where the relationship between pairs of columns (particularly, between adjacent pairs) is important; and resolving ambiguous situations that may arise in case of individual projections.

We consider different types of pair-based projections and investigate the complexity of the corresponding problems based on establishing a relation with hypergraphs, or representing the problems as systems of integer linear inequalities/constraints.

The rest of the paper is organized as follows. Subsections 1.1 and 1.2 present necessary definitions, preliminaries, and basic concepts. Main results are given in Sect. 2. Section 3 provides conclusions and intentions the vision for possible future research.

1.1 Projections, Convexity, Connectivity

Let $G_{m,n}$ denote a two-dimensional grid with m rows and n columns. A vertex set T in $G_{m,n}$ can be identified with binary matrix $A = \{A_{i,j}\}$ of m rows and n columns, where 1 s of the matrix correspond to the points of T. Let

$R = (r_1, \cdots, r_m)$, $C = (c_1, \cdots, c_n)$ and $D = (d_1, \cdots, d_{m+n-1})$ denote the row, column and diagonal (from top left to bottom right) sum vectors of A respectively, given as:

$$r_i = \sum_{j=1}^{n} a_{i,j}, i = 1, \cdots, m$$

$$c_j = \sum_{i=1}^{m} a_{i,j}, j = 1, \cdots, n$$

$$d_k = \begin{cases} \sum_{i-j=n-k} a_{i,j}, \text{if } 1 \leq k \leq n \\ \sum_{j-i=k-n} a_{i,j}, \text{if } n+1 \leq k \leq m+n-1. \end{cases}$$

Anti-diagonal (from bottom left to top right) sum vector $D^a = (d_1^a, \cdots, d_{m+n-1}^a)$ can be defined correspondingly:

$$d_k^a = \sum_{i+j=k+1} a_{i,j}, k = 1, \cdots, m+n-1.$$

Then, R, C, D and D^a are projections of T along the horizontal, vertical, diagonal and anti-diagonal directions.

Matrix $A = \{a_{i,j}\}$, and object T corresponding to it is horizontally convex (h-convex) (vertically convex (v-convex), diagonally convex (d-convex)) if 1 s in every row (in every column, in every diagonal line) of A are consecutive. $A = \{a_{i,j}\}$ is hv-convex if it is both h-convex and v-convex, and, is hd-convex if it is both h-convex and d-convex [7]. $A = \{a_{i,j}\}$ is hv-connected if the graph, obtained by representing every 1 by a vertex and by putting an edge between every pair of vertices representing horizontally or vertically adjacent 1 s, is connected. $A = \{a_{i,j}\}$ is hd-connected if the graph, obtained by representing every 1 by a vertex and by putting an edge between every pair of vertices representing horizontally or diagonally adjacent 1 s, is connected [9].

1.2 Hypergraph Degree Sequences

The projection of objects in discrete spaces along a given ray highlights some special subset of object points. When the ray is fixed, subsets of the interval of points in this direction are obtained and this is conveniently corrected with the notion of an edge in the corresponding hypergraph and with the degree of vertices of such hypergraphs. This mainly defines the graph-theoretic methodology [12, 13] chosen in this paper. Let us start with some definitions.

A hypergraph H is an ordered pair $H = (V, E)$ consisting of a finite nonempty set V, called the vertex set, and a finite multiset E, called the edge/hyperedge set, which is a family of nonempty subsets of V. H is simple if it has no repeated edges (i.e., E is a set). H is r-uniform if each edge of H has cardinality r. Note that simple 2-uniform hypergraphs are just simple graphs.

Degree of a vertex $v \in V$ is the number of hyperedges of H, containing v. Degree sequence of $H = (V, E)$ is a sequence consisting of degrees of its vertices

(usually reordered in a non-increasing order) [17,18].
Let it is given a hypergraph $H = (V, E)$, where $V = \{v_1, v_2, \cdots, v_n\}$, $E = \{e_1, e_2, \cdots, e_m\}$. The incidence matrix $X = \{x_{i,j}\}$ of H is an $m \times n$ binary matrix such that $x_{i,j} = 1$ if and only if $v_j \in e_i$.

The vertical projection of a binary matrix X corresponds to the degree sequence of a hypergraph having X as its incidence matrix, and the horizontal projection corresponds to the sizes of its hyperedges.

Let $H = (V, E)$ be a k-uniform hypergraph and $v \in V$. In [8] the concept of "subsuming" into a hypergraph was considered in the following way:

H subsumes the collection of hypergraphs $\{H_v | v \in V\}$, where H_v is a $(k-1)$-uniform hypergraph with the vertex set $V(H_v) = V - v$, and edge set $E(H_v)$ consisting of all edges $e - v$, i.e., all edges $e \in E$ for which $v \in e$, but without the vertex v. Hypergraph H is simple if and only if all the subsumed hypergraphs are simple.

Subsuming process itself by the set of subgraphs shows a high similarity to the process of object reconstructing by its projections. The following problems are introduced in [8], and their NP-completeness is proven.

DEG_SEQ [8]
Given *DegSeq*(g_i), $i = 1, 2, \cdots, n$, the degree sequences of n graphs $g_1, g_2, \cdots, g_n$. Is there a 3-uniform subsuming hypergraph G such that the subsumed graphs $G_1, G_2, \cdots, G_n$ satisfy *DegSeq*(G_i) = *DegSeq*(g_i) for $i = 1, 2, \cdots, n$?

DEG_SEQ_S (for simple hypergraphs) [8])
Given *DegSeq*$(g_i), i = 1, 2, \cdots, n$, the degree sequences of n simple graphs $g_1, g_2, \cdots, g_n$. Is there a simple 3-uniform subsuming hypergraph G such that the subsumed graphs $G_1, G_2, \cdots, G_n$ satisfy *DegSeq*(G_i) = *DegSeq*(g_i) for $i = 1, 2, \cdots, n$?

2 Discrete Tomography Problems with Pair-Based Projections

In this section we investigate a class of discrete tomography problems with pair-based projections. Vertical pair is a pair of columns in the binary matrix, horizontal pair is a pair of rows. We define 3 types of vertical pairs, and corresponding projections: unordered pairs of all columns; unordered pairs of neighboring columns; unordered pairs of columns with one fixed column. Horizontal pairs can be defined similarly.

2.1 Problems

Definition 1. *(1) $S = (s_1, s_2, \cdots, s_{C_n^2})$ is called A_VP projection of a binary matrix $X = \{x_{i,j}\}$ of size $m \times n$, if s_k is the number of intersecting 1 s in the k-th pair of columns, for $k = 1, \cdots, C_n^2$ (supposed that the pairs are enumerated beforehand, e.g. in the lexicographic order).*

(2) $S = (s_1, s_2, \cdots, s_{n-1})$ *is called* N_VP *projection of a binary matrix* $X = \{x_{i,j}\}$ *of size* $m \times n$, *if* s_k *is the number of intersecting 1 s of the k-th and* $(k+1)$*-th columns of* X, *for* $k = 1, \cdots, n-1$.

(3) $S = (s_1, s_2, \cdots, s_{n-1})$ *is called* k_VP *projection of a binary matrix* $X = \{x_{i,j}\}$ *of size* $m \times n$, *if* s_j *is the number of intersecting 1 s of the k-th and j-th columns of* X, $j = 1, \cdots, n,\ j \neq k$.

An example is given in Table 1.

Table 1. A_VP=(4,1,6,6,2,6,5,2,1,7), N_VP=(4,2,2,7) , 1_VP=(4,1,6,6)

1	1	0	1	1
1	1	0	1	1
0	1	0	1	1
1	0	0	1	1
1	1	0	1	1
1	0	0	1	1
1	1	1	1	1
0	1	1	1	0

We formulate the following problems based on the defined projections (1)–(3).

HV_AVP. Given non-negative integer vectors $R = (r_1, r_2, \cdots, r_m)$, $C = (c_1, c_2, \cdots, c_n)$ and $S = (s_1, s_2, \cdots, s_{C_n^2})$. Is there a binary matrix $X = \{x_{i,j}\}$ of size $m \times n$, such that R, C and S are its horizontal, vertical and A_VP projections, respectively?

HV_NVP. Given non-negative integer vectors $R = (r_1, r_2, \cdots, r_m)$, $C = (c_1, c_2, \cdots, c_n)$ and $S = (s_1, \cdots, s_{n-1})$. Is there a binary matrix $X = \{x_{i,j}\}$ of size $m \times n$, such that R, C and S are its horizontal, vertical and N_VP projections, respectively?

HV_ΣNVP. Given non-negative integer vectors $R = (r_1, r_2, \cdots, r_m)$, $C = (c_1, c_2, \cdots, c_n)$ and a natural number k. Is there a binary matrix $X = \{x_{i,j}\}$ of size $m \times n$, such that R, C are its horizontal and vertical projections respectively, and the sum of components of N_VP projection equals k?

HV_k_VP. Given non-negative integer vectors $R = (r_1, r_2, \cdots, r_m)$, $C = (c_1, c_2, \cdots, c_n)$ and $S = (s_1, \cdots, s_{n-1})$. Is there a binary matrix $X = \{x_{i,j}\}$ of size $m \times n$, such that R, C and S are its horizontal, vertical and k_VP projections, respectively?

2.2 Relation with Hypergraphs

While vertical projections of binary matrices correspond to degree sequences of hypergraphs, vertical pair-based projections correspond to the generalized degree sequence concept. Let H be a hypergraph with the vertex set $[n] = \{1, 2, \cdots, n\}$.

Definition 2. *Degree of a pair of vertices (i, j) of H is the number of hyperedges of H, that contain both vertices i and j.*

In matrix terms, the degree of a pair of vertices (i, j) is the number of intersecting 1 s in the i-th and j-th columns of the adjacency matrix of H.
We consider generalized degree sequences coming from A_VP, N_VP, k_VP projections:

- generalized degree sequence of a hypergraph consisting of the degrees of all unordered pairs of vertices;
- generalized degree sequence of a hypergraph consisting of the degrees of all pairs of consecutive vertices;
- generalized degree sequence consisting of the degrees of pairs of vertices $(k,1),(k,2),\cdots,(k,k-1),(k,k+1),\cdots,(k,n)$, $1 \le k \le n$.

2.3 Complexity of HV_AVP

Theorem 1. *HV_AVP is NP-complete.*

Proof. We will show that $DEG\_SEQ \propto HV\_AVP$.
First, we reformulate DEG_SEQ in an equivalent form in terms of binary matrices.

DEG_SEQ_M : Given $V(M_i) = (v_1^{M_i}, v_2^{M_i}, \cdots, v_{n-1}^{M_i})$, $i = 1, 2, \cdots, n$, and $H(M_i) = (2, 2, \cdots, 2)$, vertical and horizontal projections of some binary matrices $M_1, M_2, \cdots, M_n$; each of them has $(n-1)$ columns. Is there a binary matrix M with n columns, for which the horizontal projection is $H(M) = (3, 3, \cdots, 3)$ and i_VP projections are equal to $V(M_i)$ for $i = 1, 2, \cdots, n$?

We note that the following information also can be derived:

- the number of rows of the matrices M_i, these are $m_i = \frac{1}{2}\sum_{j=1}^{n-1} v_j^{M_i}$, $i = 1, 2, \cdots, n$
- the vertical projection of M, this is $V(M) = (\frac{1}{2}\sum_{j=1}^{n-1} v_j^{M_1}, \frac{1}{2}\sum_{j=1}^{n-1} v_j^{M_2}, \cdots, \frac{1}{2}\sum_{j=1}^{n-1} v_j^{M_n})$
- the number of rows of M, this is $m = \frac{1}{3}(\frac{1}{2}\sum_{j=1}^{n-1} v_j^{M_1}, \frac{1}{2}\sum_{j=1}^{n-1} v_j^{M_2}, \cdots, \frac{1}{2}\sum_{j=1}^{n-1} v_j^{M_n})$.

Thus, DEG_SEQ_M can be reformulated as:

DEG_SEQ_M: Given $V(M_i) = (v_1^{M_i}, v_2^{M_i}, \cdots, v_{n-1}^{M_i})$, and $H(M_i) = (2, 2, \cdots, 2)$, vertical and horizontal projections of binary matrices M_i of size $m_i \times (n-1)$, $i = 1, 2, \cdots, n$. Is there a binary matrix M of size $m \times n$, with the horizontal projection $H(M) = (3, 3, \cdots, 3)$, vertical projection $V(M) = (m_1, m_2, \cdots, m_n)$, and i_VP projections equal to $V(M_i)$ for $i = 1, 2, \cdots, n$, where $m_i = \frac{1}{2}\sum_{j=1}^{n-1} v_j^{M_i}$ and $m = \frac{1}{3}(\frac{1}{2}\sum_{j=1}^{n-1} v_j^{M_1} + \frac{1}{2}\sum_{j=1}^{n-1} v_j^{M_2} + \cdots + \frac{1}{2}\sum_{j=1}^{n-1} v_j^{M_n})$?

We will show that $DEG\_SEQ\_M \propto HV\_AVP$.
Suppose that $I(DEG\_SEQ\_M)$ is an arbitrary instance of DEG_SEQ_M.

$I(DEG\_SEQ\_M)$: Given $V(M_i) = (v_1^{M_i}, v_2^{M_i}, \cdots, v_{n-1}^{M_i})$, $H(M_i) = (2, 2, \cdots, 2)$, – the vertical and horizontal projections of some binary matrices M_i of size $m_i \times (n-1)$, where $m_i = \frac{1}{2}\sum_{j=1}^{n-1} v_j^{M_i}$ for $i = 1, 2, \cdots, n$.

We compose the following instance of HV_AVP:
$I(HV\_AVP)$: $C = (c_1, c_2, \cdots, c_n)$, where $c_i = m_i = \frac{1}{2}\sum_{j=1}^{n-1} v_j^{M_i}$;
$R = (3, 3, \cdots, 3)$ of length m, where $m = \frac{\sum_i c_i}{3}$; and $S = (s_1, s_2, \cdots, s_{C_n^2})$ composed in the following way:

$$S = (v_1^{M_1}, v_2^{M_1}, \cdots, v_{n-1}^{M_1}, v_2^{M_2}, \cdots, v_{n-1}^{M_2}, v_3^{M_3}, \cdots, v_{n-1}^{M_3}, \cdots, v_{n-2}^{M_{n-2}}, v_{n-1}^{M_{n-2}}, v_{n-1}^{M_{n-1}}).$$

We prove that $I(HV\_AVP)$ is a positive instance of HV_AVP if and only if $I(DEG\_SEQ\_M)$ is a positive instance of DEG_SEQ_M.

Suppose that $I(HV\_AVP)$ is a positive instance, i.e. there exists a binary matrix M of size $m \times n$, such that R, C and S are its horizontal, vertical and A_VP projections (recall that pairs of columns of M are ordered lexicographically). This means that:

The first $n-1$ components of S, that is $v_1^{M_1}, v_2^{M_1}, \cdots, v_{n-1}^{M_1}$, form 1_VP projections of M. Then, the next $n-2$ components, added $v_1^{M_1}$ at the beginning, that is, $v_1^{M_1}, v_2^{M_2}, \cdots, v_{n-1}^{M_2}$ will be 2_VP projections of M (because $v_1^{M_1} = v_1^{M_2}$, since $v_1^{M_1}$ is generalized degree of the pair of vertices $(1, 2)$, and $v_1^{M_2}$ is generalized degree of the pair of vertices $(2, 1)$).

The next $n-3$ components, added $v_2^{M_1}$ and $v_1^{M_2}$ at the beginning, that is, $v_2^{M_1}, v_1^{M_2}, v_3^{M_3}, \cdots, v_{n-1}^{M_3}$ will be 3_VP projections of M (because $v_2^{M_1}$ is generalized degree of the pair of vertices $(1, 3)$, the same for $(3, 1)$, which is $v_1^{M_3}$, and $v_1^{M_2}$ is generalized degree of the pair of vertices $(2, 3)$, the same for $(3, 2)$, which is $v_2^{M_3}$, and so on.

Since i_VP is the column sum vector of M_i then $I(DEG\_SEQ\_M)$ is a positive instance.

Conversely, suppose that $I(DEG\_SEQ\_M)$ is a positive instance, i.e. there exists a binary matrix M with n columns, with the horizontal projection $H(M) = (3, 3, \cdots, 3)$ such that $V(M_i)$ are i_VP projections of M. It is easy to check that $I(HV\_AVP)$ is a positive instance.

2.4 Complexity of HV_k_VP

HV_k_VP. Given non-negative integer vectors $R = (r_1, r_2, \cdots, r_m)$, $C = (c_1, c_2, \cdots, c_n)$ and $S = (s_1, \cdots, s_{n-1})$, and integer $k, 1 \leq k \leq n$. Is there a binary matrix $X = x_{i,j}$ of size m × n, such that R, C and S are its horizontal, vertical and k_VP projections.

Note that for any permutation/rearrangement/of the rows of X, the vectors $C = (c_1, c_2, \cdots, c_n)$ and $S = (s_1, \cdots, s_{n-1})$ will remain unchanged. Therefore we claim that $X = \{x_{i,j}\}$ is a solution of HV_k_VP if and only if $X' = \{x'_{i,j}\}$ is a solution, where X' is obtained from X by rearranging the rows in such a way that 1 s in the k-th column are on the first rows.

Theorem 2. *HV_k_VP has polynomial complexity.*

Proof. Consider the following 2 problems.
HV_1. Given non-negative integer vectors $R' = (r_1, r_2, \cdots, r_{s_k})$, $C = (c_1, c_2, \cdots, c_n)$, and $S' = (s_1, \cdots, s_{k-1}, s_{k+1}, \cdots, s_n)$. Is there a binary matrix $Y = \{y_{i,j}\}$ of size $c_k \times (n-1)$, such that $(r_1 - 1, r_2 - 1, \cdots, r_{s_k} - 1)$ and S' are its horizontal and vertical projections, respectively.
HV_2. Given non-negative integer vectors $R'' = (r_{s_k+1}, r_{s_k+2}, \cdots, r_m)$, $S' = (s'_1, \cdots, s'_{k-1}, s'_{k+1}, \cdots, s'_n)$, and $C = (c_1, c_2, \cdots, c_n)$. Is there a binary matrix $Z = \{z_{i,j}\}$ of size $(m - c_k) \times (n-1)$ such that R'' and $S'' = ((c_1 - s'_1), \cdots, (c_{k-1} - s'_{k-1}), (c_{k+1} - s'_{k+1}), \cdots, (c_n - s'_n))$ are its horizontal and vertical projections, respectively.

Obviously, HV_k_VP has a solution if and only if both HV_1 and HV_2 have solutions. Then, HV_k_VP has polynomial complexity because of polynomial complexities of HV_1 and HV_2 [10, 16].

2.5 Presentation by Systems of Integer-Valued Constraints

We present the considered problems as systems of integer constraints.
HV_AVP: is the following system compatible?

$$\begin{cases} (1)\ \sum_{i=1}^{m} x_{i,j} = c_j, j = 1, \cdots, n \\ (2)\ \sum_{j=1}^{n} x_{i,j} = r_i, i = 1, \cdots, m \\ (3)\ \sum_{i=1}^{m} min(x_{i,j'}, x_{i,j''}) = s_{j',j''}, 1 \le j' < j'' \le n \\ x_{i,j} \in \{0,1\} \end{cases} \quad \text{(1-HV\_AVP)}$$

$HV\_\Sigma NVP$: is the following system compatible?

$$\begin{cases} (1)\ \sum_{i=1}^{m} x_{i,j} = c_j, j = 1, \cdots, n \\ (2)\ \sum_{j=1}^{n} x_{i,j} = r_i, i = 1, \cdots, m \\ (3)\ \sum_{j=1}^{n-1} \sum_{i=1}^{m} min(x_{i,j}, x_{i,j+1}) = k \\ x_{i,j} \in \{0,1\} \end{cases} \quad (2 - HV\_\Sigma NVP)$$

HV_NVP: is the following system compatible?

$$\begin{cases} (1)\ \sum_{i=1}^{m} x_{i,j} = c_j, j = 1, \cdots, n \\ (2)\ \sum_{j=1}^{n} x_{i,j} = r_i, i = 1, \cdots, m \\ (3)\ \sum_{i=1}^{m} min(x_{i,j}, x_{i,j+1}) = s_j, j = 1, \cdots, n-1 \\ x_{i,j} \in \{0, 1\} \end{cases} \quad \text{(3-HV\_NVP)}$$

We can also add $s_j \leq min(c_j, c_{j+1})$.
In all three systems (1) provides the vertical projection and (2) provides the horizontal projection. (3) provides the *A_VP* projection; the sum of the components of the *N_VP* projection; the *N_VP* projection of $X = \{x_{i,j}\}$, respectively.

2.6 Complexity of *HV_ΣNVP*

Theorem 3. *HV_ΣNVP is NP-complete.*

Proof. If we exchange the summands in the last equation of $(2 - HV\_\Sigma NVP)$ we will get the following equivalent system:

$$\begin{cases} (1)\ \sum_{i=1}^{m} x_{i,j} = c_j, j = 1, \cdots, n \\ (2)\ \sum_{j=1}^{n} x_{i,j} = r_i, i = 1, \cdots, m \\ (3)\ \sum_{i=1}^{m}\sum_{j=1}^{n-1} min(x_{i,j}, x_{i,j+1}) = k \\ x_{i,j} \in \{0, 1\} \end{cases} \quad (2' - HV\_\Sigma NVP)$$

As $min(x_{i,j}, x_{i,j+1}) = 1$ if and only if both $x_{i,j}$ and $x_{i,j+1}$ are 1, then $\sum_{j=1}^{n-1} min(x_{i,j}, x_{i,j+1})$ in (3-HV_NVP) counts the number of piece-wise consecutive 1 s in the i-th row $i = 1 \cdots, m$.

A particular case of $(2' - HV\_\Sigma NVP)$, when $k = \sum_{i=1}^{m}(r_i - 1)$ asks for existence of binary matrix $X = \{x_{i,j}\}$ with given row and column sums such that the number of consecutive 1 s in each rows is one less than the number of its 1 s, but this is possible if and only if the matrix is h-convex.

$$\begin{cases} \sum_{i=1}^{m} x_{i,j} = c_j, j = 1, \cdots, n \\ \sum_{j=1}^{n} x_{i,j} = r_i, i = 1, \cdots, m \\ \sum_{i=1}^{m}\sum_{j=1}^{n-1} min(x_{i,j}, x_{i,j+1}) = \sum_{i=1}^{m}(r_i - 1) \\ x_{i,j} \in \{0,1\} \end{cases} \quad (4 - H\_CONVEX)$$

Thus, $(4 - H\_CONVEX)$ is equivalent to the problem of existence of h-convex binary matrix with given horizontal and vertical projections $R = (r_1, r_2, \cdots, r_m)$ and $C = (c_1, c_2, \cdots, c_n)$, and hence, $HV\_\Sigma NVP$ is NP-complete.

2.7 Complexity of HV_NVP

In frame of equation $(3 - HV\_NVP)$ consider the sum of all $s_j, j = 1, \cdots, n-1$:

$$\sum_{j=1}^{n-1}\sum_{i=1}^{m} min(x_{i,j}, x_{i,j+1}) = \sum_{i=1}^{m}\sum_{j=1}^{n-1} min(x_{i,j}, x_{i,j+1}),$$

which is equal to $\sum_{i=1}^{m}(r_i - 1)$ if and only if the matrix is h-convex. It follows that $\sum_{j=1}^{n-1} s_j = \sum_{i=1}^{m}(r_i - 1)$ if and only if the matrix is h-convex.
If we can reduce polynomially the NP-complete problem given by (4-H_CONVEX) to HV_NVP $(3 - HV\_NVP)$ then NP-completeness of HV_NVP will be proved.

Let $I(H\_CONVEX)$ with $R = (r_1, r_2, \cdots, r_m)$, $C = (c_1, c_2, \cdots, c_n)$ be an arbitrary instance of H_CONVEX.
Compose the following series of HV_NVP instances:

$$\{I(HV\_NVP)\} : R = (r_1, r_2, \cdots, r_m), C = (c_1, c_2, \cdots, c_n),$$

and

$$A = \{S = (s_1, \cdots, s_{n-1}) \; with \; \sum_{j=1}^{n-1} s_j = \sum_{i=1}^{m}(r_i - 1)\}.$$

Our goal is to show that if $I(H\_CONVEX)$ is a positive instance of H_CONVEX if and only if there is a positive instance of this problem in $\{I(HV\_NVP)\}$, i.e. if there exists $S = (s_1, \cdots, s'_{n-1})$ in A such that $R = (r_1, r_2, \cdots, r_m)$, $C = (c_1, c_2, \cdots, c_n)$, and $S = (s'_1, \cdots, s'_{n-1})$ is a positive instance of HV_NVP.

Suppose that $I(H\_CONVEX)$ is positive, i.e. there exists $X = \{x_{i,j}\}$ with $R = (r_1, r_2, \cdots, r_m)$, $C = (c_1, c_2, \cdots, c_n)$ and with $r_i - 1$ consecutive 1 s in the i-th row. Thus,

$$\sum_{i=1}^{m}(r_i - 1) = \sum_{i=1}^{m}\sum_{j=1}^{n-1} min(x_{i,j}, x_{i,j+1}) = \sum_{j=1}^{n-1}\sum_{i=1}^{m} min(x_{i,j}, x_{i,j+1}).$$

Let $s'_j = \sum_{i=1}^{m} min(x_{i,j}, x_{i,j+1})$ for $j = 1, \cdots, n-1$. Then, $S' = (s'_1, \cdots, s'_{n-1})$ belongs to A, and thus, C, R, S' is a positive instance of HV_NVP.

The converse is obvious because if $R = (r_1, r_2, \cdots, r_m)$, $C = (c_1, c_2, \cdots, c_n)$, $S' = (s'_1, \cdots, s'_{n-1})$, where $\sum_{j=1}^{n-1} s'_j = \sum_{i=})^{m}(r_i - 1)$, - is a positive instance of HV_NVP, then the matrix must be h-convex.

For the polynomiality of the reduction we need to count the cardinality of

$$A = \{S = (s_1, \cdots, s_{n-1}) | \sum_{j=1}^{n-1} s_j = \sum_{i=1}^{m}(r_i - 1)\}.$$

We have $s_j \leq min(c_j, c_{j+1})$ for $j = 1, \cdots, n-1$, where from it follows that $\sum_{j=1}^{n-1} s_j \leq \sum_{j=2}^{n} c_j$, and this means that the difference $\sum_{j=2}^{n} c_j - \sum_{j=1}^{n-1} s_j$ can be presented as

$$\sum_{j=2}^{n} c_j - \sum_{j=1}^{n-1} s_j = x_1 + x_2 + \cdots + x_{n-1},$$

where $x_j = c_{j+1} - s_j$, $j = 1, \cdots, n-1$.
Thus,

$$\sum_{j=2}^{n} c_j - \sum_{j=1}^{n-1} s_j = \sum_{j=1}^{n} c_j - c_1 - \sum_{j=1}^{n-1} s_j =$$

$$\sum_{i=1}^{m} r_i - c_1 - \sum_{j=1}^{n-1} s_j = \sum_{i=1}^{m}(r_i - 1) + m - c_1 - \sum_{j=1}^{n-1} s_j = m - c_1.$$

So, the cardinality of A is the number of integer non-negative solutions of equation $x_1 + x_2 + \cdots + x_{n-1} = m - c_1$.

Similarly, we could state that $\sum_{j=1}^{n-1} s_j \leq \sum_{j=1}^{n-1} c_j$, and present it as $\sum_{j=1}^{n-1} c_j - \sum_{j=1}^{n-1} s_j = x_1 + x_2 + \cdots + x_{n-1}$, where $x_j = c_j - s_j$, $j = 1, \cdots, n-1$, which will lead to $x_1 + x_2 + \cdots + x_{n-1} = m - c_n$.

In a particular case when $m - c_1$ (or $m - c_n$) is constant, then $|A|$ is polynomial depending on m and n.

This means only, that if the particular case of H_CONVEX when $m - c_1$ (or $m - c_n$) is constant, is NP-complete, then this particular case of HV_NVP is NP-complete as well.

Hence the question of investigating the complexity of H_CONVEX problem for this special case reminds open.

Table 2 summarizes the problems considered and their complexities.

Table 2. Problems and complexities

HV_AVP	HV_kVP	H_CONVEX_1	HV_NVP	$HV\_\Sigma NVP$
NP-complete	P	P	?	NP-complete

3 Concluding Remarks

In this paper we studied discrete tomography problems with horizontal, vertical, and paired projections, where paired projections reflect the conditions of generalized degree sequences of hypergraphs. Introducing these projections aimed at involving more information about the object under consideration in order to narrow down the class of all possible solutions when a solution exists. On the other hand, the consideration of pairwise projections successfully complements our previous research devoted to the study of structures with different classes. In this paper we considered three types of pairwise projections: all pairs of columns, all pairs with one fixed column, and all pairs of neighboring columns. We investigated the complexity characteristics of the corresponding problems; and proved either NP-completeness or polynomiality of the problem in most cases. For the case of neighboring columns (HV_NVP) the complexity question remains open. In future studies we will continue to try to establish the complexity of the HV_NVP. We also will address the consistency and reconstruction problems with paired projections involving diagonal/antidiagonal projections as well.

References

1. Aslanyan, L.: The discrete isoperimetry problem and related extremal problems for discrete spaces. Problemy kibernetiki **36**, 85–128 (1979)
2. Aslanyan, L., Sahakyan, H.: The splitting technique in monotone recognition. Discret. Appl. Math. **216**, 502–512 (2017)
3. Aslanyan, L., Sahakyan, H.: Discrete tomography with distinct rows: relaxation. In: International Scientific and Technical Conference on Computer Sciences and Information Technologies, 2018-March, pp. 117-120 (2017)
4. Barcucci, E., Del Lungo, A., Nivat, M., Pinzani, R.: Reconstructing convex polyominoes from horizontal and vertical projections. Theor. Comput. Sci. **155**, 321–347 (1996)

5. Barcucci, E., Brunetti, S., Del Lungo, A., Nivat, M.: Reconstruction of lattice sets from their horizontal, vertical and diagonal X-rays. Discret. Math. **241**, 65–78 (2001)
6. Bollobas, B.: Modern Graph Theory, vol. 184. Springer, Cham (1998)
7. Chrobak, M., Durr, C.: Reconstructing HV-convex polyominoes from orthogonal projections. Inf. Process. Lett. **69**(6), 283–289 (1999)
8. Colbourn, C.J., Kocay, W.L., Stinson, D.R.: Some NP-complete problems for hypergraph degree sequences. Discr. Appl. Math. **14**, 239–254 (1986.Comput., 1999)
9. Herman, G.T., Kuba, A. (eds.): Discrete Tomography: Foundations, Algorithms, and Applications. Springer, Cham (2012)
10. Gale, D.: A theorem on flows in networks. Pac. J. Math. **7**, 1073–1082 (1957)
11. Gardner, R.J., Gritzmann, P., Prangenberg, D.: On the computational complexity of reconstructing lattice sets from their X-rays. Discret. Math. **202**, 45–71 (1999)
12. Godsil, C., Royle, G.: Algebraic Graph Theory, vol. 207. Springer, Cham (2001)
13. Harary, F.: Graph Theory. Addison-Wesley, Reading (1969)
14. Kuba, A.: Reconstruction in different classes of 2D discrete sets. In: Bertrand, G., Couprie, M., Perroton, L. (eds.) DGCI 1999. LNCS, vol. 1568, pp. 153–163. Springer, Heidelberg (1999). https://doi.org/10.1007/3-540-49126-0_12
15. Del Lungo, A.: Polyominoes defined by two vectors. Theoret. Comput. Sci. **127**, 187–198 (1994)
16. Ryser, H.J.: Combinatorial properties of matrices of zeros and ones. Canad. J. Math. **9**, 371–377 (1957)
17. Sahakyan, H., Aslanyan, L.: Numerical characterization of N-cube subset partitioning. Electr. Notes Discr. Math. **27**, 3–4 (2006)
18. Sahakyan, H.: Numerical characterisation of N-cube subset partitioning. Discret. Appl. Math. **157**(9), 2191–2197 (2009)
19. Sahakyan, H., Aslanyan, L., Ryazanov, V.: On the hypercube subset partitioning varieties. In 2019 Computer Science and Information Technologies (CSIT), pp. 83-88. IEEE (2019)
20. Sahakyan, H., Aslanyan, L.: Connected and weighted HD-convex tomography reconstruction. In: AIP Conference Proceedings, vol. 2757, no. 1. AIP Publishing (2023)
21. Srivastana, T., Verma, S.K.: Uniqueness algorithm with diagonal and anti-diagonal projections. Int. J. Tomogr. Simul. **23**, 22–31 (2013)
22. Woeginger, G.J.: The reconstruction of polyominoes from their orthogonal projections. Inform. Process. Lett. **77**, 225–229 (2001)
23. Zhuravlev, Yu.: Selected scientific works. Moscow, Magistr (1998)
24. Zhuravlev, Yu., Aslanyan, L., Ryazanov, V., Sahakyan, H.: Application driven inverse type constraint satisfaction problems. Pattern Recognit Image Anal. **27**(3), 418–425 (2017)

Several Edge-Disjoint Spanning Trees with Given Diameter in a Graph with Random Discrete Edge Weights

E. Kh. Gimadi(✉)

Sobolev Institute of Mathematics, prosp. Akad. Koptyuga, 4, Novosibirsk, Russia
gimadi@math.nsc.ru

Abstract. We consider the problem of finding several edge-disjoint minimum total weight spanning trees of a given diameter in an undirected Graph with random edge weights, uniformly distributed on a discrete segment. We analyze $\mathcal{O}(n^2)$-time approximation algorithm and provide some sufficient conditions for this algorithm to be asymptotically optimal.

Keywords: Several spanning trees · Given-diameter · Edge-disjoint · Minimum total weight · Approximation algorithm · Probabilistic analysis · Asymptotic optimality · Discrete edge weights

1 Introduction

The Minimum Spanning Tree (MST) problem is one of the well-known discrete optimization problems. It consists of finding a spanning tree (connected acyclic subgraph, which covers all the vertices) of a minimal weight in a given undirected weighted graph $G = (V, E)$. The problem is solvable in polynomial time, for example, using the classic algorithms by Boruvka (1926), Kruskal (1956) and Prim (1957). These algorithms have tine complexities $O(u \log n)$, $O(u \log u)$ and $O(n^2)$, where $u = |E|$ and $n = |V|$. Also interested reader may refer to [1,2].

One of the actual generalizations of this problem may be the problem of finding a bounded diameter MST problem. The diameter of a graph is the maximum number of edges in the simple path between all pairs of vertices.

This problem is as follows. Given an undirected graph and given a number $d = d_n$, the goal is to find in the graph a spanning tree $\mathcal{T}$ of minimal total weight having its diameter bounded above or from below to given number d. Both problems are NP-hard in general [3].

Earlier it was studied an asymptotically optimal approach to a problem with a diameter bounded from below, from above, or equal to a given number (see, for example, a history in introductions in the papers [5,6]).

The study was carried out within the framework of the state contract of the Sobolev Institute of Mathematics (project FWNF-2022-0019).

N. Olenev et al. (Eds.): OPTIMA 2023, CCIS 1913, pp. 281–292, 2024.
https://doi.org/10.1007/978-3-031-48751-4_21

This problem models network design applications, where each "vertex" communicates with others vertices at a minimum cost, subject to a given quality requirement (subject to fixed diameter of MST in network). In the previous works, it was assumed that the input data (edge weights of the original graph) are random independent variables with a continuous distribution.

In current paper we consider the problem of finding m edge-disjoint MST's with a given-diameter d in the complete undirected graph (m-d-UMST) on random discrete data. We analyze a polynomial-time approximation algorithm to solve this problem and provide conditions for this algorithm to be asymptotically optimal. A probabilistic analysis is carried out under conditions that edge weights of given graph are identically independent distributed random variables on a discrete segment. Note, that for the first time, an asymptotically optimal approach was implemented in the work [8] more than half a century ago as applied to the classical traveling salesman problem precisely on random input data of a discrete type.

Our algorithm can be transformed to solve the problem of finding m edge-disjoint MSTs with bounded diameter from below or above. So all the applications for these problems are valid for m-d-UMST (see, for example, [7]).

2 Statement of the Problem and Description of the Analyzed Algorithm

Given a complete weighted undirected n-vertex graph $G = (V, E)$ and positive integers $m \geq 2$, d such that $m(d+1) \leq n$, the problem is to find m edge-disjoint spanning trees $\mathcal{T}_1, \ldots, \mathcal{T}_1, \ldots, \mathcal{T}_m$ with a given diameter $d = d_n \leq \frac{n}{m} - 1$ of minimum total weight. The tree containing the vertex v will be denoted by$\mathcal{T}(v)$ (Fig. 1).

Description of the Approximation Algorithm $\mathcal{A}$

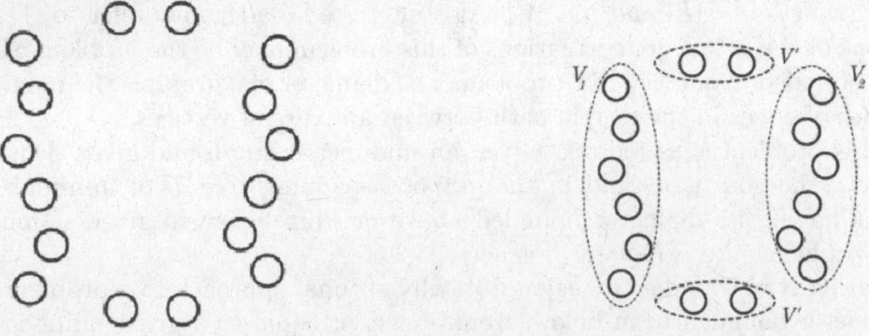

Fig. 1. Initial vertices of the graph and Stage 0 of the work of the Algorithm $\mathcal{A}$ on 16-vertex complete graph, $d = 5$.

Stage 1. Construction of m edge-disjoint paths.

For each $s = 1, \ldots, m$, starting at arbitrary vertex in the subgraph $G(V_s)$, construct in it a Hamiltonian path P_s of length $d = d_n$, using the approach "Go to the nearest unvisited vertex". After the construction of entire path $P_s = \{v_{s0}, v_{s1}, \ldots, v_{sd}\}$ we put $\mathcal{T}_s = P_s$.

All the vertices of the paths, except for the terminal ones, will by called as inner.

Stage 2. Each pair of paths P_i and P_j, $1 \leq i < j \leq m$, we connect them in a special way by the set E_{ij} of $2(d+1)$ edges, so that the constructed subgraph was composed of two $2(d+1)$-vertex edge-disjoint subtrees with a diameter equals d.

We construct the set of connecting edges as follows.

2.1. Connect each inner vertex u of P_i^1 (respectively, P_j^1, P_i^2, P_j^2) by the shortest edge to the inner vertex v of P_j^1 (respectively, P_i^2, P_j^2, P_i^1), and we add this edge to $\mathcal{T}(v)$.

2.2. Connect each end vertex of the path P_i (respectively, P_j) by the shortest edge to the inner vertex of the path P_j (respectively, P_i.). We add this edge to $\mathcal{T}_j$ (respectively, to $\mathcal{T}_i$).

Stage 3. For $s = 1, \ldots, m$ each vertex of the subgraph $G(V')$ connect by the shortest edge to the inner vertex of the path P_s. and add this edge to $\mathcal{T}_s$.

The construction of all m edge-disjoint spanning trees $\mathcal{T}_1, \ldots, \mathcal{T}_m$ is completed (Fig. 2).

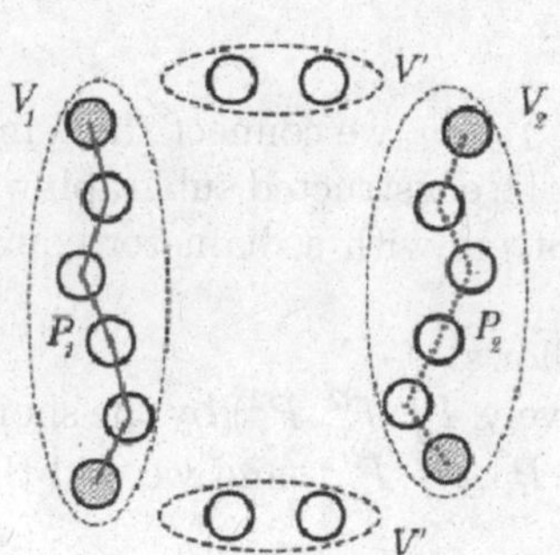

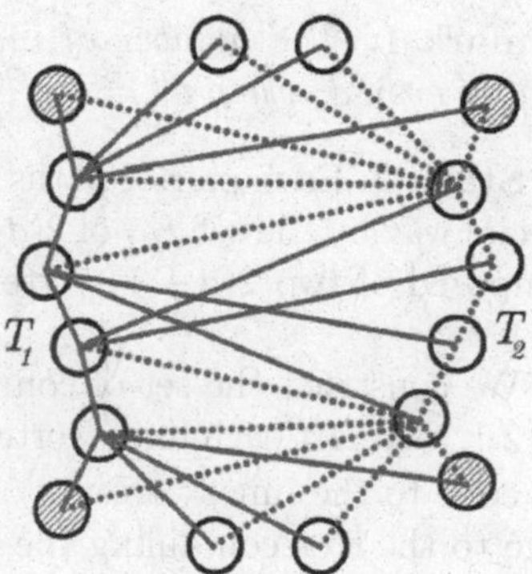

Fig. 2. The work of the Algorithm $\mathcal{A}$ on the Stage 1 and Stages 2–3 at 16-vertex complete graph, $d = 5$.

Denote by $W_{\mathcal{A}}$ the total weight of all trees $\mathcal{T}_1, \ldots, \mathcal{T}_m$ constructed by Algorithm $\mathcal{A}$. Denoting summary weights of edges, obtained on Steps 1, 2 and 3 by W_1, W_2 and W_3, we have $W_{\mathcal{A}} = W_1 + W_2 + W_3$.

Let us formulate two statements concerning Algorithm $\mathcal{A}$.

Statement 1. *Algorithm $\mathcal{A}$ constructs a feasible solution for the m-d-UMST.*

Proof. Each of the edge-disjoint constructions consists of n vertices and $n-1$ edges since we firstly create the tree as the path on $d+1$ vertices during Stage 1 and then we add edges to the tree by connecting all other vertices to the vertices in path on Stages 2–3, totally we obtain m such constructions, and we indeed get feasible solution for the m-d-UMST.

Statement 2. *Running time of Algorithm $\mathcal{A}$ is $\mathcal{O}(n^2)$.*

Proof. **Preliminary Stage 0.** In graph G, choose an arbitrary $(n - m(d+1))$-vertex subset V', and arbitrary split the remaining $m(d+1)$ vertices into $(d+1)$-vertex subsets $V_1, \ldots, V_m$.

Stage 1. For each $s = 1, \ldots, m$, starting at arbitrary vertex in the subgraph $G(V_s)$, construct in it a Hamiltonian path P_s of length $d = d_n$, using the approach "Go to the nearest unvisited vertex". After the construction of entire path $P_s = \{v_{s0}, v_{s1}, \ldots, v_{sd}\}$ we put $\mathcal{T}_s = P_s$.

The vertices of each path, except for the initial and terminal ones, are called inner.

Each path P_s. $s = 1, ..., m$. we represent as the following two halves:
1) the first $P_s^1 = \{v_{s0}, v_{s1}, \ldots, v_{sd'}\}$ with the number of inner vertices $d' = [d/2]$;
2) the second one $P_s^2 = \{v_{s(d'+1)}, \ldots, v_{sd}\}$ with the number of inner vertices $d'' = [(d-1)/2]$.

Remark 1. *The number of inner vertices in both halves is d', except for the case of even d, when $d'' = d' - 1$.*

Stage 2. Each pair of paths P_i and P_j, $1 \le i < j \le m$, we connect them in a special way by the set E_{ij} of $2(d+1)$ edges, so that the constructed subgraph was composed of two $2(d+1)$-vertex edge-disjoint subtrees with a diameter equals d.

We construct the set of connecting edges as follows.

2.1. Connect each inner vertex v of P_i^1 (respectively, P_j^1, P_i^2, P_j^2) by the shortest edge to the inner vertex v of P_j^1 (respectively, P_i^2, P_j^2, P_i^1), and we add this edge to the tree containing the vertex v.

2.2. Connect each end vertex of the path P_i (respectively, P_j) by the shortest edge to the inner vertex of the path P_j (respectively, P_i.). We add this edge to $\mathcal{T}_j$ (respectively, to $\mathcal{T}_i$).

Stage 3. For $s = 1, \ldots, m$ each vertex of the subgraph $G(V')$ connect by the shortest edge to the inner vertex of the path P_s. And add this edge to $\mathcal{T}_s$.

The construction of all m edge-disjoint spanning trees $\mathcal{T}_1, \ldots, \mathcal{T}_m$ is completed.

Denote by $W_{\mathcal{A}}$ the total weight of all trees $\mathcal{T}_1, \ldots, \mathcal{T}_m$ constructed by Algorithm $\mathcal{A}$. Denoting summary weights of edges, obtained on Stages 1, 2 and 3 by W_1, W_2 and W_3, we have $W_{\mathcal{A}} = W_1 + W_2 + W_3$.

Let us formulate two statements concerning Algorithm $\mathcal{A}$.

Statement 1. *Algorithm $\mathcal{A}$ constructs a feasible solution for the m-d-UMST.*

Proof. Each of the edge-disjoint constructions consists of n vertices and $n-1$ edges since we firstly create the tree as the path on $d+1$ vertices during Step 1 and then we add edges to the tree by connecting all other vertices to the vertices in path on Steps 2–3, totally we obtain m such constructions, and we indeed get feasible solution for the m-d-UMST.

Statement 2. *Running time of Algorithm $\mathcal{A}$ is $\mathcal{O}(n^2)$.*

Proof. Preliminary Stage 0 takes $\mathcal{O}(n)$ time.

At Stage 1 each path is built in $\mathcal{O}(d^2)$ time, thus, it takes $\mathcal{O}(md^2)$ or $\mathcal{O}(nd)$ time to construct all paths.

At Stage 2 each pair (P_i, P_j), $1 \leq i < j \leq m$, of paths is connected with the edge set E_{ij} in $\mathcal{O}(d^2)$ time, and for all $\frac{m(m-1)}{2}$ pairs of paths it is required $\mathcal{O}(m^2d^2)$, or (since $m(d+1) \leq n$) $\mathcal{O}(n^2)$-running time.

Stage 3 takes $\mathcal{O}(mn)$ operations since we connect $|G(V')| \leq n$ vertices by the shortest edge to the inner vertex of the path P_s for each spanning tree $\mathcal{T}_s$, $1 \leq s \leq m$.

So, the total time complexity of the Algorithm $\mathcal{A}$ is $\mathcal{O}(n^2)$.

3 A Probabilistic Analysis of Algorithm $\mathcal{A}$

We perform the probabilistic analysis under conditions that edge weights of graph are independent random variables η identically uniformly distributed in the positive integer segment $[1, r_n]$. Namely

$$\mathbf{P}\{\eta = r\} = \begin{cases} p, & \text{if } r \in [1, r_n], \\ 0, & \text{otherwise,} \end{cases}$$

where $p = 1/r_n$. We also introduce a denotation for the normalized biased variable $\xi = \frac{\eta - 1}{r_n}$.

By $W_A(I)$ and $OPT(I)$ we denote respectively the approximate (obtained by some approximation algorithm A) and the input I. An algorithm A is said to have *estimates (performance guarantees)* $(\varepsilon_N, \delta_N)$ on the set of random inputs of the N-sized problem. Describe the problem, if

$$\mathbf{P}\Big\{W_A(I) > (1 + \varepsilon_N)OPT(I)\Big\} \leq \delta_N, \tag{1}$$

where $\varepsilon_N = \varepsilon_A(N)$ is an estimation of *the relative error* of the solution obtained by algorithm A, $\delta_N = \delta_A(N)$ is an estimation of *the failure probability* of the algorithm, which is equal to the proportion of cases when the algorithm does not hold the relative error ε_n or does not produce any answer at all.

Following the paper [4], we say that an approximation algorithm A is called *asymptotically optimal* on the class of input data of the problem, if there exist such performance guarantees that for all input I

$$\varepsilon_N \to 0 \text{ and } \delta_N \to 0 \text{ as} N \to \infty.$$

We denote random variable equal to minimum over k variables from the class UNI$[1; r_n]$. by η_k. It is clear that $1 \leq \eta_k \leq r_n$.

Some notations: $p = 1/r_n$;, $u_r = r/r_n = rp$; $\mathcal{P}_r = \mathbf{P}\{\xi \leq u_r\} = rp$.

According to the description of Algorithm $\mathcal{A}$ the weight $W_{\mathcal{A}}$ of all constructed spanning trees $\mathcal{T}_1, \ldots, \mathcal{T}_m$ is a random value equal to

$$W_{\mathcal{A}} = m(n-1) + r_n W'_{\mathcal{A}},$$

where

$$W'_{\mathcal{A}} = W'_1 + W'_2 + W'_3.$$

W'_1, W'_2, W'_3 are normalized random variables for values W_1, W_2, W_3, respectively.

$W'_1 = \sum_{i=1}^{m} \sum_{k=1}^{d} \xi_k$ since we construct the path P_i consists of d edges and repeat this construction m times during Stage 1.

$W'_2 = W'_{2.1} + W'_{2.2}$, where

$W'_{2.1} = C_m^2 \Big(d'\xi_{d'} + d'\xi_{d"} + d"\xi_{d"} + d"\xi_{d'} \Big)$, $W'_{2.2} = 4C_m^2 \xi_{(d-1)}$, because on the Stage 2.1 for each pair of paths (totally, $C_m^2 = \frac{m(m-1)}{2}$ such pairs) we in a special way connect each inner vertex of one path by the shortest edge to the inner vertex of the other path.

And on the Stage 2.2 we connect each of two end vertices in each pair of paths by the shortest edge to the inner vertex of another path.

$W'_3 = m(n - m(d+1))\xi_{(d-1)}$ since we connect each of the $n - m(d+1)$ vertices from $G(V')$ to each path P_s, $1 \leq s \leq m$ by the shortest edge to its inner vertex.

Some notations: $p = 1/r_n$, $\mathcal{P}_r = \mathbf{P}\{\eta \leq r\} = rp$.

Denote by $S_k(R) = \sum_{r=1}^{R} r^k$ the sum of k-th powers of natural numbers $1, 2, \ldots, R$. The book [10] presents a rather cumbersome formula for calculating this function into sums of powers with the highest degree of R. Below we prove guaranteed estimates this function from above and from belou.

Lemma 1

$$\frac{R^{k+1}}{k+1} \leq S_k(R) \leq \frac{(R+1)^{k+1}}{k+1}$$

Proof. The terms of the sum $\sum_{r=1}^{R}$ grow with natural r. The upper and lower bounds for this sum are obtained by integrating the growing function x^k over the corresponding intervals.

$$S_k(R) = \sum_{r=1}^{R} r^k \leq \int_{r=1}^{R+1} r^k dr = \frac{(R+1)^{k+1} - 1}{k+1} \leq \frac{(R+1)^{k+1}}{k+1},$$

$$S_k(R) = \sum_{r=1}^{R} r^k \geq \int_{r=0}^{E} r^k dr = \frac{R^{k+1}}{k+1}.$$

Lemma 1 is proved.

Lemma 2. *For the mathematical expectation of a random variable* ξ_k *for all natural* k, *we have an upper estimate*

$$\mathbf{E}\xi_k \le \frac{1}{k+1}$$

Proof

$$\mathbf{E}\eta_k = \sum_{r=1}^{r_n} r\mathbf{P}\left\{\eta_k = r\right\} = \sum_{r=1}^{r_n} r\Big(\mathbf{P}\{\eta_k \ge r\} - \mathbf{P}\{\eta_k \ge r+1\}\Big) = \sum_{r=1}^{r_n} \mathbf{P}\{\eta_k \ge r\}$$

$$= 1 + \sum_{r=1}^{r_n-1} \Big(\mathbf{P}\{\eta > r\}\Big)^k = 1 + \sum_{r=1}^{r_n-1} \Big(1 - \mathbf{P}\{\eta \le r\}\Big)^k = 1 + \sum_{r=1}^{r_n-1} \big(1 - \mathcal{P}_r\big)^k. \quad (2)$$

We replace the summation index by $r_n - r$ and run the estimation of $\mathbf{E}\eta_k$, taking into account the upper bound from Lemma 1:

$$\mathbf{E}\eta_k = 1 + \sum_{r=1}^{r_n-1} \mathcal{P}_r^k = 1 + p^k \sum_{r=1}^{r_n-1} r^k \le 1 + p^k S_k(r_n - 1) \le 1 + p^k \frac{r_n^{k+1} - 1}{k+1} \le 1 + \frac{r_n}{k+1}.$$

From this, taking into account the formula $\xi_k = \frac{\eta_k - 1}{r_n}$, we obtain:

$$\mathbf{E}\xi_k = \frac{\mathbf{E}\eta_k - 1}{r_n} \le \frac{1}{k+1}.$$

Lemma 2 is proved.

Lemma 3. *For* $\mathbf{E}(W'_{\mathcal{A}})$ *such inequality is true*

$$\mathbf{E}(W'_{\mathcal{A}}) \le m \ln(d+1) + m^2 + \frac{mn}{d}.$$

Proof. Consider separately expectations of random variables for values W_1, W_2 and W_3.

$$\mathbf{E}W'_1 = m \sum_{k=1}^{d} \mathbf{E}\xi_k \le m \sum_{k=1}^{d} \frac{1}{k+1} = m\Big(\sum_{k=1}^{d+1} \frac{1}{k} - 1\Big) \le m \ln(d+1).$$

Let us show that $\mathbf{E}W'_2 \le 2m^2$, regardless of the parity of d.
Really, if d is odd then

$$\mathbf{E}W'_2 = C_m^2\Big(4d'\mathbf{E}\xi_{d'} + 4\mathbf{E}\xi_{d-1}\Big) \le 4C_m^2\Big(\frac{d'}{d'+1} + \frac{1}{d}\Big) = 4C_m^2\Big(1 - \frac{1}{d'+1} + 4\frac{1}{d}\Big) \le 2m^2$$

If d is even then

$$\mathbf{E}W'_2 = C_m^2\Big(d'\mathbf{E}\xi_{d'} + d'\mathbf{E}\xi_{d'-1} + (d'-1)\mathbf{E}\xi_{d-1} + (d'-1)\mathbf{E}\xi_d + 4\mathbf{E}\xi_{d-1}\Big)$$

$$\le C_m^2\Big(\frac{d'}{d'+1} + \frac{d'}{d'} + \frac{d'-1}{d'} + \frac{d'-1}{d'+1} + \frac{4}{d}\Big) = 4C_m^2\Big(1 - \frac{1}{d'+1} - \frac{1}{d'} - \frac{2}{d'+1} + \frac{1}{d}\Big) \le 2m^2$$

$$\mathbf{E}W_3' = m(n - m(d+1))\mathbf{E}\xi_{(d-1)} = m\frac{n - m(d+1)}{d} \le \frac{mn}{d} - m^2.$$

From the previous equations we get

$$\mathbf{E}(W_{\mathcal{A}}') = \mathbf{E}\Big(W_1' + W_2' + W_3'\Big) \le m\ln(d+1) + 2m^2 + \frac{mn}{d} - m^2 \le m\ln(d+1) + m^2 + \frac{mn}{d}.$$

Lemma 3 is proved.

Lemma 4

$$\mathbf{Var}\,\xi_k \le \frac{2}{(k+1)^2}.$$

Proof. Start with the calculation $\mathbf{E}\eta_k^2$ and $\mathbf{Var}\,\eta_k$:

$$\mathbf{E}\eta_k^2 = \sum_{r=1}^{r_n-1} r^2\,\mathbf{P}\left\{\eta_k = r\right\} = \sum_{r=1}^{r_n-1} r^2\Big(\mathbf{P}\{\eta_k \ge r\} - \mathbf{P}\{\eta_k \ge r+1\}\Big)$$

$$= 1 + \sum_{r=1}^{r_n-1}\left((r+1)^2 - r^2\right)\mathbf{P}\{\eta_k \ge r+1\} = 1 + \sum_{r=1}^{r_n-1}(2r+1)(1-\mathcal{P}_r)^k. \quad (3)$$

From the Eq. (3) for $\mathbf{E}\eta_k^2$ and the Eq. (2), where $\mathbf{E}\eta_k \ge 1$, we have

$$\mathbf{Var}\eta_k = \mathbf{E}\eta_k^2 - (\mathbf{E}\eta_k)^2 = \Big(\{1 + \sum_{r=1}^{r_n-1}(2r+1)(1-\mathcal{P}_r)^k\Big) - \Big(1 + \mathbf{E}\eta_k\sum_{r=1}^{r_n-1}(1-\mathcal{P}_r)^k\Big)$$

$$\le 2\sum_{r=1}^{r_n-1} r(1-\mathcal{P}_r)^k = 2\sum_{r=1}^{r_n-1}(r_n - r)(pr)^k = 2p^k\sum_{r=1}^{r_n-1}(r_n r^k - r^{k+1}).$$

In the last line, when passing from the first sum to the second one, the summation index was replaced. Next, we finish the estimation of $\mathbf{Var}\,\eta_k$ taking into account Lemma 1.

$$\mathbf{Var}\,\eta_k \le 2p^k\Big(r_n S_k(r_n - 1) - S_{k+1}(r_n - 1)\Big) \le 2p^k\Big(r_n\frac{r_n^{k+1}}{k+1} - \frac{r_n^{k+2}}{k+2}\Big) \le \frac{2r_n^2}{(k+1)(k+2)}.$$

Finally, for $\mathbf{Var}\,\xi_k$ we obtain:

$$\mathbf{Var}\,\xi_k = \frac{\mathbf{Var}\,\eta_k}{r_n^2} \le \frac{2}{(k+1)(k+2)} \le \frac{2}{(k+1)^2},$$

Lemma 4 is proved.

Lemma 5. *Let the value* $\widehat{\mathbf{E}W'_{\mathcal{A}}} = m\ln(d+1) + m^2 + \frac{mn}{d}$ *be taken by Lemma 3 as the upper estimate of the mathematical expectation* $\mathbf{E}W'_{\mathcal{A}}$. *Then the Algorithm* $\mathcal{A}$ *has the following performance guarantees (estimates of the relative error* ε_n *and the failure probability* δ_n*):*

$$\varepsilon_n = \frac{2r_n}{m(n-1)}\widehat{\mathbf{E}W'_{\mathcal{A}}}, \tag{4}$$

$$\delta_n = \mathbf{P}\left\{\widetilde{W}'_{\mathcal{A}} > \widehat{\mathbf{E}W'_{\mathcal{A}}}\right\}. \tag{5}$$

Proof. Let us show that (4) and (5) satisfy the definition of the algorithm $\mathcal{A}$ with these performance estimates:

$$\mathbf{P}\left\{W_{\mathcal{A}} > (1+\varepsilon_n)OPT\right\} \leq \mathbf{P}\left\{m(n-1) + r_n W'_{\mathcal{A}} > (1+\varepsilon_n)m(n-1)\right\}$$

$$= \mathbf{P}\left\{m(n-1) + r_n W'_{\mathcal{A}} > (1+\varepsilon_n)m(n-1)\right\}$$

$$= \mathbf{P}\left\{W'_{\mathcal{A}} - \mathbf{E}W'_{\mathcal{A}} > \frac{\varepsilon_n m(n-1)}{r_n} - \mathbf{E}W'_{\mathcal{A}}\right\} = \mathbf{P}\left\{\widetilde{W}'_{\mathcal{A}} > \frac{\varepsilon_n m(n-1)}{r_n} - \mathbf{E}W'_{\mathcal{A}}\right\}$$

$$\leq \mathbf{P}\left\{\widetilde{W}'_{\mathcal{A}} > \frac{\varepsilon_n m(n-1)}{r_n} - \widehat{\mathbf{E}W'_{\mathcal{A}}}\right\} = \mathbf{P}\left\{\widetilde{W}'_{\mathcal{A}} > \widehat{\mathbf{E}W'_{\mathcal{A}}}\right\} = \delta_n.$$

Lemma 5 is proved.

Further for the probabilistic analysis of Algorithm $\mathcal{A}$ we use the following probabilistic statement

Petrov's Theorem [9]. *Consider independent random variables* $X_1, \ldots, X_N$. *Let there be positive constants* T *and* $h_1, \ldots, h_N$ *such that for all* $k = 1, \ldots, N$ *and* $0 \leq t \leq T$ *the following inequalities hold:*

$$\mathbf{E}e^{tX_k} \leq e^{\frac{h_k t^2}{2}}. \tag{6}$$

Set $S = \sum_{k=1}^{n} X_k$ *and* $H = \sum_{k=1}^{n} h_k$. *Then*

$$\mathbf{P}\{S > x\} \leq \begin{cases} \exp\{-\frac{Tx}{2}\}, \text{ if } x \geq HT, \\ \exp\{-\frac{x^2}{2H}\}, \text{ if } 0 \leq x \leq HT. \end{cases} \tag{7}$$

Lemma 6. *Let* η_k *be random variable equal to minimum over* k *independent random variables from the class UNI considered, and* $\xi_k = \frac{\eta_k - 1}{r_n}$. *Given constants* $T = 1$ *and* $h_k = \frac{3}{(k+1)^2}$. *Then for variables* $\widetilde{\xi}_k = \xi_k - \mathbf{E}\xi_k$ *the condition (6)* $\mathbf{E}e^{t\widetilde{\xi}_k} \leq e^{\frac{n_k t^2}{2}}$ *of Petrov's Theorem holds for each* $t \leq 1$ *and natural* k.

Proof. Let us estimate $e^{t\widetilde{\xi}_k}$, keeping in mind that $\widetilde{\xi}_k \leq 1$ and $t \leq 1$:

$$e^{t\widetilde{\xi}_k} = \sum_{j=0}^{\infty} \frac{(t\xi_k)^j}{j!} = 1 + t\widetilde{\xi}_k + \frac{(t\widetilde{\xi}_k)^2}{2}\Big(1 + \frac{t\widetilde{\xi}_k}{3} + \frac{(t\xi_k)^2}{3\cdot 4} + \frac{(t\xi_k)^3}{3\cdot 4\cdot 5} + ...\Big)$$

$$\leq 1 + t\widetilde{\xi}_k + \frac{(t\widetilde{\xi}_k)^2}{2}\sum_{j=0}^{\infty}\Big(\frac{t\widetilde{\xi}_k}{3}\Big)^j = 1 + t\widetilde{\xi}_k + \frac{.(t\widetilde{\xi}_k)^2}{2}\frac{1}{1 - t\widetilde{\xi}_k/3} = 1 + t\xi_k + \frac{3(t\widetilde{\xi}_k)^2}{4}.$$

Hence, given the facts $\mathbf{E}\widetilde{\xi}_k = 0$ and $\mathbf{E}\widetilde{\xi}_k^2 = \mathbf{Var}\,\xi_k$, by previous Lemma we obtain:

$$\mathbf{E}e^{t\widetilde{\xi}_k} \leq 1 + t\mathbf{E}\widetilde{\xi}_k + \frac{3t^2}{4}\mathbf{E}\widetilde{\xi}_k^2 \leq 1 + \frac{3t^2}{2(k+1)^2} = 1 + \frac{h_k t^2}{2} \leq e^{\frac{h_k t^2}{2}}.$$

Lemma 6 is proved.

Lemma 7. *In the case $d > 4$ the following upper bound for the sum of constants $h_k = \frac{3}{(k+1)^2}$ holds*

$$H \leq 2m + m^2 + \frac{5mn}{d^2}.$$

Proof. The parameter H equal to the sum of H_1, H_2 and H_3 according to the Stages of Algorithm $\mathcal{A}$ number 1, 2, 3, respectively. Knowing that notation and estimates from above, we obtain

$$H_1 = m\sum_{k=1}^{d} h_k = 3m\sum_{k=1}^{d}\frac{1}{(k+1)^2} < 3\psi m \leq 2m,$$

where $\psi = 0.645$. We have used the well-known Euler's estimate for the inverse square equation $1 + \frac{1}{2^2} + \frac{1}{3^2} + \frac{1}{4^2} + \ldots = \frac{\pi^2}{6} < 1.645$

$$H_2 = 4C_m^2(d-1)/2h_{(d-1)/2} + h_{d-1}) \leq 6m^2\Big(\frac{(d-1)/2}{((d-1)/2+1)^2} + \frac{1}{d^2}\Big) = 6m^2\Big(\frac{2(d-1)}{(d+1)^2} + \frac{1}{d^2}\Big) \leq 12m^2\frac{d}{(d+1)^2}.$$

$$H_3 = m\big(n - m(d+1)\big)h_{d-1} \leq \frac{3mn}{d^2} - 3m^2\frac{d}{(d+1)^2}.$$

Summary:

$$H = H_1 + H_2 + H_3 < 2\,m + 12m^2\frac{d}{(d+1)^2} + \Big(\frac{3mn}{d^2} - 3m^2\frac{d}{(d+1)^2}\Big) \leq$$

$$= 2m + \frac{3mn}{d^2} + \frac{9m^2 d}{(d+1)^2} \leq 2m + m^2 + \frac{5mn}{d^2}.$$

Lemma 7 is proved.

Theorem 1. *Let the parameter $d = d_n \geq \ln n$. Then Algorithm for the problem considered is asymptotically optimal, if*

$$r_n = o\Big(\min\big(\frac{n}{\ln d_n}, d_n\big)\Big). \tag{8}$$

Proof. First of all, we note that in the course of the Algorithm $\mathcal{A}$ we are dealing with random variables only of the type ξ_k, $1 \leq k \leq d$. These variables satisfy the conditions of the Petrov's theorem for constants $T = 1$ and $h_k = \frac{3}{(k+1)^2}$.

Using the Lemma 4 and setting the threshold equal $x = \widehat{\mathbf{E}W'_{\mathcal{A}}} = m\ln(d+1) + m^2 + \frac{mn}{d}$, for $d > 4$ we have

$$TH \leq 2m + m^2 + \frac{5mn}{d^2} < m\ln(d+1) + m^2 + \frac{mn}{d} = x.$$

Therefore, we can use the first of the inequalities in (7). Let us first show that $\ln(d+1) + \frac{2n}{d} \geq \ln n$. Since the left part of this inequality decreases with respect to d, then for its minimum we have:

$$\ln(d+1) + m + \frac{n}{d} \geq \ln n + m + \frac{n}{n-1} \geq \ln n.$$

By Petrov's Theorem, we estimate the failure probability

$$\delta_n = \mathbf{P}\{\widetilde{W'_{\mathcal{A}}} > x\} \leq \exp\Big\{-\frac{Tx}{2}\Big\} \leq \exp\Big\{-\frac{m}{2}\Big(\ln(d+1) + m + \frac{n}{d}\Big)\Big\}$$

$$\leq \exp\Big\{-\frac{m}{2}\ln n\Big\} = \Big(\frac{1}{n}\Big)^{\frac{n}{2}} \to 0 \text{ as } n \to \infty.$$

From the formula (4) for the relative error we obtain

$$\varepsilon_n = \frac{2r_n\widehat{\mathbf{E}W'_{\mathcal{A}}}}{m(n-1)} = \frac{2r_n}{(n-1)} \cdot \Big(\ln(d+1) + m + \frac{n}{d}\Big)$$

$$\leq \frac{2r_n}{(n-1)} \cdot \Big(\ln(d+1) + \frac{2}{d}\Big) = \mathcal{O}\Big(\frac{r_n}{\min\big(\frac{n}{\ln(d+1)}, d\big)}\Big).$$

Hence, accordingly the condition (8) the relative error $\varepsilon_n \to 0$ as $n \to \infty$. So, we obtain asymptotically optimal solution for the problem m-d-UMST on n-vertex complete graph with weights of edges from discrete UNI$[1; r_n]$.

Theorem 1 is completely proved.

Remark 2. *In a particular case $d_n = n-1$, we have the Hamiltonian Path problem. The algorithm $\mathcal{A}$ with $\mathcal{O}(n^2)$ time complexity guarantees finding a solution with estimates*

$$\delta_n = \frac{1}{\sqrt{n}}, \ \varepsilon_n = \mathcal{O}\Big(\frac{r_n}{n/\ln n}\Big),$$

and is asymptotically optimal for

$$r_n = o\Big(\frac{n}{\ln n}\Big).$$

4 Conclusion

In this work, we have performed the probabilistic analysis of some approximation algorithm which finds in polynomial time several edge-disjoint given-diameter Spanning Trees of minimal total edge weights in a complete undirected graph. We also have obtained sufficient conditions of asymptotic optimality for this algorithm in the case of a uniform discrete distribution for the edges weights of the graph.

I am grateful to my graduate student Alexsandr Shtepa for his help in preparing the drawings explaining the work of the algorithm.

References

1. Angel, O., Flaxman, A.D., Wilson, D.B.: A sharp threshold for minimum bounded-depth and bounded-diameter spanning trees and Steiner trees in random networks. Combinatorica **32**, 1–33 (2012)
2. Cooper, C., Frieze, A., Ince, N., Janson, S., Spencer, J.: On the length of a random minimum spanning tree. Comb. Probab. Comput. **25**(1), 89–107 (2016)
3. Garey, M.R., Johnson, D.S.: Computers and Intractability. Freeman, San Francisco (1979). 340 p
4. Gimadi, E.Kh., Glebov, N.I., Perepelitsa, V.A.: Algorithms with estimates for discrete optimization problems. Problemy Kibernetiki **31**, 35–42 (1970). (in Russian)
5. Gimadi, E.K., Shevyakov, A.S., Shtepa, A.A.: A given diameter MST on a random graph. In: Olenev, N., Evtushenko, Y., Khachay, M., Malkova, V. (eds.) OPTIMA 2020. LNCS, vol. 12422, pp. 110–121. Springer, Cham (2020). https://doi.org/10.1007/978-3-030-62867-3_9
6. Gimadi, E.K., Shtepa, A.A.: On the asymptotically optimal approach to finding the minimum of the sum of weights of different-edge spanning trees of fixed diameter. Autom. Telemech. **7**, 146–166 (2023)
7. Gruber, M.: Exact and heuristic approaches for solving the bounded diameter minimum spanning tree problem. Vienna University of Technology, Ph.D. thesis (2008)
8. Perepelitsa, V.A., Gimadi, E.Kh.: The problem of finding a minimal Hamiltonian cycle in a weighted graph, in Discrete Analysis: Collection of Papers (Inst. Mat. SO AN SSSR, Novosibirsk, 1969), issue 15, pp. 57–65. (in Russian)
9. Petrov, V.V.: Limit Theorems of Probability Theory. Sequences of Independent Random Variables. Clarendon Press, Oxford (1995). 304 p.
10. Ryzhyk, I.M.: Tables of Integrals, Sums, Series... OGIZ, Moscow (1948). 400 p.

Author Index

N. Olenev et al. (Eds.): OPTIMA 2023, CCIS 1913, p. 293, 2024.
https://doi.org/10.1007/978-3-031-48751-4

Printed in the United States
by Baker & Taylor Publisher Services

Printed in the United States
by Baker & Taylor Publisher Services